Green Innovation in China

Contemporary Asia in the World

Contemporary Asia in the World

David C. Kang and Victor D. Cha, Editors

This series aims to address a gap in the public-policy and scholarly discussion of Asia. It seeks to promote books and studies that are on the cutting edge of their respective disciplines or in the promotion of multidisciplinary or interdisciplinary research but that are also accessible to a wider readership. The editors seek to showcase the best scholarly and public-policy arguments on Asia from any field, including politics, history, economics, and cultural studies.

Beyond the Final Score: The Politics of Sport in Asia, Victor D. Cha, 2008
The Power of the Internet in China: Citizen Activism Online, Guobin Yang, 2009
China and India: Prospects for Peace, Jonathan Holslag, 2010
India, Pakistan, and the Bomb: Debating Nuclear Stability in South Asia, Šumit Ganguly and S. Paul Kapur, 2010
Living with the Dragon: How the American Public Views the Rise of China, Benjamin I. Page and Tao Xie, 2010
East Asia Before the West: Five Centuries of Trade and Tribute, David C. Kang, 2010
Harmony and War: Confucian Culture and Chinese Power Politics, Yuan-Kang Wang, 2011
Strong Society, Smart State: The Rise of Public Opinion in China's Japan Policy, James Reilly, 2012
Asia's Space Race: National Motivations, Regional Rivalries, and International Risks, James Clay Moltz, 2012
Never Forget National Humiliation: Historical Memory in Chinese Politics and Foreign Relations, Zheng Wang, 2012

GREEN INNOVATION IN **CHINA**

China's Wind Power Industry and the Global Transition to a Low-Carbon Economy

Joanna I. Lewis

Columbia University Press   New York

Columbia University Press
Publishers Since 1893
New York   Chichester, West Sussex
cup.columbia.edu

Copyright © 2013 Joanna I. Lewis
Paperback edition, 2015
All rights reserved

Library of Congress Cataloging-in-Publication Data
Lewis, Joanna I.
Green innovation in China : China's wind power industry and the global transition to a low-carbon economy / Joanna I. Lewis.
    p.   cm. — (Contemporary Asia in the world)
Includes bibliographical references and index.
ISBN 978-0-231-15330-0 (cloth : alk. paper)—ISBN 978-0-231-15331-7 (pbk. : alk. paper)—
ISBN 978-0-231-52687-6 (e-book)
1. Wind power industry—China. 2. Wind power industry—Technological innovations. 3. Electric power production—Technology transfer—China. I. Title.
HD9502.5.W553L49   2013
333.9'20951—dc23
                                    2011051915

Columbia University Press books are printed on permanent and durable acid-free paper.
This book was printed on paper with recycled content.
Printed in the United States of America

Cover design: Rebecca Lown Design
Cover image: "Dabancheng Wind Farm" © Bob Sacha/Corbis

References to Internet Web sites (URLs) were accurate at the time of writing. Neither the author nor Columbia University Press is responsible for URLs that may have expired or changed since the manuscript was prepared.

Contents

List of Figures   vii

List of Tables   ix

Preface   xi

Abbreviations   xv

Chronology of Wind Power Development in China   xix

1. Green Innovation in China   1
2. China's Energy and Climate Challenge   5
3. China in the Global Wind Power Innovation System   26
4. The Role of Foreign Technology in China's Wind Power Industry Development   75
5. Goldwind and the Emergence of the Chinese Wind Industry   121

6. Wind Energy Leapfrogging in Emerging Economies 145

7. Engaging China on Clean Energy Cooperation 168

Notes 189

Bibliography 235

Index 265

Figures

2.1 Energy Consumption and Energy Intensity Trends in China 9
2.2 China's Greenhouse Gas Footprint 10
3.1 Major Components of a Modern Wind Turbine 27
3.2 Wind Energy R&D Expenditures in OECD Countries 29
3.3 Wind Power Leaders, Cumulative Installations 32
3.4 Global Wind Turbine Market Shares 37
3.5 Activities and Actors in China's Innovation System Under Central Planning and Since Reforms 45
3.6 Countries Leading in Wind Energy and Renewable Energy Patents 47
3.7 Energy R&D Investment in China 48
3.8 Average Size of Wind Turbines Installed Annually, Selected Countries 60
3.9 Average Wind Turbine Size and Start of Local Manufacturing, Selected Countries 61
3.10 Number and Ownership of Wind Turbine Manufacturers in China 62
3.11 Map of Wind Power Development and Wind Technology Manufacturing Facilities in China 63

3.12 Map of Chinese Wind Turbine Manufacturer Experience Overseas 64
3.13 Chinese and Foreign Wind Turbine Prices in the Chinese Market 66
4.1 Timeline of NEG Micon, Vestas, and Gamesa Company Structure in China 76
4.2 Wind Installations in China (Vestas, NEG Micon, Nordtank, Micon, NedWind, Gamesa) 78
4.3 Chinese Wind Market Shares (Vestas, NEG Micon, Nordtank, Micon, NedWind, Gamesa) 79
4.4 Timeline of Nordex Company Structure in China 88
4.5 Wind Installations in China (Nordex and Xi'an Nordex) 91
4.6 Annual Market Shares in China (Nordex and Xi'an Nordex) 92
4.7 Timeline of U.S. Wind Companies in the Chinese Market 96
4.8 U.S. Wind Turbine Models Installed in China (U.S. Windpower, Tacke, Zond, GE Wind) 102
4.9 Chinese Market Shares of U.S. Turbine Manufacturers 103
4.10 China Wind Turbine Market Shares 107
4.11 Chinese Market Share by Company Ownership Type 108
5.1 Goldwind's Installed Wind Capacity in China 122
5.2 Timeline of Goldwind's Company Structure in China 125
6.1 Role of China, India, and South Korea in Global Wind Power Development 146
6.2 Wind Turbine Market Shares in China, India, and South Korea 155
6.3 Wind Power Technology Transfer Networks in China, India, South Korea, and Beyond 160
7.1 Carbon Dioxide Emissions in the United States and China 170

Tables

2.1 China's New Strategic and Emerging Industries in the Twelfth Five-Year Plan 23
3.1 Leading World Wind Markets and National Turbine Manufacturers 30
5.1 Technology Development Models in the Chinese Wind Industry 136
7.1 Timeline of Major Events in U.S.–China Clean Energy Cooperation 175

Preface

When I first began examining China's strategies for clean energy development in the late 1990s, I discovered that although wind power appeared to have excellent potential for near-term utilization in China, little was happening in the way of wind power development. Since then I have witnessed a major change in the attention Chinese policy makers give to wind power development. The policy regime to promote renewable resources, and wind power in particular, has changed significantly with the introduction of several major policies to support wind energy development. As a result, over the past decade China has transitioned from a country with only a handful of wind turbines to the largest wind power market in the world.

This marked change has unsurprisingly caught the attention of wind energy technology companies from around the world. In 2001, when I conducted my preliminary survey of the wind turbine manufacturing companies that were active in the Chinese market, I assumed this would turn into a study of the failed attempts of technology transfers that appeared prevalent in the handful of Sino-foreign joint ventures that constituted the industry, as well as of why no Chinese manufacturers had been able to come up with a commercially viable wind turbine technology design.

But by the time I moved to China in fall 2003, the wind turbine market there was already beginning to look quite different. What had appeared to be a stagnant market in the late 1990s and early 2000s had witnessed the entrance of several new, dynamic participants, including Goldwind, the first successful domestic turbine manufacturer, and GE, the first multinational corporation to enter the wind energy business—both drawn to the new policy regime and potential of the Chinese market. It soon became evident that this research would include not just an examination of technology transfer models used by firms to bring wind turbines to China but also a study of the role of multinational firms and domestic Chinese companies in clean energy innovation and deployment throughout the world.[1]

Research for this book was conducted from 2001 to 2010. Interviews were conducted in China (primarily Beijing but also Shanghai, Nan'ao, Shantou, Guangzhou, Baoding, Hohhot, and Xining) and Washington, D.C., over this period, as well as in Austin (at the 2003 American Wind Energy Association conference); Denver (at the 2004 World Renewable Energy Congress); and New Delhi and Seoul (2010 and 2011). Additional interviews took place at the United Nations climate change negotiations in Bonn (2005–2007); Montreal (2005); Nairobi (2006), Bali (2007); Poznan (2008); Copenhagen (2009); and Tianjin (2010). Several positions and professional relationships facilitated my access to stakeholders in China's wind power industry. My initial introduction to many energy policy makers in China was a result of my employment with the China Energy Group of the U.S. Department of Energy's Lawrence Berkeley National Laboratory from 2000 to 2005 while enrolled as a graduate student in Energy and Resources at UC Berkeley. Since 2001 I have also worked with the Energy Foundation China Sustainable Energy Program on renewable energy issues. During the 2003–04 academic year Tsinghua University in Beijing hosted me as a senior visiting scholar while I was conducting dissertation research. During that year I also worked informally with several Beijing-based organizations involved in wind energy and climate change issues, including BJADC and Green Capital. From 2005 to 2011 I made several trips per year to China to conduct research for this project, often in conjunction with my work for the Pew Center on Global Climate Change, the Energy Foundation China Sustainable Energy Program, the Asia Society, and the U.S. National Academies.

My interview subjects fall into three broad categories: government officials working on energy and climate policy (particularly wind energy policy);[2] wind turbine company employees or former employees;[3]

and third-party wind experts including academics, consultants, staff of nongovernmental organizations (NGOs), and wind farm managers. In addition to conducting interviews in both English and Mandarin, I had informal discussions with participants in and observers of the industry on numerous occasions. I collected relevant information from discussions with leading academics in China working on wind energy, energy law and policy, electricity sector reform, and climate change policy, as well as from NGOs working to promote wind energy in China and organizations funding policy and research on wind industry development in China.[4] Key informants (defined as interview candidates who often have long-standing relationships with the researcher and differ from other interview participants because of their position in the field of research, connections, and wealth of knowledge of the subject) provided me with extensive information through frequent communications.[5] My research also relies heavily on government documents (such as laws, regulations, notices, and press releases) and company information (such as annual reports, press releases, industry studies, and media interviews). No proprietary industry information or other confidential information is included in this study.

There are known problems with data reliability in China that can make research there challenging. In the specific area of China's reported national energy statistics, researchers have studied trends in official data and systematic inaccuracies in an attempt to quantify and characterize their nature and magnitude.[6] Politically sensitive aspects of energy and climate change issues in China can lead to further problems with data reliability that could limit the accuracy of information collected and therefore the effectiveness of the research. Whenever possible I have attempted to validate data by cross-referencing data points across multiple sources and multiple interview subjects. When there are inherent uncertainties with some of the information presented, I have tried to characterize the nature of this uncertainty.

My research benefited greatly from the support and advice of many people. At UC Berkeley and Lawrence Berkeley National Laboratory, Dan Kammen, Jonathan Sinton, Ryan Wiser, David Fridley, Mark Levine, Jan Hamrin, Lynn Price, and Jiang Lin all provided valuable mentorship and feedback. Many faculty members at UC Berkeley played important roles in shaping my work, particularly Alex Farrell, You-Tien Hsing, Gene Rochlin, and Dick Norgaard, as well as my former classmates Jaquelin Cochran and Emily Yeh. My undergraduate advisors at Duke University, Marie Lynn

Miranda and Robert Keohane, provided me the initial encouragement to pursue research in this area.

In Beijing I am especially indebted to Zhang Xiliang of Tsinghua University, Wang Wanxing and Lu Hong of the Energy Foundation, and Wang Yang of BJADC for providing me with extensive information and numerous introductions over the past decade, as well as to many friends and colleagues from Tsinghua University and the Center for Renewable Energy Development of the Energy Research Institute. My research and experience in China would not have been nearly as enjoyable or productive without the valuable discussions I had with many China wind industry and climate policy observers, including Sebastian Meyer, Caitlin Pollack, Rembrandt Niessen, Carrie Lin, Chris Raczkowski, Alan Sides, Paulo Soares, Stephen Terry, Victoria Wang, Shi Pengfei, Yang Ailun, Yu Jie, Liang Zhipeng, Ren Dongming, Chen Rong, Steve Sawyer, Justin Wu, Debra Lew, Jean Ku, Eric Martinot, and David Kline. I am grateful to Stephanie Decker for introducing me to China and being a friendly face when I felt far from home.

This book would have never been completed without ongoing sources of inspiration that have led me to continue to pursue research in this area. For this I thank the many people I have had the opportunity to work with since I moved to Washington, particularly Elliot Diringer and Orville Schell. Jennifer Turner, Kenneth Lieberthal, and Bernice Lee have provided me with opportunities to present my research and receive valuable feedback, and Kelly Sims Gallagher has given me many helpful comments both on this manuscript and previous work. Georgetown University has offered a wonderful environment for research, and I want to thank my colleagues for their support and encouragement, particularly my brilliant and dedicated student research assistants Tian Tian and David Herron. My research and writing has benefited from financial support from the Energy Foundation China Sustainable Energy Program, the University of California, Berkeley, Georgetown University, the East West Center, and the Woodrow Wilson International Center for Scholars. Finally, I am ever grateful for my wonderful family and friends and their unwavering love and support.

Abbreviations

AMSC	American Superconductor Corporation
AWEA	American Wind Energy Association
CAS	Chinese Academy of Sciences
CCX	Chicago Climate Exchange
CDM	Clean Development Mechanism
CER	certified emissions reduction
CERC	Clean Energy Research Center
CF	capacity factor
CGC	China General Certification
CHP	combined heat and power
CNOOC	China National Offshore Oil Corporation
CNPC	China National Petroleum Corporation
CO_2	carbon dioxide
COP	Conference of the Parties (to the United Nations Framework Convention on Climate Change)
CSEP	China Sustainable Energy Program
CSP	concentrating solar power
CWEA	China Wind Energy Association
DOE	Department of Energy, United States

DRC	Development and Reform Commission
DWT	Danish Wind Technology
ECP	Energy Cooperation Program
EECI	Energy and Environment Cooperation Initiative
EPA	Environmental Protection Agency, United States
EU	European Union
EX-IM	Export-Import Bank of China
FIT	feed-in tariff
G-77	Group of 77
GBI	generation-based incentive
GDP	gross domestic product
GE	General Electric
GHG	greenhouse gas
GtC	gigatons carbon
GW	gigawatt (10^9 watts)
GWEC	Global Wind Energy Council
HEC	Harbin Electric Machinery Company
HPEC	Harbin Power Equipment Company
IEA	International Energy Agency
IFC	International Finance Corporation
IGCC	integrated gasification combined cycle
IMAR	Inner Mongolia Autonomous Region
IP	intellectual property
IPR	intellectual property rights
JCCT	US-China Joint Commission on Commerce and Trade
JUCCCE	Joint U.S.-China Collaboration on Clean Energy
JV	joint venture
KEMCO	Korea Energy Management Corporation
kW	kilowatt (10^3 watts)
kWh	kilowatt hour
LVRT	low-voltage ride-through technology
M&A	mergers and acquisitions
MEP	Ministry of Environmental Protection, People's Republic of China
MFA	Ministry of Foreign Affairs, People's Republic of China
MIIT	Ministry of Industry and Information Technology, People's Republic of China
MMS	mandatory market share
MNES	Ministry of Non-Conventional Energy Sources, India

MNRE	Ministry of New and Renewable Energy, India
MOEP	Ministry of Electric Power, People's Republic of China
MOF	Ministry of Finance, People's Republic of China
MOST	Ministry of Science and Technology, People's Republic of China
MOU	memorandum of understanding
MRV	measurement, reporting, and verification
Mtoe	million tons of oil equivalent
MW	megawatt (10^6 watts)
MWp	megawatt peak
NCCCC	China National Coordination Committee on Climate Change
NDRC	National Development and Reform Commission, People's Republic of China
NEA	National Energy Administration, People's Republic of China
NEC	National Energy Commission, People's Republic of China
NERC	National Engineering Research Center, People's Republic of China
NRDC	Natural Resources Defense Council
NREL	National Renewable Energy Laboratory, United States
OECD	Organization for Economic Co-operation and Development
OSC	Optimal Speed Controller
PPA	power purchase agreement
POSTECH	Pohang University of Science and Technology
PRC	People's Republic of China
PTC	production tax credit
PV	photovoltaics
R&D	research and development
RD&D	research, development, and demonstration
RMB	renminbi (official currency of the People's Republic of China)
S&ED	Strategic and Economic Dialogue
S&T	science and technology
S&T Agreement	US–China Agreement on Cooperation in Science and Technology
SCM	Subsidies and Countervailing Measures

SDPC	State Development and Planning Commission, People's Republic of China
SED	Strategic Economic Dialogue
SEPA	State Environmental Protection Agency, People's Republic of China
SERC	State Electricity Regulatory Commission, People's Republic of China
SETC	State Economic and Trade Commission, People's Republic of China
SOE	state-owned enterprise
SPC	State Planning Commission, People's Republic of China
SSTC	State Science and Technology Commission, People's Republic of China
TCE	tons of coal equivalent
TEDA	Tianjin Economic-Technological Development Area
TCX	Tianjin Climate Exchange
TWh	terawatt hour
TYF	Ten-Year Framework for Cooperation on Energy and Environment
UN	United Nations
UNFCCC	United Nations Framework Convention on Climate Change
U.S. OTA	United States Congress, Office of Technology Assessment
USTR	United States Trade Representative
USW	United Steelworkers
WTG	wind turbine generators
WTO	World Trade Organization
XWEC	Xinjiang Wind Energy Company

Chronology of Wind Power Development in China

1986   First imported utility-scale wind turbine is installed in China.
2000   First wind turbine manufactured by a Sino-foreign joint venture is installed in China.
2001   First wind turbine made with Chinese-owned intellectual property is installed in China.
2005   China National People's Congress passes the Renewable Energy Law of the People's Republic of China.
2006   China adopts its first national energy intensity target as part of the Eleventh Five-Year Plan.
2007   First megawatt-scale wind turbine made by a Chinese company is installed in China.
       China becomes largest national carbon dioxide emitter.
2008   First domestically produced multimegawatt wind turbine is installed in China.
       First Chinese-developed wind turbine is exported to the United States.
2009   China adopts first national feed-in tariff policy for wind.
       First multigigawatt wind base project begins construction in Gansu province.

China adopts its first national carbon target.
For first time China installs more wind power than any other country in a single year.

2010 China announces a pilot domestic carbon trading program.
China completes its first offshore wind farm.
China becomes world's largest national energy consumer.
China becomes country with most installed wind power capacity in the world.

Green Innovation in China

1
Green Innovation in China

China's energy system has significant global implications. Now the world's largest emitter of greenhouse gases, China has become the focus of scrutiny as the imperative to address climate change has gained international support. At the core of the climate change challenge is China's energy sector, the world's single largest source of emissions.

As the biggest coal-consuming and coal-producing nation in the world, China is perhaps an unlikely place to find a burgeoning wind power industry. Yet today China is the biggest wind power market in the world and builds almost all its wind turbines at home. China's wind power capacity has increased over a hundredfold in the past decade (from 344 MW in 2000 to 44,733 MW in 2010).[1] Just a decade ago the country had only a handful of wind turbines in operation—all imported from Europe and the United States.

The story of China's rapid rise in the global wind power industry provides valuable insight into the country's domestic energy strategies and global positioning, and into the evolving nature of technology transfer and diffusion around the world. While China's advances in clean energy may now be heightening global trade tensions, they also may in time help the world address the challenge of climate change. Alongside other green, low-emissions technologies, wind power offers an affordable option for reducing the carbon footprint of the electricity sector.[2] While there are some technical challenges to dramatically increased deployment of this technology, many of which China is experiencing firsthand, none is insurmountable.

This book examines how China developed a world-class domestic wind power industry. Elucidating China's innovative ability in a strategic, global, and green industry, it finds that China is now doing more than just replicating technologies innovated elsewhere and manufacturing them inexpensively; it is in fact investing in and succeeding at green innovation. China is beginning to serve as a center for global technological innovation—innovation achieved by both domestic and foreign firms—and green innovation from China could play a crucial role in the global transition to a low-carbon economy.

China's emergence as a green energy leader comes at a time when energy is at the top of most national security agendas, and when climate change is being linked to extreme weather events and disasters.[3] Given global concern about the impact of China's rapidly increasing energy needs on global supplies and its ability almost singlehandedly to change the global climate system, it is worthwhile to understand how and why Beijing has embraced green innovation. Renewable energy technology development is now positioned at the core of China's overarching national economic plan and supported by its industrial policy.[4] Cooperating with China to bring green innovations to market presents an opportunity for nations to build partnerships with an emerging global superpower.

China's wind industry provides a compelling example of technological leapfrogging. It also demonstrates new models of technology transfer, the movement of technology, intellectual property rights (IPR), and knowledge across borders.[5] Until recently the predominant model for foreign firms operating in China was to establish a jointly owned enterprise with a Chinese partner, creating a clear pathway for technology transfer to occur. But in the past decade a loosening of Chinese government restrictions on the ownership of foreign firms has resulted in far fewer joint ventures. Instead technology transfers increasingly are occurring between companies located in different countries with synergies along the technology development continuum; for example, German engineering design firms with little manufacturing ability and Chinese manufacturing dynamos with little innovative facility. China increasingly is not only a recipient of technology from industrialized nations but also the source of technology being transferred to other developing nations. While such technology transfers are commonly facilitated via licensing agreements, models of technology transfer in which the recipient of the transfer plays a far more active role, including mergers and acquisitions (M&A) and joint development, are prevalent.[6]

Access to networks for learning and innovation has also played a rarely examined but crucial role in the development of China's wind industry.[7] Beijing's widespread use of policies that all but forced localization, including a government-mandated domestic content requirement for wind turbines and other financial incentives for locally based manufacturers, resulted in the shift of overseas wind power technology manufacturing to China without necessarily transferring IPR to Chinese firms. But as foreign firms relocated to China, they helped create a learning network within the Chinese wind industry. This allowed for the transfer of technical know-how to Chinese firms through several indirect but highly effective channels, such as the movement of skilled personnel between foreign and Chinese firms—both across borders and within the Chinese wind power community. Chinese firms have relied heavily on this learning network within China, in contrast to other emerging Asian wind technology firms that accessed global knowledge by constructing their own network of facilities and partnerships overseas. Many emerging wind power firms, however, have gained access to technology transfers through common knowledge sources, creating a network of IP being transferred throughout Asia and beyond.

While many seek to understand whether China is innovating or merely imitating, China's capacity to innovate can be difficult to measure. The confluence of political, institutional, and economic characteristics that is unique to China[8] creates a system for promoting and rewarding innovation that is inherently different from that of market economies.[9] As a result, traditional science and technology (S&T) metrics, such as research funds invested or patents produced, are limited in their ability to capture the nature of the learning that is taking place in China's emerging technology industries. The company case studies presented in this volume begin to fill in the gaps of our understanding about innovation in China. Additionally, they offer more nuanced characterizations of learning than traditional metrics provide.

These findings build on existing theories of the role of the state in innovation[10] and promise to shape our broader understanding of China's S&T capability and its innovative "niche" in the global economy.[11]

The book begins with a close look at China's energy system in order to understand the challenges it faces in transitioning to a low-carbon economy. It examines how China's political leadership is addressing climate change, both at home, with domestic energy efficiency and renewable energy policies, and abroad, through participation in global climate change negotiations. It also analyzes how the emergence of domestic policy support for low-carbon development has contributed to a shift in China's international positioning (chapter 2).

The next several chapters focus on one crucial low-carbon technology being developed in China: large-scale wind turbines. Chapter 3 explores the global origins of wind power technology and looks at how China's national innovation system shaped the development of its wind industry. Expanding on the nature of China's innovation system for wind energy, chapter 4 examines how foreign-owned firms brought their wind power technologies to China, and how Chinese companies assimilated foreign technology over the past two decades through both successful and failed partnerships. Chapter 5 provides an in-depth look at one leading Chinese wind company, Goldwind, and how it has been able to develop its own wind power technology. It also examines other models for technology development used by Chinese wind turbine manufacturers, as well as the ongoing challenges faced by this emerging industry.

Looking beyond China, chapter 6 compares the development of its wind power industry with that of the two other Asian emerging economy wind industry leaders, India and South Korea, focusing on the technology acquisition strategies among firms and the common sources of knowledge across the three countries. A comparison across these three economies—all of which are rapidly emerging as central players in the global wind industry—illustrates how late-comers to the industry have been able to use different models of technology transfer to build up their own technological expertise with varied success.

The final chapter turns to the relationship at the center of global discussions about clean energy and climate change—that of the United States and China. These two countries not only are the two largest economies, energy consumers, and greenhouse gas emitters but also are currently the two largest renewable energy (and wind energy) markets in the world. They share many similarities with respect to renewable energy resource endowments and development goals and face similar technical, political, and social barriers to scaling up the use of renewable energy. As a result, U.S.-China cooperation in clean energy technologies may play a crucial role in solving our global climate change challenge and improving the U.S.-China bilateral relationship. Chapter 7 recommends areas for expanded cooperation.

By chronicling the history of wind power technology development around the world and China's rise within the industry, this book provides a lens into China's role in green innovation, Beijing's evolving political positioning around energy and climate change, and the outlook for a global transition to a low-carbon economy.

2
China's Energy and Climate Challenge

China has made unprecedented achievements in the past three decades.[1] Its economic growth rates have exceeded those of any other country in the world, enabling a tenfold increase in per capita income and lifting an estimated 400 million people out of poverty. While rising energy consumption fueled this rapid economic growth, the overall energy intensity of the economy has decreased, making China three times more energy efficient.

Yet these figures only begin to tell the story of China's rise and its impact on the environment. While economic output has increased rapidly, the share contributed by energy-intensive products has also increased, causing the economy to become less energy efficient for the first time in decades. Rapid economic growth has also come at a great toll to the local environment. China's environmental trends include deteriorating water quality, increasing water scarcity, escalating air pollution, increasing land degradation, and growing desertification. These environmental challenges affect the health and welfare of the current population, threaten the prospects for future generations, and challenge China's ability to sustain economic growth rates in coming decades. Most sources of pollution in China can be traced back to energy use, and particularly to the country's reliance on coal at the core of its energy system.

China must now decide whether it can continue to rely on its economic growth strategy of the past, which has come at the expense of its own environment. As the country has grown, its environmental challenges are no longer localized—they have global reach. While most greenhouse gas (GHG) emissions do not have direct local environmental or health impacts, the sources of these emissions—such as power plants, vehicles, and industrial facilities—also emit the air pollutants that cause more localized impacts. As a result, reducing the emissions of greenhouse gases can have both global and local benefits. As the impacts of rising global GHG emissions are more comprehensively understood, it is becoming increasingly evident that climate change will exacerbate many existing environmental problems and bring some new ones as well.

The Climate Change Impacts Facing China

We now know that human activity is altering the climate. Driven primarily by a century and a half of fossil fuel combustion, carbon dioxide (CO_2) concentrations in the atmosphere reached 390 parts per million in 2011, 39 percent higher than preindustrial levels.[2] Average global temperatures have risen by 0.76 degrees Celsius since the late 1800s, and the effects are evident in extreme weather events, changed weather patterns, floods, droughts, glacial and Arctic ice melt, rising sea levels, and reduced biodiversity.[3] Average temperatures are projected to increase by another three degrees Celsius upon a doubling of carbon dioxide concentrations.[4] In China the observed data show that the nationwide mean surface temperature has risen by 1.38 degrees Celsius over the past fifty years and is projected to further increase by 3–4 degrees Celsius by the end of this century.[5] Even if all emissions were to stop today, the greenhouse gases already accumulated in the atmosphere will remain there for decades to come, resulting in more warming and stronger climate impacts.

China's first National Climate Change Assessment Report, compiled by leading climate change scientists, stated: "It is very likely that future climate change would cause significant adverse impacts on the ecosystems, agriculture, water resources, and coastal zones in China."[6] Impacts already being observed in China include extended drought in the North, extreme weather events and flooding in the South, glacial melting in the Himalayas, declining crop yields, and rising seas along heavily populated coastlines.[7] China's leaders increasingly acknowledge such reports. For example,

China's former special ambassador for climate change, Yu Qingtai, stated that "climate change . . . is in fact a comprehensive question with scientific, environmental and development implications and involves the security of agriculture and food, water resource, energy, ecology and public health and economic competitiveness," and "if the climate changes dramatically, the survival of mankind and the future of earth might be impacted."[8]

Science is ever more clearly showing that climate change is expected to wreak havoc on China. For example, studies predict that precipitation may decline by as much as 30 percent in the Huai, Liao, and Hai River regions in the second half of this century.[9] Climate change could decrease river runoff in northern China, where water scarcity is already a problem, and increase it in southern China, where flooding and heavy rains are already a problem. Declining runoff to the six largest rivers in China has been observed since the 1950s, with some rivers even facing intermittent flow.[10] Decreased precipitation is also tied to increased desertification. In China it is estimated that about 150 square kilometers of cultivated land is lost annually as a result of desertification.[11]

Climate change is expected to increase the frequency and severity of storm surges, droughts, and other extreme climate events, particularly in coastal areas. Flooding events previously expected to occur once in fifty years are expected to occur once in five to twenty years by 2050.[12] The Yellow River Delta, the Yangtze River Delta, and the Pearl River Delta currently are the most vulnerable coastal regions in China.

Since the 1950s sea-level rise has been observed along China's coastline at the rate of 0.0025 meters per year; 0.0039 meters per year along the southern China coastline.[13] Estimates of future sea-level rise along China's coastline range between 0.08 and 0.13 meters by 2030, and between 0.4 and 1 meters by 2050.[14] Higher sea levels magnify the possibility of flooding and intensified storm surges and exacerbate coastal erosion and saltwater intrusion. A 1-meter rise in sea level would submerge an area the size of Portugal along China's eastern seaboard; the majority of Shanghai—China's largest city—is less than 2 meters above sea level.[15] China's twelve coastal provinces contain about 43 percent of its population and contribute about 65 percent of its GDP.[16]

The impact of climate change on the glaciers in China's Tibetan Plateau will have severe repercussions for China's lakes and river systems. The total area of China's western glaciers is projected to decrease 27.2 percent by 2050; with a measured retreat from 1951 to 2009 of 7.8 meters per year.[17] Over the same period glacial thawing is projected to increase water discharge by 20–30 percent per year until water levels peak between 2030 and 2050, after which time water discharge will then decline as the glaciers disappear. The mountain and highland lakes that rely on inland glaciers for recharge, such as the lakes on Tibetan and Pamir Plateaus, could initially enlarge as a result of glacial melting but will eventually shrink as the glaciers are reduced over time. As a result, the Yellow and Yangtze Rivers, which support the richest agricultural regions of the country and derive much of their water from Tibetan glaciers, will initially experience floods as the glaciers melt, and then drought once the glacial runoff is gone.

Climate change could also increase the frequency and intensity of heat waves, resulting in higher mortality and morbidity from heat-related weather events. Hotter temperatures over time could increase the occurrence and the transmission of infectious diseases, including malaria and dengue fever. Malaria is still one of the most significant vector-borne diseases in parts of southern China, and climate change could expand its geographical range into the temperate and arid parts of the country.[18]

As a result of all these trends, China's agricultural system, trade system, economic development engines, and human livelihood will all face new threats in a warming world.

China's Emissions Challenge

Although the attention given to climate change by China's leadership has increased in recent years, climate change has not surpassed economic development as a national policy priority. Yet the causes of climate change—namely, GHG emissions from fossil fuels and land use—are inherently linked to economic development in the Chinese context. As the vast majority of China's emissions stem from energy consumption and industrial activity, China's climate change strategy must be examined in the context of its overall energy development strategy, which is driven by its overall economic development goals. Continued growth in the prosperity of the population is viewed as fundamental to maintaining political stability, and progress to date in this regard has been impressive. Although China quadrupled its GDP between 1980 and 2000, it did so while merely doubling the amount of energy it consumed over that period, marking a dramatic achievement in energy-intensity gains not paralleled in any other country at a similar stage of industrialization. This allowed China's energy intensity (ratio of energy consumption to GDP) to decline. Without this reduction in

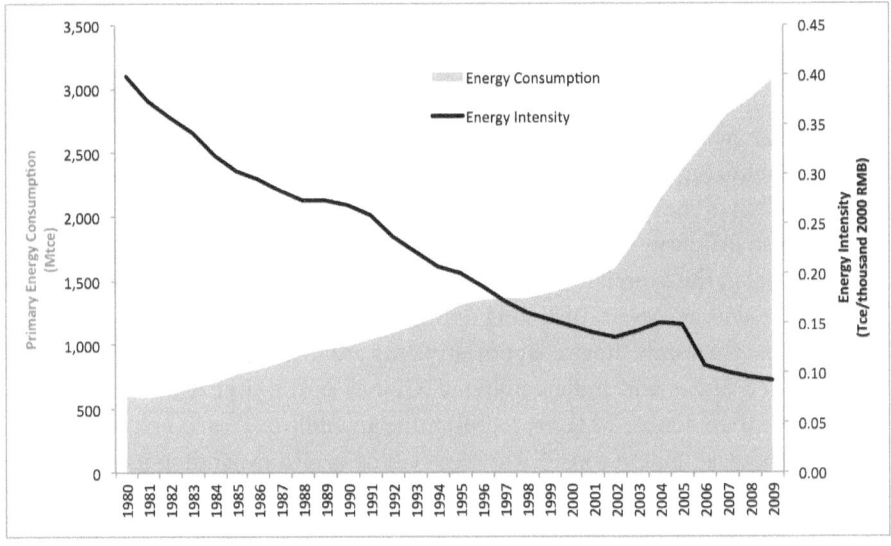

FIGURE 2.1 Energy Consumption and Energy Intensity Trends in China
National Bureau of Statistics (2010); China Energy Group (2008).

energy intensity, China would have used more than three times the energy that it did during this period to produce the same level of economic output. This energy savings meant less reliance on imported energy resources, and less carbon dioxide emissions, over this time period.

Between 2002 and 2005, however, this trend reversed, and energy growth surpassed economic growth for the first time in decades. As its energy demand has boomed, China's emissions have soared, and the country became the world's largest annual emitter of CO_2, surpassing the United States, around 2007. As recently as 2004, experts were projecting that China's CO_2 emissions would not surpass those of the United States until after 2030.[19] In 2006 this date was revised to 2013,[20] but new estimates released in 2007 demonstrated that China had already reached the number one spot, catching many (including the Chinese government) off guard.[21] Looking ahead, recent projections put China's emissions in 2030 in the range of 400 to 600 percent above 1990 levels.[22] Globally this translates to almost 50 percent of all new energy-related CO_2 emissions between now and 2030. If China's emissions were to continue to grow at the rate of 8 percent per year (the average annual growth rate between 2005 and 2010), by 2030 it would be emitting as much CO_2 as the entire world is today. In contrast, U.S. emissions are expected to increase by 22 percent above 1990 levels by 2030.[23] In historical terms, the United States is by far the largest contributor to the greenhouse gases now in the atmosphere, responsible for 27 percent of energy-related CO_2 emissions since 1900, while China accounts for only about 10 percent of these cumulative emissions.[24]

China's increase in energy-related emissions in recent years has been driven primarily by industrial energy use, fueled by an increased percentage of coal in the overall fuel mix. China relies on coal for more than two-thirds of its energy needs, including approximately 80 percent of its electricity needs. Currently more coal power plants are installed in China than in the United States and India combined. China's coal power use is expected to more than double by 2030, representing an additional carbon commitment of about 86 billion tons.[25] Although China is also expanding its utilization of nuclear power and nonhydroelectric renewables, these sources comprise 2 percent and 0.8 percent of China's electricity generation, respectively. China's share of electricity generated by renewable energy rises to 16 percent if hydropower is included.[26]

China's overall economic development statistics reveal that, despite the emergence of modern cities and a growing middle class, China is still largely a developing country. Although rapid economic growth has made China the second-largest economy in the world, its GDP per capita is still below the world average. While GDP per capita in rural China lags behind that of urban areas, 2011 marked the first time in history that more than half of China's population lives in urban China.[27] The gap between the best avail-

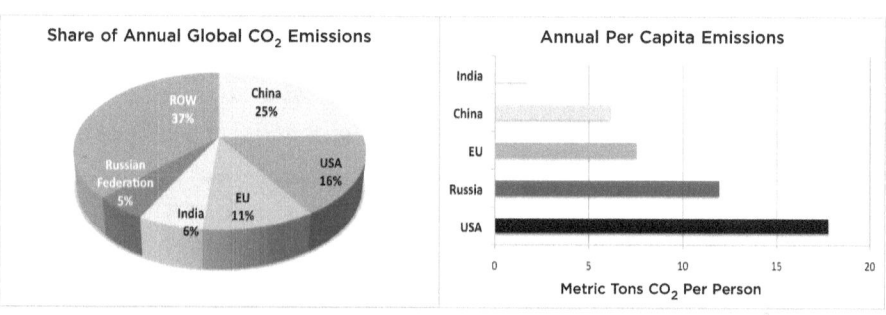

FIGURE 2.2 China's Greenhouse Gas Footprint
Data are for 2010. Includes CO2 emissions from fossil fuels only.

CDIAC (2011); World Bank (2011).

able technologies worldwide and what exists in China is still large, although advanced energy technology is increasingly available and in many cases being developed indigenously. China's per capita GHG emissions are below the world average and approximately one-third those of the United States.

All these factors shape the climate challenge faced by China's leadership. It is increasingly difficult for the country to rein in its GHG emissions growth as investment surges continue in heavy industry. Changing China's emissions trajectory will require either a substantial shift away from coal or massive investments in capturing the CO_2 emissions from coal-based energy sources. Simultaneously China must further increase the efficiency with which it uses energy resources to minimize the environmental impacts of meeting the future economic development needs of its population.

China's power sector is responsible for about one-half of energy-related CO_2 emissions, and the vast majority (about 98 percent) of power-sector CO_2 emissions come from coal use.[28] Given China's substantial domestic coal reserves and its heavy investment in coal-fired power plants over the past few decades, coal will likely remain an inescapable foundation of its economy for years to come. To render coal a climate-friendly energy source, however, will require significant advances and sustained investment in new technologies to burn it more efficiently as well as to capture and sequester the resulting GHG emissions.

China's increase in energy-related emissions in the past few years has been driven primarily by demand from its booming industrial sector. Industry consumes about 70 percent of China's energy, and China's industrial base supplies much of the world. For example, China in 2010 produced about 44 percent of the world's steel and 66 percent of aluminum.[29] Studies have estimated the CO_2 emissions embedded in China's domestic production of goods for export represent about 30 percent of the country's total annual CO_2 emissions.[30]

Climate Politics in China

China's energy challenges are shaping the way its leadership is approaching climate mitigation at the domestic level, as well as its positioning in international climate negotiations. In addition, as the possible impacts of climate change are increasingly understood, they are contributing to growing concern across the government about how to address the problem. Both the country's energy challenges and its susceptibility to climate impacts may serve as the driving forces behind any future climate change mitigation strategy the country adopts.

A look at the institutions that have been responsible for climate change policy is one way to understand how the government has approached this issue over time. Starting in the 1980s, China treated climate change as a primarily scientific issue and gave the State Meteorological Administration responsibility for advising the government on policy options in international climate negotiations, including those surrounding the 1992 United Nations Framework Convention on Climate Change (UNFCCC). As political awareness and sensitivity surrounding climate change increased in the late 1990s, this role shifted to the more powerful State Development and Planning Commission (formerly the State Planning Commission), which has since evolved into the National Development and Reform Commission (NDRC). The move indicated a shift in the relative value given to the issue, as well as perhaps a shift in perspective, where climate change was seen not only as a scientific issue but increasingly as an issue of economic development.[31] The NDRC also serves as the primary energy policy decision-making authority in China, and this move may have reflected the clear need for climate priorities to be coordinated better with energy decisions. The NDRC was also home to the former National Coordination Committee on Climate Change (NCCCC), which oversaw climate activities across the NDRC, the Ministry of Foreign Affairs (MFA), the Ministry of Science and Technology (MOST), and the former State Environmental Protection Administration (SEPA).

Further institutional change came with the release of China's national climate change plan in June 2007, announcing a high-level leading group on climate change chaired by Premier Wen Jiabao and reporting to the State Council. The National Leading Committee on Climate Change replaced the NCCCC and now included representatives from twenty government agencies. Subsequently the Foreign Ministry announced that it had also established a leading group in charge of international work on climate change, headed by Foreign Minister Yang Jiechi. Later that year Ambassador Yu Qingtai was appointed to be China's new special representative of the Foreign Ministry for climate change negotiations. The role of the special representative is to help implement China's domestic action plan to respond to climate change and to demonstrate "the government's active participation in international cooperation on responding to climate change."[32] In addition, former MFA negotiator Su Wei was appointed to be the director general of the NDRC's Department of Climate Change.

The NDRC and MFA are currently responsible for formulating China's international negotiation positions, although the Chinese delegation at the UN climate negotiations includes representatives from many other agencies and organizations, including MOST, the Ministry of Finance (MOF), the Ministry of Environmental Protection (MEP), as well as representatives from the forestry, transportation, meteorological, and agriculture agencies. The delegation also frequently includes academic experts from a variety of research institutions, including the Chinese Academy of Sciences, the Chinese Academy of Social Sciences, Tsinghua University, Renmin University, and the Energy Research Institute.

Leading Chinese research organizations that often provide analytical input to shape government policy decisions have significantly scaled up their work on climate change over the past decade.[33] The government released its first National Assessment Report on Climate Change in late 2006, conducted as a collaborative effort among more than twenty government departments and taking four years to complete.[34] Structured similarly to the Intergovernmental Panel on Climate Change reports, the Chinese assessment consists of three parts: climate change history and trends, impacts and adaptation, and mitigation and socioeconomic evaluation.

China released its much-anticipated National Climate Change Program report on June 4, 2007.[35] This plan provided a comprehensive synthesis of the current Chinese policies that were serving to moderate GHG emissions growth and to help the country adapt to climate impacts. The majority of the policies and programs mentioned in the plan were not climate change policies per se, but policies implemented throughout the economy, particularly in the energy sector, that have the effect of reducing GHG emissions. Many of these policies have been enacted to help the country meet its broader economic development strategies and, if implemented effectively, will also serve to mitigate emissions.

China's Second National Communication on Climate Change is being prepared for submission to the UNFCCC. The first National Communication was submitted over a decade ago. As of late 2011 the second communication was still under review, with NDRC reporting that it will be released in 2012.

THE G-77 AND CHINA

Historically China's position in the international climate negotiations has rarely deviated from that of the rest of the developing world, as collectively articulated by the Group of 77 (G-77), a group of 131 (formerly 77) developing countries. Developing countries have used their solidarity strategically to influence the climate change negotiations, despite the growing economic differentiation and often disparate climate policy interests within the developing world. Aware of their limited weight if acting in isolation, developing countries attempt to build common positions in the coalition of the G-77, the largest intergovernmental organization of developing states in the United Nations.[36] The G-77 provides a means for these countries to articulate and promote their collective economic interests and enhance their joint negotiating capacity on all major issues within the UN system.

China has frequently associated itself with the G-77 despite not having the problem of limited weight in acting alone. It has used its alliance with the G-77 block as protection against being singled out as the largest developing-country emitter. Yet its size also allows it to take a leadership role in formulating the positions of the G-77. China has a hand in crafting the group's position while ensuring that a large contingent of countries will stand at its side when it is presented before the world. The G-77 has consistently emphasized the historical responsibility that the industrialized world brings to the climate change problem and the disparity between per capita emissions that persists between the developed and developing world, resisting any commitments to reduce the group's own GHG emissions. In recent years China's alliance with the G-77 has not waned. China has, however, taken a far more proactive role in articulating its own national climate change plans to the international community.

In June 2005 then SEPA director Xie Zhenhua, now a vice minister of the NDRC and frequent head of the Chinese delegation to the UNFCCC, stated that he hoped "that some countries would, according to the obligations which are provided for in the Kyoto Protocol, implement in a substantive way their obligations and take up their commitments" and that, "on the Chinese side, the Chinese government would make its own decision after making some assessments of the implementation by other countries."[37] In this statement Xie was signaling that China was waiting to see whether the developed countries would follow through on their own UNFCCC obligations before it announced its own mitigation commitments. This position was subsequently reinforced by Chinese foreign minister Yang Jiechi, who said that developed countries should "continue to take the lead in reducing emissions after 2012."[38] In November 2009, however, the Chinese leadership announced its intention to implement its first-ever domestic carbon-intensity target (based on the ratio of carbon emissions to GDP), consisting of a 40 to 45 percent reduction below 2005 levels by 2020.[39] This target came within hours of President Obama's announcement that the United States would reduce its carbon emissions "in the range of 17 percent" from 2005 levels by 2020, and that the president himself would attend the UN international climate change negotiations in Copenhagen.[40] These closely timed announcements illustrate that China did indeed continue its wait-and-see approach, in that it waited for the U.S. announcement before making its own. There is little doubt, however, that the Chinese target had been under development for quite a while prior to the announcement.[41]

Countries within the G-77 are beginning to diverge somewhat in their positions, which could leave China in a more isolated negotiating position. Some countries with tropical forests, including Brazil and a coalition of thirty-two rain-forest countries such as Costa Rica and Papua New Guinea, stated a willingness to take on voluntary avoided-deforestation targets in return for compensation in the months leading up to the Bali negotiations.[42] Historically the G-77 position had not included voluntary international targets of any form. Another subgroup of G-77 member countries, referred to as the "BASIC" negotiating block, emerged around the time of the Copenhagen negotiations. BASIC, the acronym given to the large, emerging economy emitters of Brazil, South Africa, India and China, has served primarily as a platform for mutual assurance, ensuring that no single one of these countries so crucial to the overall politics of the negotiations steps too far out of line with the negotiating positions of the others.

China historically has been consistent in its position that, as a developing country, it will not take on any binding international commitments to reduce its GHG emissions, although it is willing to take voluntary actions.[43] Some of China's hesitancy to make international commitments stems from reasonable concerns about energy data quality and transparency. In developing countries, where resource constraints result in limited data quality, inventories of national greenhouse gas emissions are notoriously inexact.[44] Having in place a national emissions inventory system will likely be a crucial step in enabling the adoption and enforcement of any binding emissions reduction policies, whether enacted nationally or internationally. At the Conference of the Parties to the UNFCCC (COP 16) in Cancun, developing parties agreed to increase the frequency of their national communications to at least every four years, and of national emissions inventories to every two years, as well as agreeing to the international measurement, reporting, and verification of both inventories and mitigation actions.[45] China has not publicly released any official carbon dioxide emissions inventories that date past 1994, although an updated inventory is expected to be part of its next national communication.

THE CLEAN DEVELOPMENT MECHANISM

China has ratified the primary international accords on climate change—the UNFCCC and the Kyoto Protocol—but as a developing country it has no binding emissions limits under either accord.[46] It has, however, been an active participant in the Clean Development Mechanism (CDM) established under the protocol, which grants certified emissions reduction (CER) credits for verified reductions in developing countries, which developing countries can then use to help meet their Kyoto targets.

The Chinese government initially approached the CDM somewhat cautiously, taking a more involved role in the project approval process than did other developing countries, which resulted in a relatively late start to the Chinese carbon market. Although CDM projects became eligible for crediting in 2000 (five years before the Kyoto Protocol entered into force), China did not ratify the treaty until August 2002, its designated national authority overseeing CDM projects was not established until June 2004, and the State Council did not adopt rules for the management of CDM projects until October 2005.[47] China was initially skeptical about the introduction of the Kyoto mechanisms under the UNFCCC, not only viewing the CDM as a way for developed countries to avoid their own responsibilities to reduce emissions but also expressing concern about the potential for foreign exploitation of rights to ownership of emissions credits.[48] China has long had protectionist tendencies and resisted foreign involvement in various sectors and activities, particularly industries deemed to have an impact on national economic security.

China's position toward the CDM changed dramatically over time, however, as the country began to realize the economic and political benefits that the CDM could provide. The CDM has become a vehicle for China to help stimulate investment in projects that mitigate GHG emissions and to help cover the incremental cost of higher-efficiency or low-carbon technology. Another benefit of China's leadership in the CDM is that it provides a way in which China can be viewed internationally as being proactive on the climate issue. Now the world leader in terms of CDM-induced greenhouse gas reduction credits in the CDM pipeline, China has learned how to use the CDM to its advantage.

Concerns about foreign involvement in Chinese CDM projects have not waned. The rules governing the CDM in China are viewed as "carefully crafted . . . to heavily favor Chinese interests and control, and to ensure Chinese 'resources' are protected," and have become a cause for complaint by many potential foreign investors—particularly the stipulation that only majority-owned Chinese enterprises may serve as project owners.[49] Despite these restrictions and complaints, China has emerged as the leading CDM host country, with 64 percent of expected annual CERs registered as of January 20, 2012.[50] International concern over China's potential abuse of the CDM has likely stemmed from its leading role, although many concerns are likely legitimate.[51] For example, many of China's wind energy projects that had applied for CDM credits were denied when the CDM executive board raised questions over whether the projects could be considered real (or "additional") emissions reductions that would not have occurred in the absence of CDM credits.[52]

The future of the CDM is linked to ongoing negotiations about whether to renew the Kyoto Protocol after its first commitment period expires in 2012. Despite international agreement in December 2011 at COP 17 in Durban to extend the Kyoto Protocol, China's ability to remain the largest CDM market is questionable. As China's emissions continue to soar, industrialized countries will increasingly expect it to unilaterally take more mitigation actions and will likely be less willing to help subsidize these actions through the CDM.

A NEW NEGOTIATING POSITION?

The December 2011 climate negotiations in Durban, South Africa, saw two potentially significant changes to long-held elements of China's negotiating position. The first was China's willingness to adopt legally binding commitments, rather than just voluntary commitments, as part of a future climate change agreement. This was reflected in China's support of the Durban Platform in which UNFCCC parties agreed "to launch a process to develop a protocol, another legal instrument or an agreed outcome with legal force under the Convention applicable to all Parties."[53] The second was a new openness toward discussing absolute GHG emissions targets, rather than just intensity targets. These changes are very likely a result of the programs that have been implemented domestically in the wake of China's carbon-intensity target and since the Copenhagen negotiations to both measure and monitor domestic emissions and to implement domestic carbon trading programs. Since early 2011 China's National Energy Administration (NEA) has discussed implementing a cap on total domestic energy consumption by the year 2015, which could certainly pave the way for an absolute emissions target as well.[54]

Another notable change in China's approach to the international climate negotiations that was on display at the Durban climate talks was a new level of openness, illustrated through a series of public events hosted by the Chinese delegation and featuring Chinese negotiators, researchers, and NGOs that far outnumbered those convened by China in previous negotiations. The Chinese delegation rarely spoke about domestic energy and climate change initiatives at the UNFCCC meetings until 2007, and the first official UNFCCC side event hosted by a Chinese NGO took place only as recently as 2009. This shift toward more proactive engagement, both with other country delegations and with the civil society community, points to an increased willingness on the part of China to both explain its domestic climate and energy challenges and articulate its accomplishments.

Clean Energy and Climate Action in China

China has begun to implement many national policies and programs to address its increasing GHG emissions and reliance on fossil fuels. The energy-intensity increases between 2002 and 2005 served as the impetus for Beijing to launch a variety of energy-efficiency programs, including a nationwide energy-intensity target. Promoting both energy efficiency and renewable energy has become a fundamental part of both China's national development strategy and its climate change mitigation strategy.

ENERGY-EFFICIENCY PROGRAMS AND INTENSITY TARGETS

A suite of energy-efficiency and industrial-restructuring programs have aimed to drive down China's energy intensity. One of the core elements of the Eleventh Five-Year Plan period, spanning from 2006 to 2010, was to lower national energy intensity by 20 percent. China's Top 1,000 Program has helped to cut energy use among the biggest energy-consuming enterprises (representing 33 percent of China's overall energy consumption, 47 percent of industrial energy consumption, and 43 percent of carbon dioxide emissions).[55] The 10 Key Projects Program provides financial support to companies that implement energy-efficient technology. In addition, many inefficient power and industrial plants have been targeted for closure. Such closures helped contribute to the decline in energy intensity experienced during the Eleventh Five-Year Plan period, with a reported 72.1 GW of thermal capacity closed between 2006 and 2010—equivalent to 16 percent of the total capacity added over that period.[56] An additional 8 GW of coal plants were targeted for shutdown in 2011, with further closures no doubt on tap over the next five years.

The government has also strengthened local accountability for meeting targets by intensifying oversight and inspection. China's 2007 Energy Conservation Law required local governments to collect and report energy statistics and required companies to measure and record energy use. Each province and provincial-level city was required to help meet China's goal to cut energy intensity by 20 percent and was assigned its own target, ranging from 12 to 30 percent. Governors and mayors were held accountable for meeting their targets, with penalties ranging from forced public apologies to removal from their positions. In addition, experts from Beijing conducted annual site visits to facilities in each province to assess their progress.

The 10 Key Projects Program required selected enterprises to undergo comprehensive energy audits and offered financial rewards based on actual energy saved. The audits followed detailed government monitoring guidelines and had to be independently validated. There have also been increases in staffing and funding in key government agencies that monitor energy statistics and implement energy-efficiency programs. In 2008 alone China reportedly allocated RMB 14.8 billion of treasury bonds and central budget as well as RMB 27 billion of governmental fiscal support to energy-saving projects and emission cuts.[57]

While industrial energy consumption in China is increasing, it has been offset considerably by energy-efficiency improvements. Recent surges in energy consumption by heavy industry have caused the government to implement measures to discourage growth in energy-intensive industries compared with sectors that are less energy intensive. Beginning in November 2006 the Ministry of Finance increased export taxes on energy-intensive industries. Simultaneously, import tariffs on twenty-six energy and resource products, including coal, petroleum, aluminum, and other mineral resources, were reduced. Whereas the increased export tariffs were meant to discourage relocation of energy-intensive industries to China for export markets, the reduced import tariffs were meant to promote the utilization of energy-intensive products produced elsewhere.

Programs included in the Twelfth Five-Year Plan build directly on the Eleventh Five-Year Plan's energy-intensity target and its associated programs, including a new target to reduce energy intensity by an additional 16 percent by 2015.[58] While this may seem less ambitious than the 20 percent reduction targeted in the Eleventh Five-Year Plan, it likely represents a much more substantial challenge. The largest and least efficient enterprises have already undertaken efficiency improvements, leaving smaller, more efficient plants to be targeted in this second round. Also under preparation is a Top 10,000 Program, modeled after the Top 1,000 Program and adding an order of magnitude of companies to the mix. But as the number of plants grows, so do the challenges of collecting accurate data and enforcing targets. While the country likely fell just short of meeting the energy-intensity target of 20 percent in the Eleventh Five-Year Plan (the government reported a 19.1 percent decline), there is no doubt that much was learned though efforts to improve efficiency nationwide.

While estimates have been made of the potential carbon emissions savings that could accompany the 20 percent target, China did not put forth any targets that explicitly quantified its carbon emissions until late 2009.[59] In the lead-up to the Copenhagen climate negotiations in the fall of 2009, the Chinese government pledged a 40–45 percent reduction in national carbon intensity from 2005 levels by 2020.[60] To achieve this 2020 target, the Twelfth Five-Year Plan set an interim target of reducing carbon intensity 17 percent from 2010 levels by 2015. Whether this target will result in a deviation from expected carbon emissions over this time period depends on the corresponding GDP growth, but many studies have found that the target will be challenging to achieve without additional, aggressive policies to promote low-carbon energy development.[61]

Also promised in the Twelfth Five-Year Plan is an improved system for monitoring GHG emissions, which will be needed to assess compliance with the carbon-intensity target and to prepare the national GHG inventories that, under the 2010 Cancun Agreements, are to be reported more frequently to the UNFCCC and undergo international assessment. Apart from the policies contained in the Twelfth Five-Year Plan, there have been numerous reports of new policies that may be introduced to help China achieve these carbon-intensity reductions, including carbon-trading programs and carbon taxes.

A handful of provinces have announced the beginnings of pilot carbon-trading schemes. In July 2010 the NDRC announced the selection of pilot low-carbon provinces and cities, including the provinces of Guangdong, Liaoning, Hubei, Shaanxi, and Yunnan and the cities of Tianjin, Chongqing, Shenzhen, Xiamen, Hangzhou, Nanchang, Guiyang, and Baoding. The mandate for these low-carbon pilot cities and provinces was somewhat vague, however, and the program fell short of mandating the implementation of carbon trading programs. An October 2011 NDRC notice included the announcement of the provinces and municipalities selected to pilot a cap-and-trade program for carbon dioxide: Guangdong, Hubei, Beijing, Tianjin, Shanghai, Chongqing, and Shenzhen.[62] The Tianjin Climate Exchange, partially owned by the founders of the Chicago Climate Exchange, is positioning itself to be the clearinghouse for any future carbon-trading program, although several other exchanges have been established around the country.[63]

Implementation of a carbon-trading scheme in China, even on a small scale or on a pilot basis, will not be without significant challenges. Both domestic and foreign-owned enterprises operating in China have already raised concerns about how the regulation could affect their bottom lines. But the key challenge is likely technical, resulting from the minimal capacity currently in place to measure and monitor carbon emissions in China. China's first-ever carbon target will require an important change in the country's data collection and transparency practices. The government has announced that the national-level carbon-intensity target will be allocated across each province, municipality, and economic sector and enforced with new monitoring rules.[64] Measuring and enforcing these subnational targets will require periodic inventories of GHG emissions and a significantly improved statistical monitoring and assessment system to ensure that goals are met; Beijing has indicated that these systems are indeed under development.[65]

There is no question that China's announcement of its first carbon intensity target and pilot cap-and-trade programs represents a monumental change in the country's approach to global climate change. It is also important to recognize, however, that even with this target in place, absolute emissions could continue to increase rapidly. A meaningful reduction of emissions encouraged by a carbon-intensity target hinges on future economic growth rates and the evolving structure of the Chinese economy, as well as on the types of energy resources utilized and the deployment rates of various technologies, among other factors. Carbon intensity, like energy intensity, has declined substantially over the past two decades. Between 1990 and 2005 China reduced its carbon intensity by 44 percent. It is also projected to reduce its carbon intensity 46 percent from 2005 levels by 2020, even while its emissions increase by 73 percent during this same period.[66] This has sparked much debate over whether this domestic policy target is sufficient based on China's role in the global climate challenge.

If implemented effectively, however, a carbon-intensity target will not only accelerate the energy-efficiency improvements already taking place in response to the energy-intensity target but will also further promote the development of low-carbon energy sources like nuclear power, hydropower, and renewables. In addition, implementing a carbon policy through a domestic carbon-trading program, or through other financial incentives like a carbon tax, would be a significant step toward implementing a comprehensive climate policy in China, complementing ongoing efforts to improve energy efficiency and promote low-carbon energy sources. And as recent government reports have mentioned, these programs may well pave the way for an absolute target on energy or on carbon emissions in the coming years.

RENEWABLE ENERGY

One key element of China's energy strategy, as well as its low-carbon development strategy, is the promotion of renewable energy technologies. This effort was kick-started with the passage of the Renewable Energy Law of the People's Republic of China, which became effective on January 1, 2006.[67] The law created a framework for regulating renewable energy and was hailed at the time as a breakthrough in the development of renewable energy in China. It established a national renewable-energy target, a mandatory connection and purchase policy, a feed-in tariff system, and a cost-sharing mechanism, including a special fund for renewable-energy development.[68]

The government has also set a target of producing 15 percent of its primary energy from nonfossil sources by 2020, which includes renewable energy and nuclear power. Although increases in wind power in particular have been impressive in recent years, this energy source is still dwarfed by large-scale hydropower. Hydropower capacity is projected to more than double by 2020, requiring a new dam roughly the size of the Three Gorges Project to be constructed every two years between now and then. Power companies have mandatory renewable-energy targets for both their generation portfolios and annual electricity production that they must meet. In December 2009 amendments to the Renewable Energy Law were passed, further strengthening the process through which renewable-electricity projects are connected to the grid and dispatched efficiently.[69]

Policies to promote renewable energy also include mandates and incentives to support the development of domestic technologies and industries—for instance, by requiring the use of domestically manufactured components. China invested fifty billion dollars in 2010 in renewable-energy development, far more than any other country in the world.[70] It also reached the top of Ernst and Young's renewable energy "country attractiveness" index, which examines the domestic environment for investment in renewables.[71]

The Twelfth Five-Year Plan includes a target to increase nonfossil energy sources (including hydro, nuclear, and renewable energy) to 11.4 percent of total energy use (up from 8.3 percent in 2010).[72] It also includes many new industrial policies to support clean energy industries and related technologies. Industries targeted include the nuclear, solar, wind, and biomass energy technology industries, as well as hybrid and electric vehicles and energy savings and environmental protection technology industries.[73] These "strategic and emerging" industries are being promoted to replace the "old" strategic industries such as coal and telecom (often referred to as China's pillar industries), which are largely state-owned and have long benefited from government support. (Over 70 percent of the assets and profits of state-owned enterprises, or SOEs, are concentrated in the pillar industries.[74]) This move to rebrand China's strategic industries likely signals the start of

TABLE 2.1 China's New Strategic and Emerging Industries in the Twelfth Five-Year Plan

Old Pillar Industries	New Strategic and Emerging Industries
National defense	Energy saving and environmental protection
Telecom	Next generation information technology
Electricity	Biotechnology
Oil	High-end manufacturing (e.g., aeronautics, high-speed rail)
Coal	New energy (nuclear, solar, wind, biomass)
Airlines	New materials (special and high-performance composites)
Marine shipping	Clean-energy vehicles (PHEVs* and electric cars)

Sources: Government of the PRC, "Decision on Speeding Up the Cultivation and Development of Emerging Strategic Industries"; HSBC, "China's Next 5-Year Plan: What It Means for Equity Markets," HSBC, October 2010.

*Plug-in hybrid electric vehicles

a new wave of industrial policy support for the new strategic industries, which may include access to dedicated state industrial funds, increased access to private capital, or industrial policy support through access to preferential loans or R&D funds. The State Council intends to guide RMB 10 trillion of domestic and foreign investment into these sectors by the end of the Twelfth Five-Year Plan. Other Twelfth Five-Year Plan targets encourage increased innovative activity, including a target for R&D expenditure to account for 2.2 percent of GDP, and for 3.3 patents per 10,000 people. During the Eleventh Five-Year Plan period, an estimated 15.3 percent of government stimulus funding was directed toward innovation, energy conservation, ecological improvements, and industrial restructuring.[75]

China possesses the ingenuity and institutional capital to meet certain elements of the climate challenge better than other elements. While it still faces significant challenges in the enforcement of regulations, it has technical, engineering, and innovation capacity that many other developing countries lack. China serves to gain from developing many of the technologies that will be crucial to dealing with climate change, from renewable energy to carbon capture and sequestration technologies, which would allow the country to continue to rely on fossil fuels while mitigating some of the most severe impacts of climate change.

China is poised to become the world leader in renewable energy technology development. It is already a leading global manufacturer of wind power technology and solar photovoltaic technology, as well as biomass power technologies, hydropower technology, and solar hot water heaters. China is also involved in several efforts to develop and demonstrate low- or zero-emission coal technologies. These include GreenGen, a 400-MW-scale integrated gasification combined cycle (IGCC) plant to which carbon capture and sequestration (CCS) technologies will be added by 2020,[76] the Near Zero Emission Coal partnership between China, the EU, and the UK, the goal of which is to have a coal plant with CCS online by 2020,[77] and a postcombustion CCS project between China's Huaneng Power Company and the Australian Commonwealth Scientific and Industrial Research Organization, among others.[78]

It is at least possible, then, that China may develop sufficient expertise in and commitment to next-generation low-carbon energy technologies to be able to benefit from climate change mitigation rather than merely tolerating it. Such a scenario would meet the precondition of China's political leaders, who have stated that climate change "can only be solved through development."[79] This is one of the few scenarios in which China is a winner in the climate change challenge, and in which the rest of the world stands to benefit as well.

If China were to stake its future exclusively on a policy of high growth fueled by high emissions, showing only defiance or indifference in the face of deteriorating environmental conditions both locally and around the world, climate change and its associated environmental impacts could pose an increasingly dire threat to the stability of the Chinese state. Yet China is not without options. It is already well poised to become a leader in the low-carbon technology revolution. It has also come to perceive itself, and wishes to be perceived by others, as one of the world's leading nations, and not simply its biggest. Although it is unrealistic to expect China to sacrifice its drive for economic growth in order to mitigate the impact of climate change in the near term, it is not impossible that these two goals will converge.

3
China in the Global Wind Power Innovation System

People have been harnessing energy from the wind for thousands of years. Windmills were developed to aid in the grinding of grain and the pumping, irrigation and drainage of water, with the first simple windmills believed to have been used in China as early as 200 B.C. Merchants and crusaders are thought to have carried the idea for the technology back to Europe, where it came to be used for a variety of industrial purposes in the centuries preceding the Industrial Revolution. Windmills were first used to generate electricity in the United States in the late nineteenth century, when the emergence of central power stations relegated smaller electricity-generating windmills to the more isolated farms of the Great Plains until widespread rural electrification made them all but obsolete.[1] This chapter examines the global origins of the modern wind turbine industry and explores the wind innovation system that has emerged in China.

Origins of Modern Wind Turbine Technology

Industrialization in Europe and America led to the steam engine and electricity technologies replacing most windmills that had been used for water

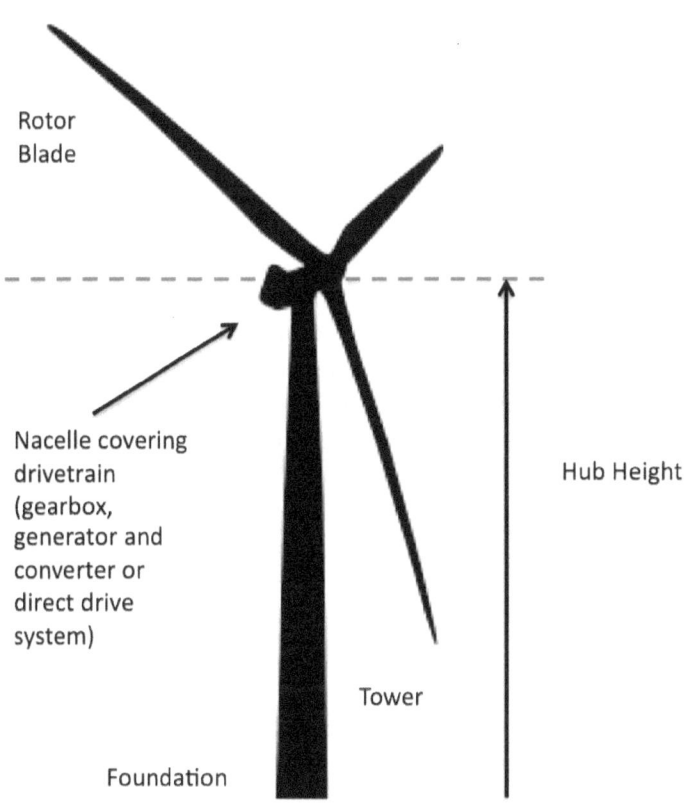

FIGURE 3.1 Major Components of a Modern Wind Turbine
Illustration by author using wind turbine graphic by Erland Howden from the Open Clip Art Library (http://www.openclipart.org).

pumping and electricity generation. The development of wind turbine technology continued, however, with larger windmills for electricity generation appearing in Denmark around 1890, and the first MW-scale wind turbine put into operation in Vermont during World War II.[2] It was not until the oil embargoes of the 1970s, however, that the technology gained interest worldwide, triggering an explosion in wind power technology research and development (R&D) emanating from Europe (primarily Germany, Sweden, the Netherlands, and Denmark) and the United States.

Early wind turbines were still quite small, in the range of less than 100 kW, with rotor-blade diameters of less than 15 meters. Although the origins of much of today's modern utility-scale wind turbine technology are connected to those of small wind turbines, these two technology pathways have since diverged, and there is now little relationship between the small and the large machines now that the large machines have become highly computerized.

Many companies from different backgrounds entered the wind industry during the R&D boom of the mid-1970s to mid-1980s. Common company backgrounds included shipbuilding, agricultural machinery, and aerospace, where there were expected to be useful knowledge spillovers to the wind power industry. Much of the R&D during this period focused on the theoretical physics of wind power, as well as competing technology designs of vertical versus horizontal axis machines, and turbines using one to four blades.[3]

In the late 1980s the technology started to mature, primarily because of the increased opportunity for demonstration projects that came with electricity policy reforms in California and Europe, which supported the construction of many early wind farms.[4] Demonstration turbines gradually became larger as the technology became more advanced, with rotor-blade diameters in the range of 15–30 meters.[5] During this period the first design codes and national standards for wind turbine technology were developed to facilitate quality control and interconnection with the electric grid. The companies that had entered the wind industry during the boom period began to see their technologies either succeed or fail, and the industry began to consolidate through bankruptcies, mergers, and acquisitions.

By the early 1990s the technology had matured, and several commercially successful companies achieved mass production of 500 kW and 600 kW wind turbines with rotor-blade diameters in the range of 30–50 meters.[6] R&D continued to focus on the design of larger and larger turbines with increased efficiencies and lower costs per unit.

Today the majority of wind turbine models being deployed around the world exceed 1 MW in capacity (1,000 kW) and have a rotor-blade length of well over 50 meters. Modern wind R&D is focused on continued design improvements to increase the resilience and the efficiency of the turbines, as well as on improved power electronics that facilitate smoother integration with the power grid. As countries begin to pursue offshore wind power development, R&D is also focused on developing even larger wind turbines designed specifically for offshore environments.

R&D conducted in Europe and the United States, and to a lesser extent in Canada and Japan, has resulted in significant cost reductions and technology improvements, allowing today's wind power technology to produce electricity at a cost that is comparable to that of power from conventional electricity generation in many parts of the world. The United States has invested the most in wind energy R&D measured cumulatively from 1974 to 2009, followed by Germany, the Netherlands, the United Kingdom, and Demark.[7] But if the size of the countries is taken into account, their relative contribution is rather different. Of the other four countries in the top five, three (Germany, the Netherlands, and Denmark) have invested a larger proportion of their GDP than has the United States.[8] In fact, using the GDP-adjusted numbers, Denmark invested over eight times more than the United States, and the Netherlands invested

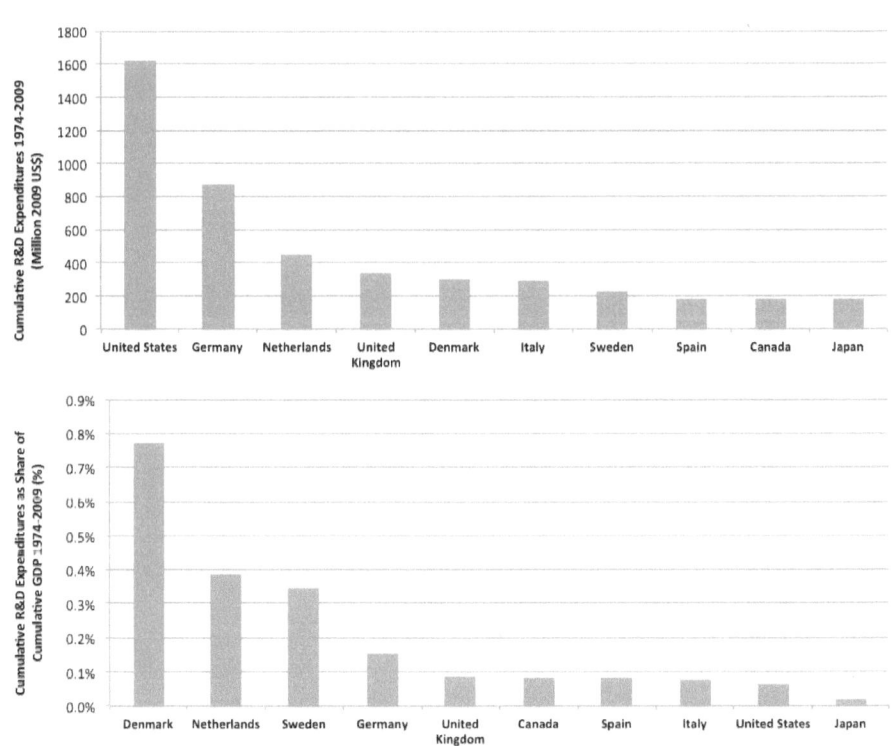

FIGURE 3.2 Wind Energy R&D Expenditures in OECD Countries

IEA (2010)

TABLE 3.1 Leading World Wind Markets and National Turbine Manufacturers

Country	Cumulative wind capacity, end of 2009 (MW)	Annual wind installations, 2009 (MW)	Cumulative government wind energy R&D expenditure 1974–2009 (millions of 2009 U.S. dollars)	Global market share of major domestic companies, 2009 (total; and for individual companies)	Domestic market share of major domestic companies, 2009 (total; and for individual companies)
United States	35,159	9,922	$1,621.82	13%; GE 12.6% Clipper <1%	46.4%: GE (40.3%) Clipper 6.1%
Germany	25,777	1,917	$874.17	21%; Enercon 8.5% REpower 3.4% Nordex 2.5%	75.8%: Enercon 60.2% REpower (8.8%). Fuhrlander 4.9% Nordex 1.9%
China	25,104	13,000	—	26.1%; Sinovel 9.3% DEC 5.4% Goldwind 7.3%, Mingyang 2% United Power 2.1%	78%; Sinovel 25.4%, Goldwind 19.8% DEC 14.6% United 5.6% Minyang 5.4% XEMC 3.3% Sewind 2% Windey 1.9%
Spain	19,149	2,459	$186.02	7.9%; Gamesa 6%, Acciona 1.9%	43.3%; Gamesa 37.1%, Acciona 6.2%
India	10,926	1,271	—	5.9%; Suzlon 5.9%	72.3%; Suzlon 55.1% Vestas 9.4%, SWL 4.3%, Pioneer Wincon 3.5%
Italy	4,850	1,114	$293.44	<1%	Leitwind 0.5%
France	4,492	1,088	$39.70	<1%	Alstom 1.7%
UK	4,051	1,077	$342.88	—	—
Portugal	3,535	673	$7.04	—	—
Denmark	3,465	334	$304.54	14.5%; Vestas 21.1% Siemens 6.6%	Siemens 100%

TABLE 3.1 *(continued)*

Sources: BTM Consult, *International Wind Energy Development: World Market Update 2009* (Denmark: BTM/Navigant Consulting, 2010); GWEC, *Global Wind 2009 Report* (Brussels: GWEC, 2009), IEA/OECD, "IEA Energy Statistics OECD R&D Database" (OECD/IEA, 2010), http://www.iea.org/stats/rd.asp.

Notes: Only leading wind turbine manufacturers are listed, not all manufacturers, so company shares may not add up to the total national market share. Nationality of manufacturer is determined by the country where the company was originally founded. In the case of Siemens, the majority of its wind power technology was obtained by its acquisition of Bonus, a Danish company, in late 2004, so for the purposes of this table Siemens is classified as Danish company even though it is technically German.

nearly six times the amount of the United States. Of the top ten OECD countries with consistent data on wind energy research between 1974 and 2009 (United States, Germany, Netherlands, United Kingdom, Denmark, Italy, Sweden, Spain, Canada, and Japan), the United States ranks in the bottom half, ahead of just Japan, Spain, the United Kingdom, and Italy.[9] In the United States wind energy R&D increased annually from 1974 and peaked in 1981, with R&D levels yet to return to the amount invested that year.[10]

The countries that invested the most in R&D were not necessarily the ones that produced successful wind turbine manufacturing companies. Other factors were likely just as important as R&D in determining which countries became the birthplace for the leading wind power technology manufacturers, including the nature of the domestic policy environment and the learning opportunities and networks that supported the emerging companies.[11] Together, these factors determined the national innovation systems for wind energy in these countries.

Understanding National Innovation Systems

Vestas, arguably the most successful wind turbine manufacturing company in the world, hails from Denmark. While Denmark invested far less in wind energy R&D than countries like the United States and Germany over the past few decades, it managed to produce the company that has dominated the global wind power technology market for almost three decades.[12] More than half the wind turbines on the international market in the 1990s were produced by the prospering Danish wind turbine industry.[13] Denmark has also been able to achieve the highest wind power penetration of any country in the world, with 26.2 percent of Denmark's power coming from wind energy in 2010.[14]

Denmark's success is in stark contrast to the experience of the Netherlands, a country with centuries of experience in building windmills, but one that never managed to meet its goals of either producing a successful turbine manufacturing company or developing a sizable domestic source of wind electricity.[15] A similar comparison can be made between Denmark and the United States. Although the United States had put much more money into R&D than Denmark, much early U.S. R&D for wind turbine development was allocated to the aerospace industry, and U.S. aerospace companies were generally less successful than lower-tech Danish firms throughout the 1980s and 1990s.[16] The Danish firms, as early movers in the industry, also benefited greatly from the on-the-ground experience they obtained with their demonstration projects built in California in the 1980s. General Electric (GE), the only globally successful U.S. wind turbine company, acquired some of the technological knowledge of the early American wind companies.[17]

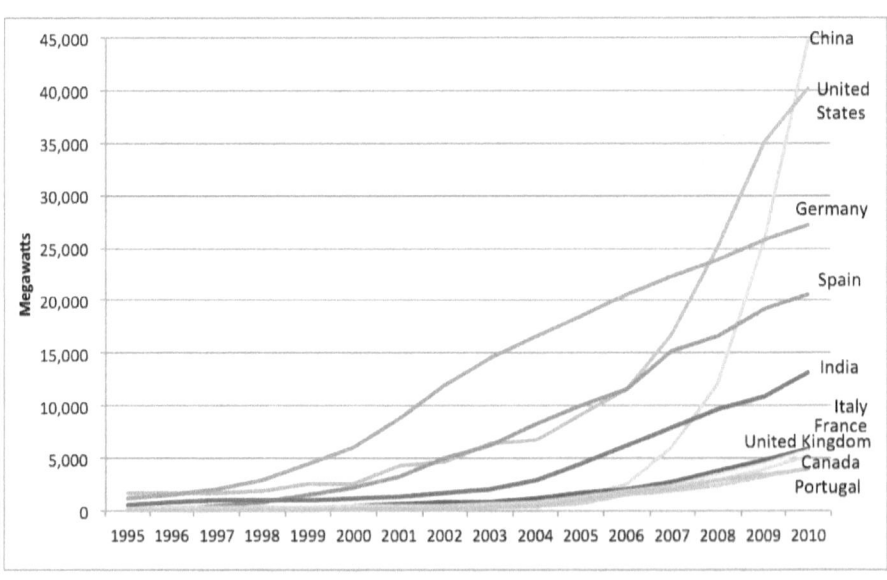

FIGURE 3.3 Wind Power Leaders, Cumulative Installations

Author's own database compiled from national wind energy associations, global wind energy association reports, and Windpower Monthly *"Windicator" statistics.*

The differing amounts of success achieved by these countries in producing a commercial technology provider are frequently attributed to the differences in the national systems of innovation in which the emerging companies operated. The national innovation system includes "the network of public and private institutions that fund and perform R&D, translate the results of R&D into commercial innovations and effect the diffusion of new technologies"[18] and is influenced by domestic policy or other factors that encourage innovative activity.[19]

The organization and distribution of innovation-related activities often differ among countries and regions. Studies have emphasized such differences between developed and developing countries,[20] and among countries falling within those broad categories, or as distinguished by regional characteristics with some similarities in innovation models among the Asian "late industrializing countries."[21] Additionally, organizational categories such as "research institute," "firm," or "government" can differ significantly in meaning and in terms of the range of activities undertaken by these organizations across different national or industrial contexts. In the case of China in particular, the distinction between public and private organizations and the type of activities an organization undertakes can be very different from that of economies that are not centrally planned.[22]

As the technology development strategies of transnational firms become increasingly global in scope—for example, as companies establish R&D centers in multiple countries with scientists who have been trained around the world—national innovation systems may eventually become obsolete.[23] In addition, the rise of multinational corporations with a global presence has created a new model for innovation through the global generation of technology. Multinational firms now take advantage of global experience to shape their innovative activity and conduct technological innovation within a global network.[24] As a result, there are clear limits to understanding the success of wind power technology firms based exclusively on the national innovation systems in which they operate. The presence of these companies in different international markets, the frequency with which they look globally to pursue forms of technology development or acquisition outside their national borders, and the clear linkages between the origins of technological know-how among companies in different countries point to the need for a more global model of innovation systems, which take into account the specific characteristics of different technologies and sectors.[25] Such frameworks can aid in our understanding of the innovation dynamics of even national wind industries, focus on the firm as the unit of analysis, encompass the various technologies within a particular sector, and account for globally based learning activities.

LEARNING IN NETWORKS

The concept of learning encompasses many ways to acquire knowledge that are relevant to thinking about technological innovation. One author defines learning as "the way firms build, supplement and organize knowledge and routines around their activities and within their cultures, and adapt and develop organizational efficiency by improving the use of the broad skills of their workforces."[26] While learning often benefits firms, it is of course the individuals within a firm that do the actual learning.[27]

What is being learned? Learning may include embodied or disembodied knowledge, tacit or formalized knowledge, experience-based knowledge, or knowledge based on R&D.[28] Put another way, learning can include know-how, know-why, and know-what, with know-how acquired by doing, know-why by searching, and know-what by using.[29] While R&D is aimed primarily at the generation of knowledge, these other kinds of learning are more frequently by-products of activities that are performed for other purposes. For example, early wind turbine design was a bottom-up process, based on practical experience and learning by doing, such as the early experience obtained by Danish firms through experimental projects in California and Denmark in the 1980s.

Networks are a way to think about how information is often learned or transferred between people or between firms. While formal learning networks might include company training programs or research partnerships, informal learning networks can also play a very important role in the exchange of ideas. Regional and global learning networks have likely played a large role in the development of wind turbine technology over time. The wind industry—characterized by its small number of firms, highly specialized technology, and geographically specific hubs of innovation (often near wind development locations)—is likely to exhibit many of the characteristics of the regional learning networks that have been observed in other industries and locales.[30] Studies have hypothesized that learning networks are a crucial determinant of the ability of wind power firms to gain success with a new technology.[31] Just as the early wind development in Denmark and the United States provided a crucial learning ground in the 1970s and 1980s, the emerging wind markets of India, China, and South Korea are serving as valuable regional learning networks for newer firms.[32] The increasingly global reach even of new firms, often facilitated by technology transfer partnerships with overseas firms, has also provided a valuable means of accessing global learning networks of knowledge and innovation.

TECHNOLOGICAL LATECOMERS

Countries that were not part of the group of early wind turbine innovators have used different strategies to foster the development of their own domestic large wind turbine manufacturing companies. These nations all attempted some form of "technological catch-up," a concept that in the most dramatic of cases is referred to as "technological leapfrogging," which has been documented across industries and technologies.[33]

A common strategy among latecomer firms has been to obtain, through a technology transfer, a technology from a company that has already developed advanced wind turbine technology. Technology transfers can occur through many different models.[34] One model is through a licensing agreement that gives the licensing firm access to a certain wind turbine model, often with some restrictions on where it can be sold. Another model includes establishing joint-venture partnerships between two companies, either to share a license or for collaborative research and development. Firms also can opt to collaborate to jointly develop a new technology design (joint development) and then share the associated intellectual property, which can be done without forming a new company or a joint-venture enterprise. If a firm has the capacity and means, it can also obtain access to technology through the purchase of ownership rights in a company with the desired technology or other forms of mergers and acquisitions (M&A).

A technology transfer may or may not include technological know-how associated with the development of the technology itself. The physical transfer of technology alone is likely insufficient to ensure the transfer of the technological knowledge that recipient companies would need to produce comparable wind technology domestically and to ensure its continued operation and maintenance in the field. Cases have shown that the transfer of technology without supplemental know-how—also referred to as the software needed to accompany the hardware—may detract from the lasting effectiveness of the technology transfer.[35] For example, a purchase of the blueprints and license to produce one model of wind turbine will likely be less valuable than an arrangement that also includes on-site training of the workers in the purchasing company by the transferring company.

A complete and successful technology transfer would mean that innovation could now be achieved at the receiver's side. The ability for a complete technology transfer to occur is dependent on the recipient company's ability to adopt an externally sourced technology and apply it internally (often called absorptive capacity), which is in part determined by the enabling conditions for technological innovation nationally (the national innovation system).

Although the acquisition of technology from overseas companies is one of the easiest ways for a new wind company to quickly obtain advanced technology and begin manufacturing turbines, there is a disincentive for leading wind turbine manufacturers to license proprietary information to companies that could become competitors. An example of this fear has been realized by Vestas, which licensed its turbine technology to Gamesa and now views the company as a major competitor in the global market.[36] This is increasingly also true for technology transferred from developed to developing countries, where a similar technology potentially could be manufactured in a developing-country setting with less-expensive labor and materials and result in an identical but cheaper product. For this reason developing country wind power manufacturers often obtain technology from second- or third-tier wind power companies that have less to lose in terms of international competition, and more to gain in fees paid from the license.[37]

Practicing at Home

Historically it has been common for wind turbine manufacturers to get their start in their home-country markets. Home-market experience was important for many of today's leading wind turbine manufacturers, with few exceptions. All of today's top wind turbine manufacturers got their start in their home markets, including Vestas (Denmark); Siemens (which acquired Bonus—originally from Denmark); GE (USA); Enercon, Nordex and REpower (Germany); Gamesa (Spain); Suzlon (India); Sinovel, Goldwind, Dongfang Electric Corporation (DEC), United Power, Mingyang, Sewind, and XEMC (China). In addition, all continue to be dominant suppliers in their home markets, with the exception of the Danish manufacturers, as there is little remaining potential for onshore wind development in Denmark.

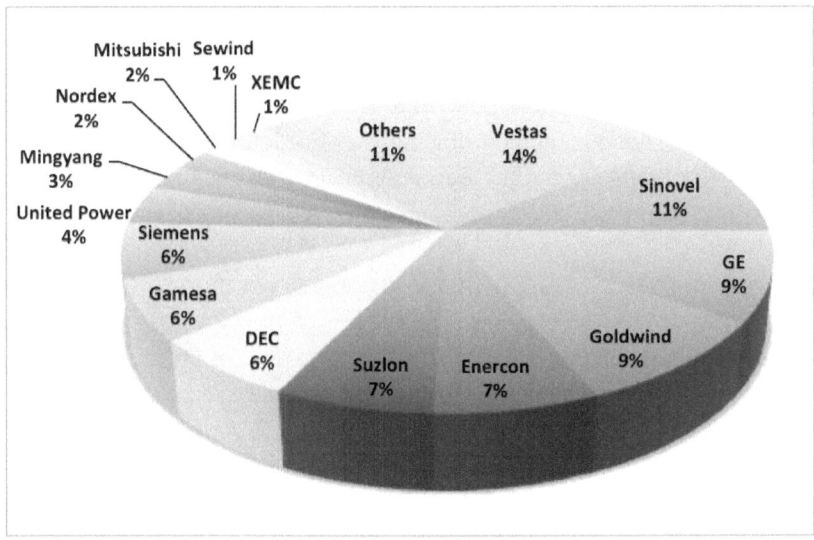

FIGURE 3.4 Global Wind Turbine Market Shares

Shares are for 2010. BTM Consult (2011).

There are few leading global wind turbine companies that did not primarily rely on their home market in the early stages of their technology development, including Mitsubishi of Japan; this will likely also be true of the emerging Korean manufacturers. By beginning their industry experience in a home market, companies can benefit from national government R&D support or policy support for demonstration projects.[38] It is particularly important that government policies be used to create a sizable, stable annual demand for wind power to give companies the long-term planning horizon necessary to allow for investing in the future. It is estimated that a minimum annual demand of 150–200 MW for three or more years is crucial to developing a nascent local manufacturing industry, while a more capable and aggressive local industry is likely to require at least 500 MW a year.[39]

Studies have highlighted the important role of the domestic market in supporting the growth of a domestic industry,[40] as a country can most easily act to support its own economic and social interests in its own home market.[41] To build a new internationally competitive industry such as wind turbine manufacturing, it is necessary to put policies in place that ensure a stable demand if one does not already exist.[42] Feed-in tariffs, which provide long-term price support for wind energy, are a common way that countries have promoted sizable and stable markets for wind electricity. Many countries have also implemented policies to specifically promote the development of a local wind manufacturing industry in addition to just the deployment of wind energy, including Germany, Spain, Canada, Brazil, and China.

THE DECISION TO LOCALIZE A WIND INDUSTRY

There are many policy mechanisms that can be used to promote the utilization of wind power, but such use does not necessarily lead to nor require the development of a local wind power industry. Wind power technology can be imported from abroad, and it often makes economic sense to do so until a large enough domestic demand has been established. Wind technology companies must decide, upon making a decision to enter a new national wind power market, whether it makes more sense for them to import their turbines or to shift their own manufacturing facilities into that country. So too must governments decide, in drafting their wind energy research, development, and demonstration (RD&D) plans and support policies, what it is that they are trying to achieve with respect to both domestic industry development and technology deployment.

A local wind industry may aspire to manufacture complete wind turbine systems; to manufacture certain components and import others; or perhaps just to serve as an assembly base for wind turbines imported from abroad. Each of these approaches implies different goals for manufacturing, degrees of localization and technology ownership, and policy incentives at work. Each model requires a different degree of localization ranging from partial to full—the turbine is only fully localized when it is being completely manufactured in the target country.

When assessing the benefits of developing a local wind technology manufacturing industry, it is important to recognize that in each of the models listed above, much of the know-how and intellectual property associated with the turbines may remain in the hands of foreign firms, with no technology transfer or in-country innovation taking place. This is because the models do not specify company ownership. If the above activities are not performed by a locally owned firm, but rather by a foreign firm, can they still be called "localized"? For example, if a foreign wind turbine manufacturer chooses to set up a factory in another country to manufacture its turbines but maintains control of the intellectual property rights associated with the technology, can the technology being manufactured locally actually be claimed as local?

Another way to examine localization is by the extent to which local manufacturing benefits the local economy. If a foreign company locally manufactures a turbine with imported labor and materials, little economic benefit from that factory's presence will flow to the town where the factory is located. This could occur under a scenario where highly skilled labor is needed and it is less expensive to import laborers than to train new staff. Alternatively, even if local laborers are used in the manufacturing process, they could be subject to strict nondisclosure agreements that prevent them from taking the information they learned on the job and bringing their expertise to other employers or using it to start their own companies. The purchase of local materials such as the steel used in wind turbines could benefit the local economy financially but still would not require the transfer of any of the knowledge associated with incorporating the steel in the turbine design.

Consequently it is useful for governments hoping to promote local manufacturing within a region to be very clear whether the goals of creating this industry are to create jobs and a demand for raw materials or to facilitate the transfer of advanced wind power technology and the associated know-how required to develop a domestic wind turbine manufacturing company within its borders. Policy incentives to support these varied goals may be quite different.

THE BENEFITS OF LOCAL MANUFACTURING

The potential benefits of local wind turbine manufacturing include economic development opportunities through job creation and sales of new products, opportunities for the export of domestically made wind turbines to international markets, and cost savings that result in lower-cost wind turbine equipment, lower cost of wind-generated electricity, and higher growth rates in domestic wind capacity additions. Another, less tangible benefit to wind technology localization, but clearly a motivating factor for several countries, is a desire for national achievement in what is viewed as a leading green technology industry with a sizable potential for growth.

The development of any new industry, including wind turbine manufacturing, creates new domestic job opportunities. Less clear is the extent to which these new jobs may replace other jobs. Wind development is often credited with creating more jobs per dollar invested and per kWh generated than fossil fuel power generation. One study estimates that wind power creates 27 percent more jobs than the same amount of energy produced by a coal plant and 66 percent more jobs than a natural gas combined-cycle power plant,[43] although such estimates are somewhat difficult to substantiate. Direct jobs are typically created in three areas: manufacturing of wind power equipment, constructing and installing the wind farm, and operating and maintaining the farm over its lifetime. Approximately two-thirds of the labor requirements are in the manufacturing of the wind power equipment, which includes turbines, blades, towers, and other components, while the remaining one-third is accounted for by installation, services, transport, and development. Of these components, rotor blades are the most labor-intensive and therefore are a crucial element of local manufacturing of wind turbines since they generally bring the most jobs.[44]

Local manufacturing of wind turbines or wind turbine components can potentially reduce costs through a reduction in labor costs, raw materials costs, and transportation costs. The improved servicing and response times that come from locally based manufacturing may further reduce costs and improve operations.[45] The cost of the wind turbine itself is estimated to be about 70–75 percent of the total installed cost for an onshore project or 40–50 percent for an offshore project, although this will vary substantially from project to project. The remaining costs primarily include construction costs (foundations, grid connection, roads, and sea cables), development and legal costs, and land acquisition costs, and there will be variation in these remaining expenditures depending on the location of the wind farm site.

A less tangible benefit to entering the wind industry can stem from a national goal to develop domestic wind power technology companies as a source of both national pride and national technological achievement. Most countries favor domestically manufactured products when given a choice between domestic and imported products if quality is perceived to be equivalent. Wind turbines can serve as a symbol of national technological success in engineering a cutting-edge, green technology that can be displayed to the nation.

Countries with lower wage rates, such as India and China, expect to be able to realize cost savings through domestic manufacturing of wind turbines compared with their European and American counterparts. This cost reduction is potentially significant for those turbine components that are particularly labor intensive. Rotor-blade manufacturing, for example, is labor intensive and could thus benefit from lower labor costs.[46]

Cost savings from in-country production could also be realized if a country is dependent on importing foreign turbines from overseas and shipping costs are high. Transportation costs can be particularly severe for sizable, heavy equipment. As a result, towers are often the first component to be manufactured in a local market (towers are also not as technically sophisticated as other components). The Canadian Wind Energy Association estimated that transport costs for wind turbines, composed of both overseas shipping expenditures and on-land freight transport, represent 5–10 percent of the entire system cost for imported turbines and 3–5 percent for domestically made turbines.[47] Reduced delivery lead times for wind turbines and components are another cost-saving factor in local manufacturing.[48] Better customer service and faster access to customer service staff and technical staff as well as spare components in case of mechanical problems may further reduce costs or improve project operations. The actual cost reduction that can be realized through localizing production, however, is a calculation that will vary greatly from country to country depending on the availability of local components and the local cost of labor and materials.

BARRIERS TO ENTRY

While there are many potential benefits to local wind manufacturing, there are also challenges to developing a new industry. There are significant barriers to entry into what has become a relatively mature industry, particularly as turbine sizes grow larger and the technology becomes more complex. Many companies have decades of experience in wind power technology R&D, and the leading turbine manufacturers are becoming larger and encompassing more global market share through mergers and acquisitions. Limited indigenous technical capacity and quality control makes technology development in new markets difficult, particularly when policies mandate local manufacturing. National standards requiring the use of advanced technology can initially shut out emerging firms with inferior technology. In addition, there are limited global locales possessing a skilled labor force in wind power, with Denmark still being a leading location for both skilled laborers and an experienced network of key components suppliers to support turbine manufacturers, even as its dominance in the industry has waned.

Current goals for technological innovation in the wind industry include cost reduction, efficiency improvement, and increasingly sophisticated grid interaction technology for both onshore and offshore wind turbines. Continuous advancements create a barrier to new entrants that may struggle to catch up to the best available technology. Firms looking to enter will have to decide whether to compete with another model of a currently popular turbine and risk it being outdated in the near future or to develop a larger turbine that does not yet have a commercial application in the hope that it soon will, or they will have to find another competitive edge such as producing a popular turbine type at a lower cost.

Certain wind turbine components are technically sophisticated and must last for years with little maintenance. Quality control is therefore of primary importance in the wind industry. Many technologically advanced countries have been able to enter the wind market at a late stage without much prior experience in wind turbine manufacturing owing to their relatively developed technical knowledge base. Countries with less indigenous technical capacity will have a harder time attempting to develop new technologies, particularly wind turbine technology, where experience in other industries has been shown to result in spillovers that can be an asset in wind technology development.[49] Even the perception of poor quality can severely limit market growth.

When establishing a new wind turbine manufacturing facility in a country with minimal experience with wind power development, the burden lies on the pioneering company to either manufacture or import all necessary components for the wind system. In addition, spare parts must be kept on hand in case repairs are needed, along with skilled maintenance technicians. A company that is starting from scratch in a new market will need to either import skilled labor from its home country, if it was previously established elsewhere, or train local workers to manufacture, sell, and service its turbines.

Established wind power markets with a history of wind companies within their borders, including Denmark, Germany, and the United States, have trained a skilled labor force in the wind industry. In addition, the presence of large turbine manufacturers leads to the creation of supporting technology industries, and these countries also have a relatively established network of wind turbine components suppliers on hand. There are examples of new wind turbine manufacturers locating to these well-established markets with labor and components readily available. For example, India-based Suzlon decided to base its international headquarters in Denmark, even though it was unlikely to sell its turbines to the Danish market.[50] Another emerging issue surrounds the availability of raw materials in the country where manufacturing is occurring. For example, China is estimated to currently control 97 percent of the world supply of rare earth metals that are needed in wind turbines using permanent magnet generators.[51]

One of the largest barriers to entry in the wind industry remains the use of tariff or nontariff trade barriers, despite the fact that the World Trade Organization (WTO) has established stringent trade regulations among member countries that prevent the use of trade barriers. The WTO Technical Barriers to Trade Agreement "tries to ensure that regulations, standards, testing and certification procedures do not create unnecessary obstacles" to trade and "discourages any methods that would give domestically produced goods an unfair advantage," and many other WTO provisions include similar guidelines.[52] To this end, policies that tax the importation of wind turbines, or even policies that require the use of domestically produced turbines, could be construed as "protectionist" and barriers to trade. Protectionist policies that may differentially support local industries like wind turbine manufacturing have increasingly become the focus of trade disputes.[53]

Characterizing China's National Wind Energy Innovation System

There was almost never a time when China considered pursuing wind power utilization without simultaneously pursuing the development of an indigenous wind power technology industry. There are two parts to the concept of localization as discussed throughout this chapter: locally making or manufacturing something, and adapting the locally manufactured product to meet the demands of the local context or market.[54] In discussing the localization of wind turbine manufacturing in China, of primary importance is the ability to locally manufacture the wind system in-country, since specific technical adaptations for the Chinese market are relatively minor.[55] As discussed above, localization of manufacturing can mean shifting foreign production to the location of the target market without the transfer of any technology to domestic firms. China always intended for its wind industry to ultimately consist of Chinese-owned firms and Chinese-owned technology, and it established a national innovation system for the wind sector that left little to chance in ensuring that this goal was met.

SCIENCE AND TECHNOLOGY DEVELOPMENT IN CHINA

China has faced challenges in building a strong national system of innovation. Going back thousands of years, there have been periods where China led the world in science and technology (S&T) development before falling behind Europe and North America during the industrial revolution.[56] China has been playing catch-up ever since, causing its leadership to experiment with policies to promote industrialization and institutional reforms aimed in part at establishing a national system of innovation that would eventually secure China's position at the global technological frontier.[57]

The defining characteristics of China's innovation system under central planning included activities being distributed among organizations by function; decision making that exhibited a multicentric, fragmented authoritarianism; and the dominant output criterion being quantity, not quality.[58] Overall this meant that there was little incentive for proactive innovation, except in designated research institutes. Additionally, there was little opportunity for linkage or interaction between actors, and these sorts of linkages have been shown to be vital to technological innovation and diffusion.[59]

In the wake of the Cultural Revolution, which all but quelled any innovative activities, the government, led by Deng Xiaoping, embarked on a reform program with a strong emphasis on S&T development, beginning with the National Conference on Science and Technology of 1978.[60] The Sixth to Eighth Five-Year Plans (1981–1995) included strong commitments to revitalized science and technology initiatives.[61] The 1985 Decision on Reform of the Science and Technology Management System aimed to wean research institutes from government funding and encouraged collaborations with enterprises, stating, "we should promote the commercialization of technological achievements and exploit the technology market so as to suit the needs of the socialist commodity economy," and that "in restructuring the science and technology system, emphasis should be placed on encouraging partnership between research, educational and designing institutions on the one hand and production units on the other and on strengthening the enterprises' capability for technology absorption and development."[62]

The 863 Program, launched in March 1986, awards competitive grants for applied research in key sectors, including energy, and remains an important part of China's national S&T program today. High-technology development zones were also established around the country to provide incentives to attract innovative enterprises, with over fifty such zones in place today. In addition, the May 1995 Decision on Accelerating Scientific and Technological Progress promoted the fundamental idea that S&T is a primary productive force in all fields.⁶³

During this reform period, it is estimated that the primary state R&D institution, the Chinese Academy of Sciences (CAS), spun off hundreds of enterprises based on its developments.⁶⁴ The process of reform has been ongoing within CAS as it seeks to redefine the role of its extensive system of over one hundred research institutes. During the 1980s CAS was almost fully financed by an annual budget appropriation, but it has since had to diversify its revenue sources to include support from a variety of central and local government sources and the National Natural Science Foundation of China, as well as private enterprises.⁶⁵ According to one study, "CAS has evolved away from an old Soviet model of an isolated set of basic-research laboratories . . . into a system of national laboratories designed to provide a national base of basic-research competence.⁶⁶ In an effort to spur domestic technological innovation and to diffuse applied technologies across government, industry, scientific, and academic communities, China has also established numerous National Engineering Research Centers (NERCs) across the country.

While reforms were ongoing, the government also looked abroad in an attempt to modernize technology and seed innovation. Many students were

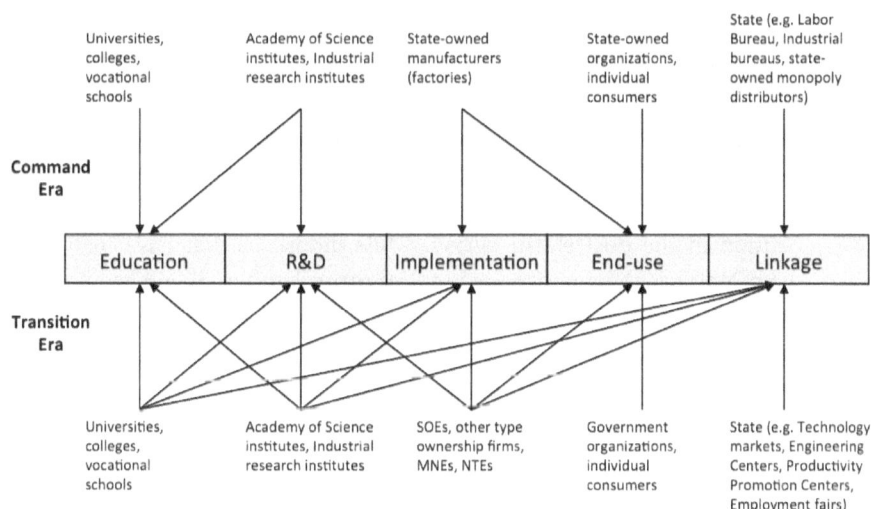

FIGURE 3.5 Activities and Actors in China's Innovation System Under Central Planning and Since Reforms

Reprinted from Research Policy 30, Liu and White, Comparing Innovation Systems (2001), 1091–1114, with permission from Elsevier.

sent overseas to be educated, and World Bank loans were used to upgrade laboratory facilities.[67] The government spent an estimated $40–$70 billion between 1979 and 1993 on imported technology.[68] While the acquisition of foreign technology was an important component of the government's reform strategy in the 1980s and 1990s, a key part of China's current S&T plan aims to reduce China's dependence on foreign technologies, and there is now a concerted effort being made to foster indigenous innovation rather than to rely on foreign innovation. The Eleventh Five-Year Science and Technology Plan (spanning the 2006–10 five-year plan period) includes targets to invest 2 percent of GDP in R&D, to reduce dependence on foreign technologies by 40 percent, to increase the contribution of technologies to economic growth to 45 percent, to rank in the world's top ten countries in citations used in international science papers, to rank in the top fifteen countries in patents granted, and to have fifty million people working in the S&T field, including seven million scientists, technicians, and engineers.[69]

Since economic reforms began in China in 1978, innovation performance in Chinese energy industries has strengthened. Political and economic reforms that have most likely influenced the innovation system include the decentralization of decision making and resource allocation, the encouragement of competition between firms, the increased competition for labor and for jobs, the greater diversity within organizations and less functional specialization, and the increased public R&D funding for basic research. By 1993 more than half of China's large state-owned enterprises had established technical development centers aimed at improving production efficiency as well as increasing product quality and marketability. Innovation performance has been shown to be strongest in those industries that have experienced the most institutional transformation and increased market competition in the postreform period, while industries that have maintained the prereform characteristics of central control and weak intellectual property protection have demonstrated less innovative activity.[70] China's SOEs still dominate much of the energy sector, and industrial R&D investments remain comparatively weak as technology investments continue to be predominantly determined and financed by the central government.

By 2009 total R&D had risen to 1.7 percent of China's GDP, up from 0.76 percent a decade earlier, with enterprise spending constituting about 73 percent of that amount.[71] As recent government policies target the energy and the carbon intensity of the economy, innovation in energy-saving and low-carbon energy technologies are increasingly rewarded and encouraged.[72] Universities are playing a more and more important role in innovative activities in China, providing important research inputs to both government and industry endeavors.

As China continues to make the transition to a market economy, its national innovation system will continue to evolve as well, particularly as the marketplace further rewards innovative activity. Although innovative activity was not as valued in China under a centrally planned government system, there is a clear trend toward increased investment in R&D and innovative activity among the private sector, as well as an increase in patents being granted to enterprises. Many studies have pointed to unclear property rights and intellectual property (IP) protection, as well as weak patent law and contract law, as major remaining barriers to innovation in China, although these areas are evolving rapidly as well.

MEASURING INNOVATIVE ACTIVITY IN CHINA

China's intellectual property rights system is still very much a work in progress; as a result, standard measures of innovative activity frequently

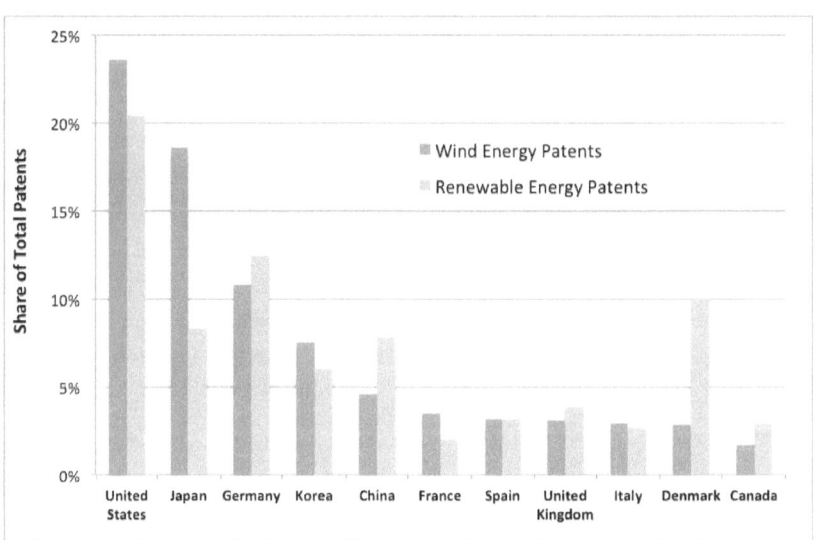

FIGURE 3.6 Countries Leading in Wind Energy and Renewable Energy Patents
Represents share of total patents filed under the Patent Co-operation Treaty by inventor's country of residence in 2009.

OECD Patent Database (2011).

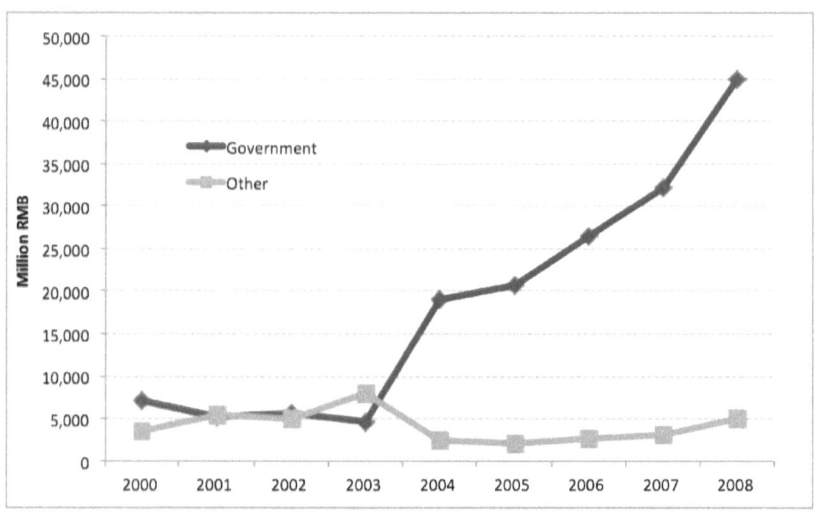

FIGURE 3.7 Energy R&D Investment in China

"Government" is estimated by including government grants for intramural R&D expenditure in R&D institutions, government grants for S&T activities in non-state-owned enterprises, and total funding for S&T activities in state-owned enterprises in the processing of petroleum, coking, and nuclear fuels, and the production and supply of electric power and heat power. It also includes government funding for S&T activities in institutions of higher education in the areas of energy technology and power and electrical engineering. "Other" is estimated by including intramural R&D expenditure in R&D institutions other than government grants, and funding for S&T activities in non-state-owned enterprises other than government grants in the following sectors in the processing of petroleum, coking, and nuclear fuels, and the production and supply of electric power and heat power. It also includes funding for S&T activities in institutions of higher education other than government grants in the areas of energy technology and power and electrical engineering. R&D reported in million RMB (nominal values) as reported annually.

Kempener et al. (2010), compiled from Zhongguo keji tongji nianjian (China Statistical Yearbook on Science and Technology) (2001–2009).

used in industrialized nations are not necessarily the most accurate metrics in a country where the concept of applying for patents is still relatively recent, or where detailed R&D accounts are not well tracked. What R&D statistics are available are often not disaggregated and do not fit clearly into standard categories of public and private, as many of China's leading innovators are in fact firms that are both partially state owned and privately owned.[73] In addition, standard outputs to innovation such as patents or R&D investment also do not distinguish clearly between technologies of domestic and international origin.[74] Foreign investment can be measured, but the technology transfer or innovative activity that results from such investments can be hard to discern. As China's own innovation policy establishes targets measured in terms like dependence on foreign technology, patents, citations used in international science papers, and numbers of engineers, its system of quantifying such metrics will no doubt improve over time.

National energy R&D statistics published by the Chinese government illustrate a strong, increasing trend in "government R&D," which includes publicly funded R&D at both public and private enterprises and national research institutes.[75] Other sources of energy R&D, including private support for research institutions and R&D funded by private enterprises, have remained relatively flat. This is not surprising since so many enterprises in the energy sector, including the wind sector, in China are still at least partially state owned. In addition, China ranks fifth globally in both wind energy and renewable energy patents generated in 2009.

China's Wind Energy Policy Regime

MAPPING GOVERNMENT INSTITUTIONS AND PLANS

China's energy policy bureaucracy frequently undergoes changes in structure and leadership. At time of writing, the National Energy Commission (NEC), formally established in January 2010, is the top administrative authority for energy policy making in China.[76] The NEC is chaired by the premier, and its General Office is run by the director of the National Development and Reform Commission (NDRC), a powerful agency with a wide-ranging set of responsibilities that includes most energy policy, planning, and pricing decisions.[77] The twenty-one NEC members include ministers from several government agencies (MOST, MOF, Ministry of Environmental Protection [MEP]), as well as other government and military representatives. It is ultimately responsible for government investment in large energy projects, as well as energy pricing decisions, energy R&D investments, and international energy cooperation.

The NEC is supported by the National Energy Administration (NEA), housed within the NDRC. The NEA absorbed several NDRC energy offices when it was established, including the Office of the former National Energy

Leading Group, and comprises nine departments, including the Department of New and Renewable Energy.[78] While there reportedly have been ongoing discussions within the Chinese leadership to establish a Ministry of Energy, there has been opposition, particularly among the large state-owned energy companies, to creating an all-powerful energy agency.[79] The State Electricity Regulatory Commission (SERC) also plays a role in renewable energy governance since it technically oversees China's power and electricity industry.

Wind energy policy in China therefore is governed by all the government institutions with jurisdiction over the energy sector. Other ministries that are not part of the traditional energy policy-making bureaucracy also frequently promulgate regulations that directly affect the wind power sector. For example, the MOF frequently administers government subsidies, and the MIIT administers industrial restructuring measures.

Many of China's energy policies are centrally planned in the context of the five-year plans issued by the NDRC, which can include specific targets for different energy sources. Separate plans have sometimes been issued for renewable energy, notably the Medium- and Long-Term Plan for Renewable Energy Development in China issued in 2007, and the Renewable Energy Development Plans that accompanied the Eleventh and Twelfth Five-Year Plans.[80] The major laws, plans, and policies relevant to promoting wind energy development in China over the past two decades are elaborated below.

WIND ENERGY POLICY TIMELINE

Various forms of government policy support have specifically targeted the wind power industry. The wind innovation system in China includes both direct support for industrial development and indirect support through the establishment of a local market for wind power. The government has used different types of policies to emphasize different areas of support over time, and the policy structure has clearly influenced firm strategies for technology development in this sector.

The first major policy to specifically support wind power in China came in 1994 when the government, led by what was then the Ministry of Electric Power, released the Provisions for Grid-Connected Wind Farm Management. The provisions mandated that grid operators facilitate interconnection of wind farms and set a purchasing price for wind power based on a pricing principle of generation cost plus repayment of loan and interest plus a "reasonable" profit.[81] In addition, the provisions stipulated that the entire grid bear any incremental cost of wind power over the average cost of conventional electricity. This "cost-plus-profit" formula persisted for several years, encouraging wind development in certain provinces but reportedly leading to huge subsidies in many instances where the profit margin was set relatively high.[82]

In 1997 the SETC launched the Double Increase Program, which aimed to double the 80 MW of wind capacity that were then installed and encouraged (but did not mandate) that a larger share of local content be incorporated in turbines used. However, the future outlook for wind power utilization in China was likely too uncertain, and 80 MW too small a quantity, to encourage local manufacturing by turbine suppliers at this stage. Additionally, local content requirements conflicted with the requirements of most foreign government loans, which were already being used to support many wind farm ventures in China. These loans were typically in the form of tied aid that came from various foreign governments (including Denmark, Germany, and the United States) to support the sales of their own domestic wind farm technology to China. This aid from foreign governments helped to subsidize the cost of early wind power development in China. About 74 MW of wind power was successfully installed under the Double Increase Program, essentially meeting the program target.[83]

In 1997 the SDPC began its Ride the Wind Program to promote a model of "demand created by the government, production by joint venture enterprise, and ordered competition in the market."[84] Two joint-venture enterprises to domestically manufacture wind turbines were established: one between the Spanish company Made and the Chinese company Yituo, part of China's Luoyang First Tractor Factory, a commercial wing of the Chinese Ministry of Machinery; and another between German wind company Nordex and Xi'an Aero Engine Corporation. The technology transfers carried out through this program started with a 20 percent local content requirement and a goal of an increase to 80 percent as learning on the Chinese side progressed.[85] The Made-Yituo joint venture focused on a 660 kW turbine transferred by Made, and the Xi'an-Nordex joint venture focused on a 600 kW turbine transferred by Nordex. The program experienced limited success, however, likely due in part to the "arranged marriages" between the Chinese and foreign partners. In addition, companies were selected from industries that were thought to be appropriate to wind technology but had little experience in manufacturing wind turbines—not unlike what occurred in the early years of the U.S. wind industry. China's target to install 1,000 MW of wind power capacity by the year 2000 was not met by a long shot, with members of the wind industry blaming the failure on unclear approval procedures and unrealistic local content requirements.[86]

A key driver of wind development between 2003 and 2007 was the wind resource concessions for government-selected sites awarded to developers selected through a competitive bidding process. Each concession project included approval to develop the selected project site, a power purchase agreement for the first thirty thousand hours of the project, guaranteed grid interconnection, financial support for grid extension and access roads, and preferential tax and loan conditions granted to the winning bidder by the central government. Five rounds of wind concessions produced eighteen wind projects ranging from 100 MW to 300 MW in size, totaling 3,350 MW of new wind installations. While the wind concession program got off to a bumpy start owing to reported gaming with the bidding system,[87] the program was ultimately successful in helping the government determine the current price for wind power in China, setting the groundwork for the establishment of national feed-in tariffs.[88] Tariffs issued under the wind concession program between 2003 and 2007 ranged from RMB 0.42 to 0.551 per kWh.[89]

The wind concession projects were also the first meaningful instance where the use of locally made wind turbines was both requested and rewarded. All wind farm projects approved by the NDRC during the Ninth Five-Year Plan (1996–2000) required that wind turbine equipment purchased for these projects contain at least 40 percent local content, but by the 2003 wind concession program that percentage had increased to first 50, and then 70 percent. Since there were very few Chinese turbine manufacturers in the market at this time, these local content requirements most directly affected the foreign wind turbine manufacturers, causing most of them to establish manufacturing facilities in China.[90]

China's local content requirement for wind turbines was further institutionalized in the 2005 NDRC Notice on the Relevant Requirements for the Administration of the Construction of Wind Farms.[91] This notice clarified the basis upon which wind projects would be approved, with the major criteria being the project's proximity to the power grid to facilitate the dispatch of electricity, and the rate of using domestically manufactured equipment. If the localization rate for the project was less than 70 percent, it would not be approved. While some components were still being imported, the Customs Administration applied import duties on any wind equipment brought into China from abroad.[92] The 2005 Requirements also clarified the procedures by which the government would approve wind farms, stating that the NDRC must approve projects greater than 50 MW, while projects under 50 MW would be approved by the provincial or local Development and Reform Commission authorities.[93]

A new era of policies to support renewable energy development began in China in 2006 with the launch of the Renewable Energy Law of the People's Republic of China.[94] The law, while not specific to wind energy, directly benefited wind power development by establishing a framework for regulating renewable energy. The law established the basis for setting national renewable energy targets informed by provincial energy plans. It also put in place a mandatory connection and purchase policy and authorized the establishment of feed-in tariffs for renewable electricity. It set the groundwork for a cost-sharing mechanism for renewables through the establishment of a special fund for renewable energy development, and it called for much needed national surveys of available renewable energy resources.

Several additional regulations were issued in order to implement the goals established in the Renewable Energy Law, including the 2006 Interim Measures on Renewable Energy Electricity Prices and Cost Sharing Management,[95] establishing a surcharge on electricity rates to help pay for the cost of renewable electricity,[96] and the 2007 Interim Measures on Revenue Allocation from the Renewable Surcharge,[97] which aimed to improve equity among provinces in bearing the costs of renewable electricity through an equalization program. Since electricity generated from renewable sources is often consumed in a different region from where it is generated, the 2007 Interim Measures required provincial grid companies to exchange their shortfall or surplus of surcharges with grid companies from other regions.[98]

In 2007 the Medium- and Long-Term Plan for Renewable Energy Development in China clarified China's renewable energy targets, stating that by 2010 China would aim to raise the share of renewable energy in total primary energy consumption to 10 percent, and by 2020, it will aim to raise this share to 15 percent.[99] The plan also established technology-specific targets, including a 5 GW target for grid-connected wind power by 2010 and 30 GW by 2020, which has since been increased to 200 GW by 2020 and 1,000 GW by 2050.[100] It also announced the first target for offshore wind power: 1 GW by 2020.[101] In addition, the plan announced a mandatory market share (MMS) for renewable energy. Somewhat similar to the Renewable Portfolio Standards of some U.S. states, the plan called for nonhydro renewable power generation in parts of the country covered by large-scale grids to reach 1 percent of total power generation by 2010 and at least 3 percent by 2020. Furthermore, all electricity-generating companies owning capacity of over 5 GW were required to expand their renewable capacity such that it comprised 3 percent of their total capacity by 2010 and over 8 percent by 2020.[102] This obligation falls on the large power companies and is one of the primarily reasons these companies have been developing large wind projects over the last few years. By the end of 2009, 57 percent of wind power in China had been developed by the so-called "big five" state-owned power companies: Guodian (Longyuan Electric Group), Datang, Huaneng, Huadian, and Guohua.[103]

It was the 2007 plan that first announced the Chinese government's strategy for developing large-scale wind power bases, with plans refined further in the March 2008 Eleventh Five-Year Renewable Energy Development Plan. Initial plans included three GW-scale wind farm bases in Jiangsu, Hebei, and Inner Mongolia to be built before 2010, plus an additional six bases in Xinjiang, Gansu, Jiangsu, Shanghai, Inner Mongolia, Hebei, and Jilin by 2020. Other regions slated for large-scale wind development included Guangdong, Fujian, Shandong, and Liaoning.[104] By 2008 it was further clarified that there were to be seven wind power bases of at least 10 GW each in Gansu, Xinjiang, Hebei, Jilin, eastern and western Inner Mongolia, and coastal Jiangsu, which together would total 138 GW by 2020. Few countries in the world are pursuing wind power development of this scale.

With such extensive wind power development planned to take place in parts of the country that currently either are not large electricity load centers or do not have an extensive power transmission infrastructure already in place comes significant technical challenges related to electricity transmission. These centrally administered and coordinated government plans for large wind bases allow for transmission planning to occur around the bases. The location of the bases also signals to manufacturers where to set up their factories, and in fact many of the provincial and local governments that are home to the wind bases are offering incentives for local manufacturing in their region. In addition, several of the wind base projects have set a minimum size for the turbines to be installed there, in the range of 1.5 MW to 2 MW, to encourage the use of advanced technology. This is causing turbine manufacturers to shift at least portions of their facilities to the region near the bases. Current wind turbine manufacturing facilities therefore tend to be located in provinces with large wind installations.

Prior to 2009 wind power prices could be determined in a variety of ways. Prices for the concession projects were established by a bidding process, and other large-scale projects received a tariff that was determined on a project-by-project basis, theoretically informed by a nearby concession tariff, as specified in the Management Rules on the Administration of Power Generation from Renewable Energy.[105] Within specific provinces, provincial governments also experimented with their own wind power pricing systems since provincial governments still had the authority to set prices for projects under 50 MW. Guangdong, one of the earliest provinces to experiment with wind development, was the first to set its own feed-in tariff for wind power in 2004 of RMB 0.528 per kWh, which was later increased to RMB 0.689 per kWh.[106] This patchwork of coexisting wind power pricing systems meant that prospective wind power developers had to negotiate a complex regulatory landscape.

This changed rather substantially with the August 2009 NDRC Notice on Improving Grid-Connected Wind Power Tariff Policy, which established a unified nationwide pricing system and fixed return on investment, thereby standardizing the development process for wind farms in China.[107] The policy set four feed-in tariff levels across the country, varying by region based on wind resource class. Category I resource areas had the best wind resources and therefore received the lowest tariff, while Category IV areas had the poorest wind resources and therefore received the highest tariff. Category I resource areas were to receive RMB 0.51 per kWh; Category II, RMB 0.54; Category III, RMB 0.58; and Category IV, RMB 0.61. Setting a higher tariff in low wind resource regions encourages wind power development despite less opportunity for electricity production.

Since 2005 the Kyoto Protocol's Clean Development Mechanism has provided another pricing subsidy for wind power development in China by putting a price on the carbon emissions offset by these projects.[108] While the value of the certified emissions reductions earned by a wind project will vary with its size and performance, the CDM has been estimated to provide a subsidy to wind projects approximately equivalent to RMB 0.1 per kWh.[109] This is a rather substantial amount, equivalent to roughly the total difference in the feed-in tariff between the best and the worst wind resource regimes across China. The ability of Chinese wind developers to continue to access carbon finance via the CDM is increasingly in question, however, owing to the impending end of the first commitment period of the Kyoto Protocol in 2012 and the increasing skepticism surrounding whether wind projects in China represent a true deviation from business-as-usual greenhouse gas emissions.[110]

In December 2009 amendments to the Renewable Energy Law were passed, further strengthening the process through which renewable electricity projects are connected to the grid and dispatched efficiently. One major change was related to the administration of the Renewable Energy Development Fund. Many grid companies found the surcharges to be insufficient to cover the higher cost of purchasing renewable power and were mismanaging or diverting the funds to other purposes. The amendments required that the funds be administered centrally, rather than collected directly by grid companies, and that grid companies seek compensation for the additional costs associated both with purchasing renewable power and with interconnection.[111]

The Ministry of Science and Technology has subsidized wind energy R&D expenditures at varied levels over time. In an effort to help Chinese turbine manufacturers develop new products and technologies, MOST funded research to develop technologies for 600 kW machines during the Ninth Five-Year Plan (1996–2000).[112] A prototype machine developed through this research was approved at the national level. MOST is now supporting the development of megawatt-size wind turbines, including technologies for variable-pitch rotors and variable-speed generators, as part of the 863 National High Tech R&D Program. The Eleventh Five-Year Development Plan of Science and Technology (2006–10) included support for the commercialization of wind turbines 2 to 3 MW in size.

In April 2008 the Ministry of Finance issued a new regulation mandating that the tax revenue from key components and raw materials for large turbines (2.5 MW and above) be returned to the state to channel the money back into the technology innovation and capacity building in the wind industry. Also that year, MOF announced the Interim Measures on the Management of Special Project Funds for the Industrialization of Wind Power Generation Equipment, which provided funding support for the commercialization of wind power generation equipment.[113] The measures specified that for all "domestic brand" wind turbines (with over 51 percent Chinese investment), the first fifty turbines with over 1 MW produced would be rewarded with RMB 600 per kW from the government. The measures further specified that the wind turbines must be tested and certified by China General Certification Center (CGC) and must have entered the market, been put into operation, and been connected to the grid. In addition, the National Energy Bureau has reportedly granted licenses to sixteen official national energy R&D centers to research topics such as blade R&D, large-scale grid-connected wind power systems, and offshore wind power equipment.[114]

Various preferential tax policies have also been placed on wind technology equipment manufacturers over the past decade. In 2001 the value-added tax on wind electricity was reduced by half.[115] In 2005 the Renewable Energy Law called for new tax benefits to be put into place to promote industrial development in renewable energy, leading to the National Guidance Catalogue for Renewable Energy Industry Development, which gave special tax status to wind power generation projects and equipment manufacturers, and the Enterprise Income Tax Law, which levied reduced income tax rates on wind manufacturers.[116] More recently, the Ministry of Finance and the State Administration of Taxation released guidelines on the taxation of imports, exempting components for wind turbines larger than 1.5 MW from customs duties and import sector value-added tax while still taxing the importation of complete wind turbines less than 3 MW. This served to further discourage the import of complete wind turbines while helping local manufacturers obtain imported components.[117]

In 2009 while U.S. Commerce Secretary Gary Locke was in China he asked for the removal of the local content requirement on wind turbines, arguing that it was a trade barrier for foreign firms. In a somewhat surprising move, the Chinese government promptly agreed and issued the November 2009 NDRC Notice on Abolishing the Localization Rate Requirement for Equipment Procurement in Wind Power Projects.[118] While the removal of the local content requirement was viewed as an achievement for foreign wind manufacturers, it likely had little impact in the Chinese wind sector where foreign firms had already established in-country manufacturing facilities, essentially rendering local content requirements obsolete. In addition, at the time the requirement was removed, there were already growing concerns at the highest levels of government about the health of the Chinese wind sector owing to reports of substantial overcapacity. In August 2009 the State Council had listed wind turbine production as an "excess capacity sector," causing the Ministry of Land and Resources to reportedly deny all applications for new wind turbine manufacturing facilities in an effort to slow down growth in the sector.[119] In early 2010 the MIIT released draft Wind Power Equipment Manufacturing Industry Access Standards, which aimed to "promote the optimization and upgrading of the industrial structure of the wind power equipment manufacturing industry, enhance enterprises' technical innovation, improve product quality, [and] restrict the introduction of redundant technology" to "guide the industry's healthy development."[120] This was to be accomplished by restricting the operation of wind turbine manufacturers that did not have the capability to produce a 2.5 MW or larger turbine, did not have at least five years of experience in a related industry, and did not meet various financial, R&D, and quality-control requirements. These stringent requirements, if implemented and enforced, would greatly restrict many firms from participating in the industry—Chinese and foreign alike.[121]

In early 2010 the NDRC introduced the first policy governing offshore wind development in China, the Measures for the Administration of Offshore Wind Power Development.[122] These measures set general guidelines for the planning and approval of offshore wind projects and stipulated that the central government approve, manage, and supervise all such projects. The decision to require central approval was likely due to the complex issues related to ocean navigation and coastline access surrounding offshore wind projects, many of which could potentially be national security concerns. Because of China's coastal geography, many of China's offshore projects are actually in intertidal zones or subtidal mudflats.

A concession program—not unlike the early wind concession program for onshore wind development—was initiated in May 2010 for four offshore projects in Jiangsu province located in Binhai (300 MW), Sheyang (300 MW), Dafeng (200 MW), and Dongtai (200 MW). One major difference in the requirements for eligible bidders for these concessions from the earlier onshore concessions was that no foreign-owned companies were permitted to apply for the offshore projects.[123] The only way that foreign-owned companies could participate was as part of a Sino-foreign joint venture where the Chinese partner held over a 50 percent controlling share in the company. Foreign-owned turbine technology was technically not excluded from the bids, but it is proving increasingly rare in China for Chinese-owned developers to partner with foreign-owned turbine manufacturers, particularly since many Chinese developers already have existing relationships with Chinese technology suppliers.[124] By the end of 2010 China had 100 MW of wind capacity installed offshore and an additional 15,100 MW already proposed, planned, or under construction.

In mid-2011 the NEA approved a series of eighteen technical standards to improve regulation of technology development in the wind sector.[125] The regulations included the Notice on Strengthening the Management of Wind Power Plant Grid Integration and Operation and the Provisional Management Methods for Wind Power Forecasting.[126] The wind forecasting regulation initiates a requirement that all wind farms now engage in forecasting to better predict power output. The new grid codes for wind farm interconnection, established as part of the Notice on Strengthening the Management of Wind Power Plant Grid Integration and Operation, are being developed in conjunction with the China National Standardization

Commission, in an attempt to prevent further wind-related disruptions to the power grid.[127]

Many of China's policies that governed wind power development over the past decade were informed by the government's understanding of China's total wind resource potential, where the best wind resource sites were located, and what wind-powered electricity actually costs. As studies increasingly elucidate better data on these three topics, however, it is becoming clear that some of the early policies may have been based on incorrect assumptions.

Official Chinese wind resource measurements have changed rather dramatically over time. Wind resource maps created in the 1980s were very low resolution and produced total national wind resource estimates of about 250 GW of onshore potential and 750 GW of offshore potential. These estimates were referenced in many of the key policy documents described above (including the 2007 Medium- and Long-Term Plan for Renewable Energy Development) and served as the basis for informing many wind policy decisions.

Recent wind resource estimates for China conducted since 2007 by both Chinese officials and experts both in and outside of China have illustrated that these early estimates were highly inaccurate. It is likely that China's onshore resources were dramatically understated, while offshore resources were substantially overstated. In December 2009 the China Meteorological Administration (the same government agency that had performed the earlier wind resource estimates) released new wind resource estimates for China of 2,380 GW onshore and 200 GW offshore, although some offshore estimates are substantially higher.[128] China's early resource assessments had no doubt influenced the setting of national and regional targets for wind power development, as well as decisions about what types of wind power technology to support for research and development. Substantial R&D funds in China have been directed towards offshore wind power technology and demonstration projects. As a result, it is not surprising that leading figures in China's wind industry have questioned the rationale for investments being made in offshore wind technology.[129]

Evidence of Learning in the Chinese Wind Industry

While the Chinese wind industry has been the subject of much attention owing to its explosive growth in recent years, there have been few attempts to assess the actual learning that has taken place within the industry.[130] As described above, there are a variety of ways to measure learning—some more appropriate for the Chinese context than others.

TECHNOLOGY SIZE

One way to assess technological progress in wind power technology is by the average size of the wind turbines being installed annually. Since the size of individual wind turbines has increased over time, and since the majority of China's wind power installed in recent years has come from Chinese technology manufacturers (over 80 percent in 2009), the size of the turbines installed is an approximate measure of the technology level of Chinese wind technology providers. China is still installing smaller wind turbines on average than other countries, even those that have fallen behind China in terms of annual installations. These countries, including Denmark, Germany, the United States, and Spain, all were earlier innovators in the industry. The local manufacturing of wind turbines in China began around 1996, about two decades after it began in Denmark. While China has clearly made strides, its companies

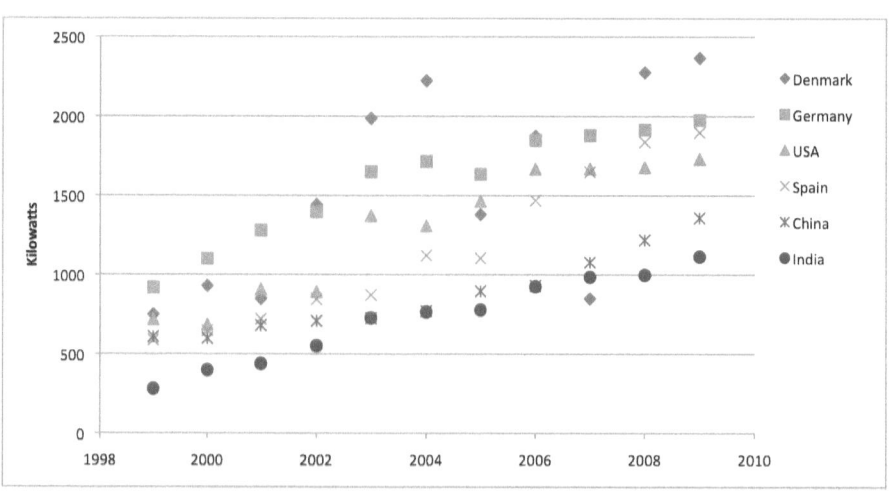

FIGURE 3.8 Average Size of Wind Turbines Installed Annually, Selected Countries

BTM Consult (2010 and previous years).

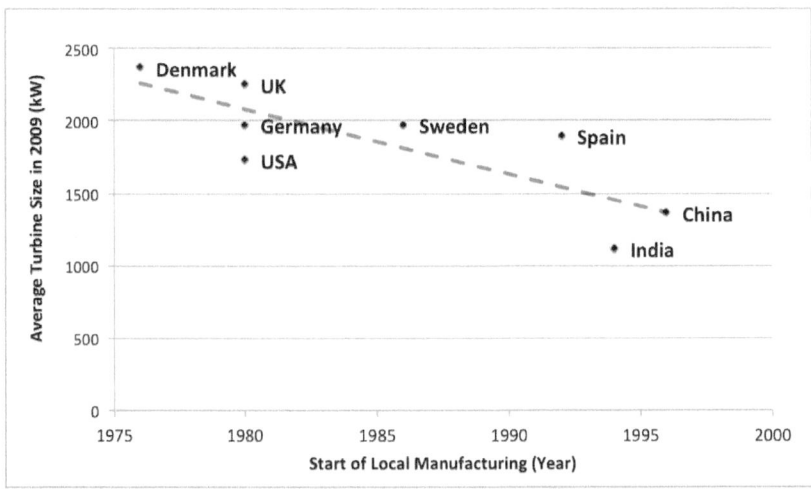

FIGURE 3.9 Average Wind Turbine Size and Start of Local Manufacturing, Selected Countries

Author's estimates and BTM Consult (2010 and various years).

are still working to catch up in terms of the level of technology they are manufacturing.

NEW ENTRANTS AND MARKET SHARES

When the first utility-scale wind turbine was installed in China in 1985, it was imported from Denmark. At that point, pioneering wind company Vestas pretty much had the Chinese wind market to itself. Over the next decades a handful of other foreign wind turbine manufacturers imported turbines to China. The mid-1990s saw the establishment of the first Sino-foreign joint ventures in wind turbine manufacturing, and the late 1990s the establishment of the first Chinese-owned wind turbine manufacturers. By the mid-2000s many new Chinese manufacturers had entered the Chinese market.[131] Between 1999 and 2009 the number of Chinese-owned companies installing at least one wind turbine in the country annually increased from two to thirty-four. The actual number of Chinese wind companies is thought to be much larger, likely in the range of eighty or more, although they are yet to participate in a commercial wind farm development in China.

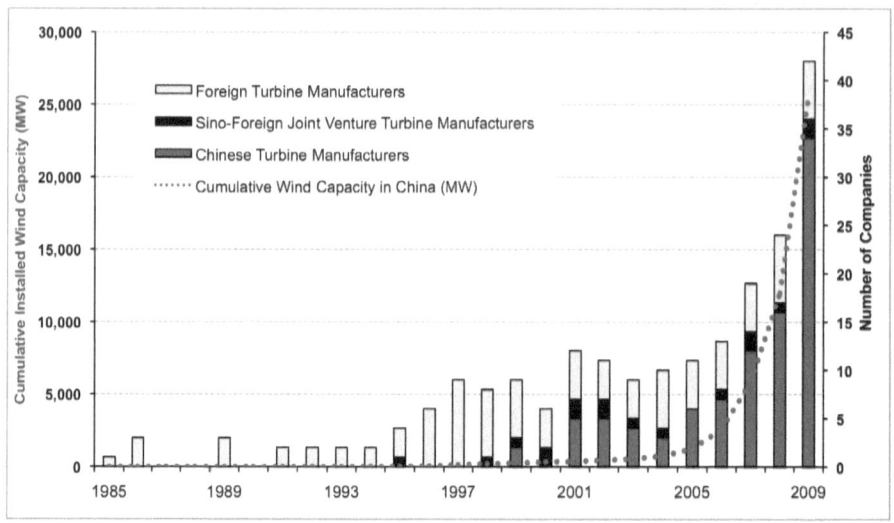

FIGURE 3.10 Number and Ownership of Wind Turbine Manufacturers in China
Author's database of wind energy manufacturers in China.

RESEARCH AND DEVELOPMENT

Many foreign-owned wind turbine manufacturers involved in the Chinese market have shifted larger shares of their total R&D expenditures into China. For example, in October 2010 Vestas announced that it had established a new wind power R&D center in Beijing, with an investment of $50 million over five years and a staff of two hundred by 2012, and Suzlon is reportedly in the process of opening a R&D center in China as well.[132] While few Chinese-owned wind turbine manufacturers provide transparent reports of their R&D expenditures, there is some evidence that levels of R&D are increasing as a percent of total revenue. For example, the Chinese company Minyang increased its R&D expenditures by 31 percent from 2009 to 2010 as it worked to develop a new 3 MW wind turbine design.[133]

ACCESS TO LEARNING NETWORKS

While China was barely on the map in terms of wind power development in the 1980s and 1990s, it was already becoming a global test bed for wind

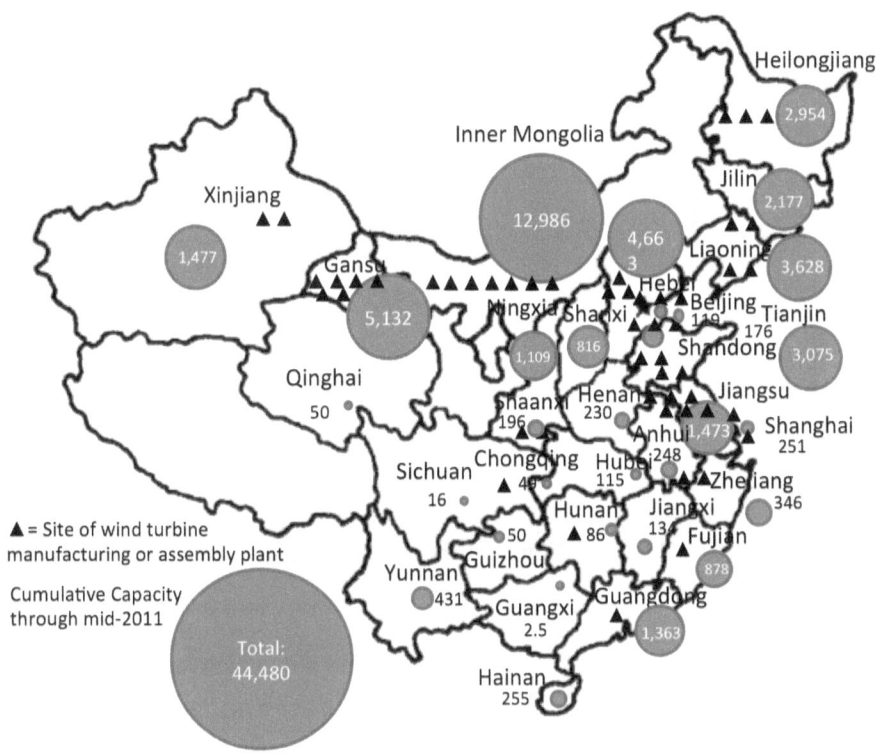

FIGURE 3.11 Map of Wind Power Development and Wind Technology Manufacturing Facilities in China

Illustration by author. Figure data from Windpower Monthly, *GWEC, and author's database.*

turbine technology. Early wind farms such as Dabancheng in Xinjiang and Nan'ao in Guangdong featured wind turbines from all over the world and wind power engineers working side by side as they tested their technology. By 2000 only 344 MW of wind power capacity had been installed in China, but these wind turbines had been manufactured by more than seventeen companies from six countries.[134] Early Chinese firms were likely able to benefit from the learning networks created by foreign firms developing wind farms in localized geographic regions.

As more and more wind turbine manufacturing facilities are being constructed throughout China, there are increasing opportunities for regional learning networks to be established. Many wind manufacturing facilities

FIGURE 3.12 Map of Chinese Wind Turbine Manufacturer Experience Overseas

Illustration by author based on reports of company activities through 2011.

are being constructed in close proximity to the government-planned large wind bases described earlier in the chapter to allow for easy access to wind farm construction projects.

As Chinese turbine manufacturers start to export their wind turbine technology overseas, they are increasingly looking to globalize their understanding of the technology for different national conditions. Chinese wind turbine manufacturers have already exported turbines to countries in Europe, Asia, and North America and South America and have announced plans to export to Africa and Australia. Technology transfers have already begun to occur from Chinese-owned companies to other countries. Some of the leading Chinese turbine manufacturers are also establishing research facilities and sales offices outside China.

LEARNING AND TECHNOLOGY COST

China has been playing an increasingly important role in global wind turbine deployment as it becomes an ever larger wind technology manufacturing base. Learning curves, frequently used to understand past cost trends and forecast future cost reductions for various energy technologies, start with the premise that increases in the cumulative production of a given technology lead to a reduction in its cost.[135] Therefore if wind power technology cost decreases over time with manufacturing experience, China surely is playing an important role in that decrease.

Many studies have calculated learning rates for wind power from 1980 to 2000, suggesting that historical cost reductions have been significant (though there is relatively little agreement on the magnitude of those reductions). In the United States, the major site of wind power development outside of China in the past decade, for example, installed costs for wind power fell by about $2,700 between 1980 and 2000, reaching a low point of about $700 per kW between 2000 and 2002.[136] Between 2002 and 2009, however, costs in the United States increased by about $800 per kW or over 100 percent, despite over 30 GW of new U.S. wind power installations representing one quarter of capacity installed globally over that period.[137] The *IPCC Special Report on Renewable Energy* suggests that a 9 percent learning rate occurred in onshore wind power in the United States from 1982 to 2009, despite increasing wind power plant investment costs from 2004 to 2009.[138]

Estimates of global learning curves are more limited, however, as are learning curve studies of wind energy costs in countries other than the United States.[139] While the indicators discussed above illustrate the learning that has taken place in the Chinese wind industry, there are challenges to trying to quantify the learning that has taken place in monetary terms. This is due in part to the fact that China has only become a major source of wind power deployment and manufacturing over the last five or so years, and cost data are limited. Some studies have attempted to use the wind concession bid prices to generate a learning curve for China's wind industry, but confirmed occurrences of gaming and price distortion in the concession projects likely invalidate such metrics and as a result call the accuracy of such estimates into question. Prices also are not always the best indicator of costs owing to a variety of market factors that influence selling price. For example, recent reports of overcapacity in the Chinese wind turbine industry likely account for at least some of the observed drop in prices.

Chinese wind farm price data do illustrate significant declines in wind turbine pricing: the average wind turbine in China is selling for about RMB 3,700 per kW in 2011, compared with an all-time high average

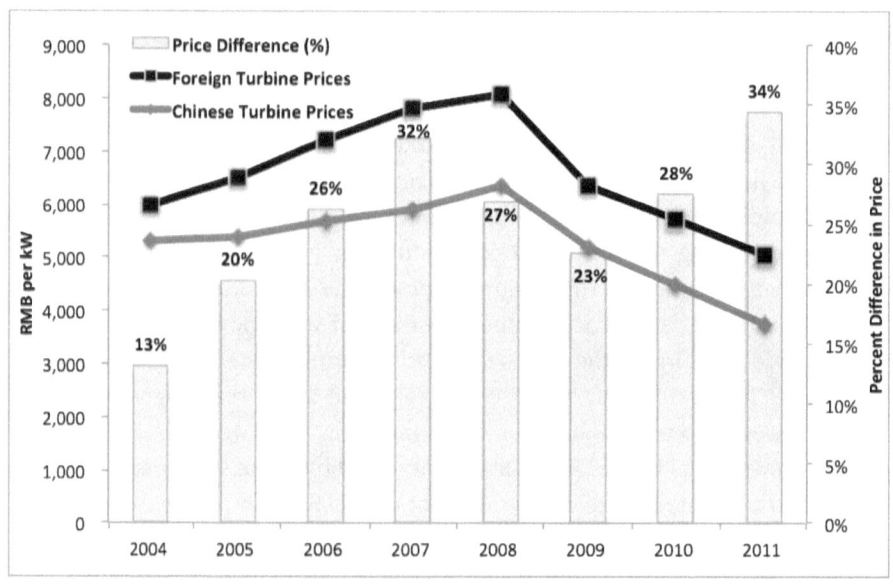

FIGURE 3.13 Chinese and Foreign Wind Turbine Prices in the Chinese Market
Based on reported prices bid to concession and wind base projects, and CWEA data.

price of RMB 6,400 per kW in 2008, representing a cost reduction of almost 50 percent over this three-year period. Chinese turbine prices are also consistently lower than the prices of turbines sold by foreign-owned manufacturers, even though these turbines are also manufactured in China. The average price difference between the foreign-owned and Chinese-owned turbines being sold in the Chinese market ranged from 13 to 34 percent between 2004 and 2011, with 2011 seeing the largest price difference.

While China has faced challenges in building a strong national system of innovation, its quest for expanded wind power utilization has gone hand in hand with its development of an indigenous wind power technology industry. This is evident in its early policies that not only supported the deployment of wind energy but also nurtured the development of a domestic wind technology industry.

Although innovative activity was not as valued in China under a centrally planned government system, there is a clear trend toward increased private-sector investment in R&D and private-sector innovative activity, as well as an increase in patents being granted to enterprises. Despite these advances, many barriers to innovation persist, including unclear property rights and intellectual property protection, and weak patent and contract law. As China continues to make the transition to a market economy, its national innovation system will continue to evolve as well. And as innovative activity is further rewarded in the marketplace, there will be an increasing incentive to protect such activity through reforms to the intellectual property regime and legal system.

China was barely on the map in terms of wind power development in the 1980s and 1990s, yet it was already becoming a global test bed for wind turbine technology. Early Chinese firms were likely able to benefit from the learning networks created by foreign firms developing wind farms in localized geographic regions within China. Today many foreign-owned wind turbine manufacturers involved in the Chinese market are shifting larger shares of their total R&D expenditures into China. As more and more wind turbine manufacturing facilities are being constructed throughout the country, there are increasing opportunities for regional learning networks to be established. Many wind manufacturing facilities are being constructed in close proximity to the government-planned large wind bases to allow for easy access to wind farm construction projects. As Chinese turbine manufacturers start to export their wind turbine technology overseas, they are increasingly looking to globalize their understanding of wind turbine technology for different national conditions and markets.

There are several indications that learning is taking place in China's wind industry. While China is still installing smaller wind turbines on average than those in other countries—even countries that have fallen behind China in terms of annual installations—the average size of wind turbines installed in China has increased by almost 1 MW over the past decade. Over the same time period, the number of Chinese owned companies installing commercially viable wind turbines has increased seventeenfold. The global reach of Chinese firms is also expanding, allowing for additional access to global learning networks.

The Chinese wind industry is still young, however, and it is early to assess its achievements. The next few years will be a telling time for the ultimate success or failure of today's emerging Chinese wind companies.

APPENDIX: TIMELINE OF CHINA'S MAIN WIND POWER POLICIES AND REGULATIONS

1994   Provisions for Grid-Connected Wind Farm Management
The former Ministry of Electric Power mandated the purchase of wind-generated electricity by power companies, allowing the additional cost to be spread over the grid. The sharing of the higher cost of wind was disputed, and eventually the program was removed when associated power-sector reforms were halted.

1996   National High Tech R&D Program (863 Program)—Ninth Five-Year Plan
The program was originally established in March 1986 (the year and month serving as the basis for its name). The continued implementation of MOST's 863 Program was approved as part of the Ninth Five-Year Plan beginning in 1996 (as well as in subsequent five-year plans). Over RMB 60 million was made available for renewable energy R&D through MOST and the former SDPC. R&D programs included a focus on developing 600 kW wind turbines with 40 percent local content for all new wind power projects. Local wind turbine manufacturers were able to use these funds in part for international technology transfer; local content requirements set the stage for more stringent future requirements.

Loans for Wind Farm Development
This program implemented by SDPC and MOST gave priority access to reduced interest domestic loans for wind farms with rates up to 50 percent lower than current commercial rates, with a preference given to projects using locally manufactured wind turbines. It was used to develop several small demonstration wind projects.

1997   Ride the Wind Program
New technology funds were allocated to two Sino-foreign joint ventures in wind turbine manufacturing facilitated by the former agencies SDPC and SETC. It promoted RD&D for wind power technology along with a local content incentive, although neither JV established under the program was ultimately successful.

Double Increase Program
A financial incentive program implemented by SDPC and SETC that promoted the doubling of current wind capacity with low-

or no-interest domestic loans and encouraged domestic wind turbine manufacturing. While the target for capacity installations for that year was met, the program failed to encourage significant local manufacturing.

1998 **Further Modifications to Import Duties on Wind Turbine Components**

From 1990 to 1995 imported wind turbines were exempted from customs duties to promote wind development. As expectations of a domestic wind turbine industry grew, in 1996 the Customs Administration raised customs duties on imported complete turbines and lowered them on imported components (to encourage local turbine manufacturing, with some use of international components). In 1998 further differentiation between the two was made when components were exempted from value-added taxes (VAT) and turbines were not, to further promote domestic wind turbine manufacturing. Customs duty regulations vary across components and are applied differently to firms with different ownership structures.

2001 **National High Tech R&D Program (863 Program)—Tenth Five-Year Plan**

The continued implementation of MOST's 863 Program as part of the Tenth Five-Year Plan included support for the development of megawatt-size wind turbines, including technologies for variable-pitch rotors and variable-speed generators, and supported the RD&D of many Chinese companies' early development of MW-scale wind turbine technology.

Value-Added Tax Reductions on Wind Electricity

The MOF and the State Administration of Taxation reduced the VAT for wind electricity from 17 percent to 8.5 percent, resulting in a reduction in the price of wind power electricity.

2003 **Wind Concession Program**

The NDRC administered this government-run competitive tender to develop preselected wind farm sites. The program initially mandated a local content requirement of 50 percent for wind turbines used in the concession projects, with the requirement increased to 70 percent in 2004. A total of 3,350 MW of wind power capacity was developed through five rounds of concessions from 2003 to 2007, with tariff prices ranging from 0.42 to 0.551 RMB per kWh.

2005 **Notice on the Relevant Requirements for the Administration of the Construction of Wind Farms**
This NDRC notice clarified the project approval process and criteria for wind projects, including proximity to the grid and the percentage of domestically manufactured equipment utilized. NDRC had to approve all projects greater than 50 MW, while provincial or local DRCs could approve those less than 50 MW. The local content requirement encouraged many foreign manufacturers to establish local manufacturing facilities in China, and the 50 MW approval cutoff encouraged many developers wanting to bypass central government approval to separate larger projects into multiple projects of 49 MW or smaller in size.

National Guidance Catalogue for Renewable Energy Industry Development
NDRC established the renewable energy technologies to be targeted for government R&D support, including wind, solar, biomass, geothermal, wave, and hydropower technologies.

Measures for Operation and Management of Clean Development Mechanism Projects in China
This State Council decision, implemented by NDRC in consultation with MOST and the MFA, set the domestic guidelines for CDM project development in China in line with China's role as a recipient of CDM-related carbon finance under the Kyoto Protocol for certified emissions reductions by projects that are shown to reduce GHG emissions. The CDM was used to support the majority of wind farm projects in China, providing a subsidy estimated at about RMB 0.1 per kWh, until the CDM Executive Board raised questions about the additionality of wind projects. The future of the CDM is uncertain even as the Kyoto Protocol has technically been extended beyond 2012 as the market for Chinese CERs is tied to the larger politics of the international climate negotiations.

2006 **Renewable Energy Law of the People's Republic of China**
Passed by the State Council in 2005 and implemented in 2006, this law serves as the primary legal framework for renewable energy development in China. The law established the basis for setting national renewable energy targets, set a mandatory

connection and purchase policy, authorized the establishment of feed-in tariffs for renewable electricity, established a special fund for renewable energy development, and called for national surveys of available renewable energy resources.

Interim Measures on Renewable Energy Electricity Prices and Cost-Sharing Management
This NDRC measure established a surcharge on electricity rates to help pay for the cost of renewable electricity. While a fund was successfully established to subsidize the cost of renewables, the allocation of the fund proved challenging.

National High Tech R&D Program (863 Program)—Eleventh Five-Year Plan
The 863 Program as updated for the Eleventh Five-Year Plan supported the development of MW-size wind turbines, including technologies for variable-pitch rotors, variable-speed generators, and commercialization of wind turbines of 2 to 3 MWs in size.

2007 Interim Measures on Revenue Allocation from the Renewable Surcharge
Aimed at improving equity among provinces in bearing the costs of renewable electricity, this NDRC measure established an equalization program requiring provincial grid companies to exchange their shortfall or surplus of surcharges with grid companies from other regions. Owing to reported mismanagement of funds, the interprovincial balancing mechanism was amended in 2009.

Medium- and Long-Term Plan for Renewable Energy Development in China
This NDRC plan established national targets for the share of renewable energy in total primary energy consumption (10 percent by 2010 and 15 percent by 2020) and total wind power installations (5 GW of grid-connected wind power by 2010 and 30 GW by 2020, including 1 GW of offshore power). The plan also introduced a mandatory market share (quota) requirement that 1 percent of total power generation by 2010 and at least 3 percent by 2020 come from renewable sources and required that all power-generation companies owning over 5 GW of power plant capacity achieve 3 percent of their total capacity from renewable energy by 2010 and over 8 percent by 2020. This requirement

encouraged China's five largest generating companies to lead the way in wind power development in order to meet their quotas. The plan was also the first announcement of the government's intention to develop large-scale wind bases.

2008 Eleventh Five-Year Renewable Energy Development Plan
This NDRC plan established seven wind power bases of at least 10 GW in size in Gansu, Xinjiang, Hebei, Jilin, eastern and western Inner Mongolia, and coastal Jiangsu, which together would total 138 GW by 2020. The plan also established a framework for transmission planning and encouraged local wind power equipment manufacturing in close proximity to the wind bases. The bases were initially developed through government tenders similar to the earlier wind concession projects but eventually switched to a more streamlined approval method with tariff prices established by the national feed-in tariffs after the FIT policy was established in 2009.

Interim Measures on the Management of Special Project Funds for the Industrialization of Wind Power Generation Equipment
This MOF program provided a direct subsidy of 600 RMB per kW to Chinese-owned wind turbine manufacturers for the first 50 wind turbines over 1 MW produced. To receive the subsidy, the wind turbines must be tested and certified by CGC and must have entered the market, been put into operation, and been connected to the grid. The program, funded in part by the tax revenue collected from wind equipment companies, provided a substantial subsidy for the demonstration of newly produced wind technologies for companies able to meet the eligibility requirements.

2009 Notice on Improving Grid-Connected Wind Power Tariff Policy
This NDRC notice established the first unified, consistent national pricing mechanism for wind power development in China. The program included four standardized national feed-in tariffs for wind power development with tariff levels varying by wind resource class: category I, RMB 0.51 per kWh; category II, RMB 0.54 per kWh; category III, RMB 0.58 per kWh; category IV, RMB 0.61 per kWh.

Renewable Energy Law Amendments

These amendments made several changes to the administration of the Renewable Energy Law to address problems that had arisen, including a requirement that the renewable energy fund be administered centrally rather than collected directly by grid companies, with grid companies seeking compensation for the additional costs associated both with purchasing renewable power and with interconnection. The amendments also strengthened the process through which renewable electricity projects are connected to the grid and dispatched to increase efficiency.

Notice on Abolishing the Localization Rate Requirement for Equipment Procurement in Wind Power Projects

This NDRC notice removed the long-standing local content requirement that favored domestic over imported wind technology in the Chinese market. While this was viewed as an achievement for foreign manufacturers, it likely had little impact in the Chinese wind turbine market, where most foreign firms had already established in-country manufacturing facilities.

2010   ### Wind Power Equipment Manufacturing Industry Access Standards

This MIIT regulation was aimed at consolidating the wind turbine manufacturing industry by restricting the operation of wind turbine manufacturers that did not have the capability to produce a 2.5 MW or larger turbine, at least five years of experience in a related industry, and the ability to meet various financial, R&D, and quality-control requirements. It was drafted in response to concerns about an overheated industry that had seen numerous new entrants in recent years in response to industry support programs and wind energy subsidies, though many industry stakeholders believed it would harm the industry by picking winners and eliminating healthy competition.

Measures for the Administration of Offshore Wind Power Development

These NDRC measures set the general guidelines for the planning and approval of offshore wind projects and stipulated that all offshore projects be approved, managed, and supervised by the central government.

Offshore Wind Concession Program

NDRC's first offshore wind concessions in China set the stage for future offshore wind power development. By the end of 2010 China had 100 MW of wind capacity installed offshore and an additional 15,100 MW already proposed, planned, or under construction. The first government tender for four offshore wind projects in Jiangsu province totaled 1,000 MW. No foreign-owned developers were permitted to participate in the bidding because of national security concerns associated with development along China's coast.

Further Modifications to Import Duties on Wind Turbine Components

This MOF provision adjusted import duties on major technical equipment and removed duties specifically for Chinese-owned wind turbine manufacturers importing wind turbine components to manufacture advanced wind turbines, modifying previous import duty measures. Eligible importers must be producing 1.5 MW or larger turbines and over 300 MW in sales per year.

2011

Notice on Strengthening the Management of Wind Power Plant Grid Integration and Operation

This notice from the NEA and the China National Standardization Commission was directed at reported problems with wind integration resulting in frequent wind power curtailments. The regulation included the establishment of new grid codes to address challenges with the interconnection of wind turbines. It also requires all wind farms to obtain NEA approval in order to receive the feed-in tariff subsidy.

Provisional Management Methods for Wind Power Forecasting

This NEA regulation, aimed at improving wind power integration by better predicting when wind power will be available to the grid, required all grid-connected wind farms to install forecasting systems.

Notes: Program and policy names are provided in English; in some cases the Chinese name may be slightly different. More details of key programs as well as citations for many of the regulations themselves (including their Chinese titles) are provided in chapter 3.

4
The Role of Foreign Technology in China's Wind Power Industry Development

As modern wind turbine technology was first developed by companies in Europe and the United States, it is not surprising that the first utility-scale wind turbines installed in China were imported from these regions. European and American wind turbine manufacturers were demonstrating their technology in China as early as the mid-1980s. These demonstrations created opportunities for learning and led to local partnerships, and eventually to a shift from technology imports to local manufacturing. Along the way, technology transfers from overseas companies to local Chinese companies, whether in the form of intellectual property, skilled personnel, or other informal means of knowledge transfer, helped lay the groundwork for the Chinese wind industry.

China's sizable wind resource potential, combined with its supportive domestic policy environment for wind power, has attracted the attention of many international wind companies. This chapter examines the companies that came to China from three of the early leaders in wind power technology development: Denmark, Germany, and the United States.

The Danish Pioneers

Vestas is one of the oldest and most successful wind turbine manufacturers in the world. A pioneer in the global industry, Vestas also played a role

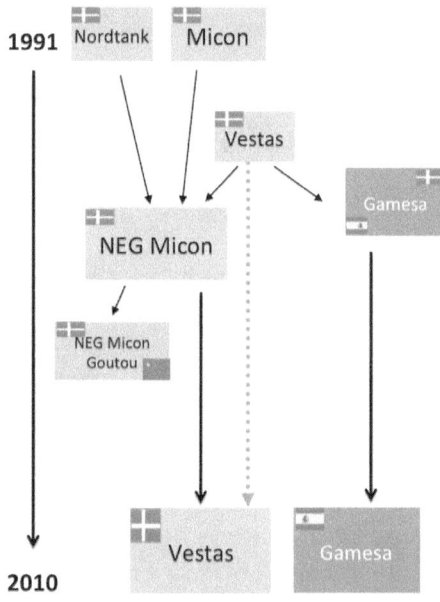

FIGURE 4.1 Timeline of NEG Micon, Vestas, and Gamesa Company Structure in China

Illustration by author.

in the early stages of the Chinese wind industry and is still active in the Chinese market today. As the Danish wind industry has evolved, a series of mergers and acquisitions have left Vestas as the only sizable Danish wind manufacturer standing. Bonus, one of Vestas's early Danish competitors, was acquired by Germany's Siemens in 2003. NEG Micon, formerly one of Vestas's competitors, was very active in the Chinese market until it was absorbed by Vestas in 2003. NEG Micon was itself the product of a merger between two Danish companies, Nordtank and Micon, in 1997. Leading Spanish wind turbine manufacturer Gamesa was founded after an early technology transfer from Vestas and subsequently entered the Chinese market as one of the company's major foreign competitors.

NEG MICON

The origins of NEG Micon are primarily rooted in two former Danish wind turbine companies, Nordtank and Micon, which merged in June

1997. The aim of this merger was "to create a company with powerful competitive advantages on the global market."[1] The Danish wind industry was doing very well at this time, but competition was fierce among several leading Danish manufacturers. The Danish headquarters of Nordtank and Micon had been located in Jutland only 50 kilometers apart. Another motivation for the merger was to combine the experience each company had gained in various overseas markets, which tended to be complementary rather than overlapping. Micon had been primarily active in South America, especially Argentina, while Nordtank had already developed extensive experience in China. The companies also complemented one another technically, with product lines coming from Micon for 600 kW, 750 kW, and 1 MW models, and Nordtank supplying a 600 kW and a 1.5 MW unit.

NEG Micon proceeded to further expand the company with a series of takeovers and acquisitions of other wind companies. In March 1998 it bought Danish company DanControl as well as UK firm Wind Energy Group and Aerolaminates, a rotor-blade manufacturer. It added Wind World to its group portfolio in August 1998, acquiring the company's gearbox expertise and its Optimal Speed Controller (OSC) technology.[2] NEG Micon took over Dutch company Nedwind in 1998, though Nedwind continued to operate as its Dutch subsidiary and independently installed its own turbines in China.[3]

NEG Micon is perhaps best known in the wind industry for a catastrophic gearbox problem identified in 1999, which led the company to carry out what has been referred to as the largest gearbox retrofit in wind turbine history—over 1,250 turbines—and almost resulted in the financial collapse of the company. The company realized that the flaw could be a problem for all its turbines that had been installed since 1996 and thus decided to replace all the gearboxes on the flawed turbines with renovated ones free of charge. The renovated gearbox came with a two-year guarantee on the replaced parts. As of this decision in 1999, only 150 of the 1,250 turbines had actually reported problems, but there was a fear that if the issue was not dealt with immediately it could seriously damage the technical reputation of the company. Although it was unclear exactly what had caused the flaw, it is believed that the bearings, in specific conditions, were not robust enough.[4]

The technical problems of NEG Micon extended to problems with Nordtank turbines as well. In 2000 the Danish wind turbine owner's association determined that the bearings on the 270 kW and 600 kW Nordtank wind turbines distributed in four European countries were guaranteed to fail within five years, and Nordtank had to do a turbine-wide retrofit. This problem was determined to be due to by the spherical roller bearings used in the Nordtank machines—a type of bearing that most manufacturers had otherwise abandoned and experts had warned against.[5] The extensive technical problems would have been the end of the company if not for a successful financial restructuring accomplished in December 1999 by a substantial purchase of new company shares by existing shareholders.

In 1998 NEG Micon began a joint-venture enterprise with a Chinese company in Beijing to form Beijing NEG Micon Goutou Wind Turbine Co. Ltd. This was primarily a sales outlet for NEG Micon's Danish turbines, which continued to be imported from Denmark. According to Goutou employees, some technology was transferred to this facility, including the nacelle cover design.[6] After Vestas took over NEG Micon, it cut back on many of NEG Micon's operations in China and let Goutou buy out the joint venture.[7] Goutou continued to operate as a fully owned Chinese company and expanded production into additional wind turbine components. In addition to its joint venture, NEG Micon also sold primarily 750 kW and 900 kW wind turbines to the Chinese market. The majority of these tur-

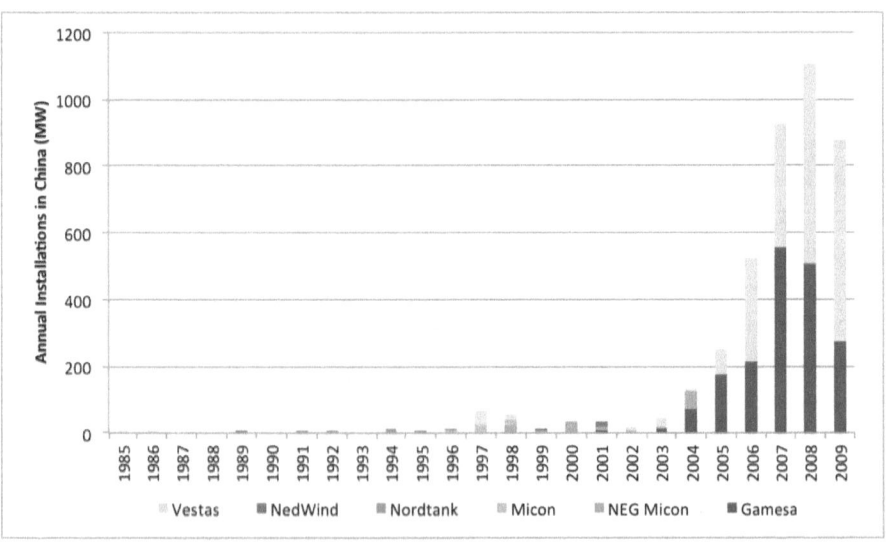

FIGURE 4.2 Wind Installations in China (Vestas, NEG Micon, Nordtank, Micon, NedWind, Gamesa)

Author's database.

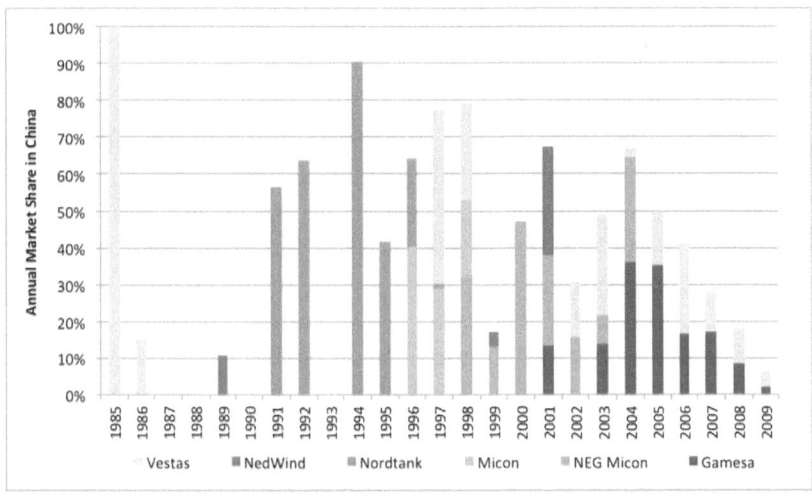

FIGURE 4.3 Chinese Wind Market Shares (Vestas, NEG Micon, Nordtank, Micon, NedWind, Gamesa)

Author's database.

bines were imported from Denmark, with some peripheral components manufactured in China.

While NEG Micon was one of the first foreign companies to transfer wind power technology to China via its joint venture, the company may have had a more lasting impact on the Chinese wind industry through the knowledge transfer associated with the demise of the company. Although NEG Micon and Vestas merged at the end of 2003, both companies already had wind farms lined up through 2004 so they essentially continued to operate as separate entities in China that year while beginning the merger process. While NEG Micon had much more of an on-the-ground presence in China than Vestas, Vestas was a global name. As a result, NEG Micon's offices in China took the Vestas name after the merger. NEG Micon had a far larger Chinese operation than Vestas, however, and while Vestas absorbed some of the staff, both foreign and Chinese, it opted to cut back China-based staff substantially. As a result, many former NEG Micon employees with extensive experience in the Chinese market opted to go work for one of the emerging Chinese wind turbine manufacturers or to start their own wind industry–related companies.[8] The last NEG Micon 900 kW turbines were installed in China in 2004 since Vestas opted to not continue to manufacture the NEG Micon turbine line for the Chinese market.

VESTAS

The history of Vestas can be traced back to 1898, but the company started to manufacture wind turbines in 1979 and became a wind turbine company exclusively in 1987. At the end of the nineteenth century, Danish blacksmith H. S. Hansen opened his first workshop in Lem, which later became an important center for the blacksmith craft. Thirty years later Hansen and his son, Peder Hansen, founded their first industrial company, Dansk Staalvindue Industri, a manufacturer of steel window frames for industrial buildings. After World War II Peder Hansen started a new company called Vestjysk Stålteknik A/S, which subsequently changed its name to Vestas. This company manufactured a variety of items, from kitchen appliances to agricultural trailers. Vestas's offices and warehouse burned down in 1960, launching several years of restructuring and consolidation. In 1968 Vestas began manufacturing hydraulic cranes for light trucks and became a successful exporter. When the two oil crises of the 1970s hit the transportation industry and caused truck crane sales to decline, Vestas was forced to look to another growth area.[9]

Inspired by the oil crises, Vestas became interested in the potential of wind turbines as an alternative source of energy. Though initial experimentations were based on the Darrieus Turbine design, the company soon opted to pursue the three-blade turbine model, which subsequently became the dominant design in the wind industry. Vestas delivered its first turbines to Danish customers in 1979. Vestas was very involved in the first wind industry boom driven by government incentives in Denmark and the United States (primarily California). It began serial production of its 55 kW wind turbines in 1980. In 1985 Vestas introduced pitch-regulation, "a major technological innovation that optimizes the energy output of a wind turbine by constantly adjusting the angle of the blades to current wind conditions," and by the end of the year it had sold 2,500 wind turbines to the United States and had 800 employees.[10]

The unstable nature of U.S. policy support for wind power ended up affecting Vestas's market performance, particularly when California tax credit legislation expired at the end of 1985. The collapse of the U.S. market forced Vestas to file for bankruptcy in October 1986, but it was able to restructure, selling off large parts of the Vestas Group and forming a new company called Vestas Wind Systems A/S in 1987. The new company was exclusively involved in wind energy, had sixty employees, and was headed by Johannes Poulsen, who remained CEO until 2001.

Vestas Wind Systems A/S began almost immediately on a path of technological innovation and international expansion. Subsidiaries were established in India (1987), Germany (1989), Sweden (1992), and the United States (1992), as well as a joint venture with Spanish company Gamesa Eólica in Spain (1994). Vestas has continued to expand geographically and by 2011 had operations in North America, South America, Europe, South Africa, Australia, New Zealand, and Asia (including China, India, Japan, South Korea, Singapore, Taiwan, and the Philippines), as well as R&D facilities in Denmark, the United Kingdom, Germany, the United States (Texas), Singapore, India, and China.

Vestas has played a unique role in contributing to innovation in wind turbine systems. While many companies that enter the wind industry do so by purchasing the rights to manufacture turbine technology that was developed by others, most of Vestas's exceptional growth during the 1980s and 1990s was internal. The company did engage in select acquisitions over the years, including Danish Wind Technology (DWT) in 1989 and Costas Computer Technology A/S, a long-standing supplier of software and components for Vestas's wind turbine control systems, in 1999.[11]

The company has continued to develop larger and larger turbine sizes. In 1981 it was selling primarily 55 kW wind turbines. It introduced several wind turbines ranging from 55 kW to 660 kW before introducing its first megawatt-scale turbine in 1999. These early designs have since been replaced by its current turbine fleet, which ranges in capacity from 850 kW to 3 MW, and R&D has begun on a 5 MW turbine for offshore use.[12] One of the early pioneers in offshore wind development, Vestas was chosen as the supplier for the first major offshore project in the North Sea, at Horns Reef, in 2000.[13]

In 2001 Vestas was the dominant global wind energy leader, but that year brought many changes in the company that would prove to be significant. Perhaps one of the biggest mistakes in Vestas's history lies in its decision to participate in a technology transfer with Spanish company Gamesa. In December 2001 "increasing strategic differences" between Vestas and Gamesa led to a sale of Vestas's 40 percent stake in the joint venture. One of the best models of a successful technology transfer in the wind industry, it resulted in the establishment of one of Vestas's biggest competitors, discussed later in the chapter.

On the morning of September 11, 2001, the company announced the resignation of the current CEO Johannes Poulsen effective April 2002. The transition to a new CEO was accompanied by news that a major competitor had entered the market: General Electric, which announced in February 2002 that it had acquired the assets of Enron Wind. GE's expertise in the conventional power business, its strong distribution channels in North America, and its financial strength made it an important new competitive threat for Vestas, particularly in the U.S. market.[14] In addition, during 2002 Vestas struggled with technical problems associated with its new flagship V80–2.0 MW turbine and faced delays at the high-profile Horns Reef offshore project. In contrast to prior years, Vestas ended 2002 with an announcement that it would lay off 495 employees, marking its first downsizing since 1986.[15] More recently the company has acquired multiple offshore contracts that have boosted company revenue, though its financial performance in recent years has been mixed.

Vestas has utilized different models for expanding into new international markets. In some cases it has chosen to form a joint venture together with a well-established partner in the target country, as was the case with Vestas RRB India Ltd. in India and Energy System Taranto S.p.a. in Italy. In other markets it has decided to enter with fully owned subsidiaries, including in the United States, Sweden, and Germany.[16] It is this fully owned subsidiary model that Vestas used as it entered the Chinese market, despite likely pressures to pursue a joint-venture arrangement.

Vestas became the first wind turbine company to enter the Chinese market when it installed the first small turbines in Shandong in 1986.[17] It built its first major wind farm in 1997 using 600 kW turbines. All the Vestas turbines used in these projects were imported to China from Denmark.

By September 2004 Vestas China had forty-five employees. Although Vestas had discussed plans to pursue local manufacturing of wind turbines in China, it did not make any official plans to do so until it had a large turbine order in hand.[18] On December 21, 2004, Vestas announced it had received an order for the delivery of fifty units of its V80 2.0 MW turbines for the 100 MW Rudong Wind Concession Project, located in Jiangsu province.[19] The concession projects (as discussed in chapter 3) were based on the condition that 70 percent of the wind turbines must be manufactured locally. Therefore Vestas announced that as a consequence of the order for the Rudong project, as well as the large potential for wind power in the Chinese market, it was planning to establish a blade factory in China. This paved the way for the company's transition from importing turbines to locally manufacturing turbines in China.[20]

Vestas has since expanded its presence in China significantly. Sales to China represented 11 percent of the company's total turbine sales (measured in megawatts installed) in 2008 and 12 percent in 2009.[21] By the end of 2009 Vestas had established four different production facilities in China, located in Tianjin, Xuzhou, Hohhot, and Shanghai, and a sales and service headquarters in Beijing. The Tianjin facility was Vestas's first in China, established in 2006, with a production scale of 1,200 blades per year and 400 employees.[22] Located in the Tianjin Economic-Technological Development Area (TEDA) near a large port on the Bohai Sea and on the transcontinental railway that links Asia and Europe, the facility was expanded in 2007 to manufacture nacelles and hubs, adding an additional 250 employees.[23] A generator-manufacturing facility was based in Tianjin that year as well, with another hundred-plus employees. The Xuzhou casting facility and Hohhot blade and nacelle facility were established in 2008 with 180 and 1,000 employees, respectively, and the Tianjin facility was expanded to include the manufacturing of control systems in 2009 with an additional 135 employees.[24] Many of the Chinese staff employed at these facilities have been brought to Denmark for training.

In 2009 Vestas developed a new wind turbine model that it designed specifically for the Chinese market, particularly the cold temperatures of Inner Mongolia. This was reportedly the first market-specific turbine ever developed by Vestas for any domestic market. While based on its existing technology, the V60–850 kW turbine was designed to be most effective in low and medium wind speeds, which Vestas estimated represented about 75 percent of unutilized onshore wind resources in China.[25] The turbine is also notably Vestas's first to be almost entirely sourced in China, with 90 percent of the components manufactured locally. The turbine was being built by the Hohhot facility in order to be close to future wind development sites in the Inner Mongolia Autonomous Region (IMAR).

Vestas claims that the design of this turbine resulted from extensive dialogues with Chinese customers and stakeholders. The former president of Vestas China, Lars Andersen, stated that "our best customers told us they wanted a kilowatt turbine that could unlock the potential of low and medium wind sites and overcome obstacles, like the tough winters in Northern Chinese areas such as IMAR. In collaboration with our Chinese partners, we designed Vestas's most advanced and productive kW turbine, and then localized it and built a factory to produce it—close to some of China's richest low and medium wind resources."[26] One problem with Vestas's V60 850 kW turbine, however, is that it is smaller than many of the turbines being installed by its competitors, and many Chinese government policies favor the use of larger, more advanced wind turbines. It is also smaller than the flagship 2 MW turbine model that the company had already been selling in China since 2004. As a result, the Chinese reception to the China-specific machine has been lukewarm. Vestas has received at least one order for the machines, but as the Chinese government looks toward industry consolidation, smaller turbine technology is being weeded out of the market.[27] Vestas opted to introduce the V112/3 MW turbine in China in June 2010, likely after realizing the smaller turbine may offer limited market opportunities. It remains to be seen whether Vestas can compete with Chinese manufacturers by expanding into areas of lower wind speed with smaller turbine sizes.

As Vestas has expanded its manufacturing presence in China, it has expanded its R&D activities based there as well. One collaboration with the State Grid Energy Research Institute in 2010 addressed challenges with wind power integration.[28] Vestas is also sponsoring wind power systems research at Tsinghua University and Xi'an Jiaotong University and has supported training efforts in China through the Sino-Danish Renewable Energy Development Program.[29] Vestas China's biggest R&D announcement to date, however, came in October 2010 when it announced the opening of a $50 million technology R&D center in Beijing, covering areas such as high-voltage engineering, aerodynamics, and material and software development. According to the company statement, Vestas plans to invest $50 million over five years and employ more than 200 engineers and technology specialists by 2012, 95 percent of whom will be recruited locally.[30] About 34 percent of the company's employees (and 6 percent of staff in the Asia Pacific region) were R&D employees in 2009, with R&D costs representing about 1.4 percent of the company's total revenue that year.[31]

GAMESA

The leading Spanish wind turbine manufacturer and among the top ten globally, Gamesa Corporación Tecnológica has had successes that can be attributed at least in part to its early partnership with Vestas. The Grupo Auxiliar Metalúrgico (Gamesa's parent company) was founded in 1976 by Tornusa, a private investment company. In the 1980s IBV (a joint venture between Spain's second largest utility, Iberdrola, and the financial institution Banco Bilboa Vizcaya Argentaria) took over 91 percent of the Gamesa Group, and in October 2000 Gamesa became publicly listed on the Spanish stock market. In the 1990s the company had two core industries, renewable energy and aeronautics, but by 2008 it had refocused its business exclusively on wind energy technology. Gamesa currently manufactures wind turbines, develops wind farms, and offers operation and maintenance services.

The Gamesa-Vestas joint venture was formed in 1994 as a way for Vestas to manufacture wind turbines in Spain. The Spanish market became particularly lucrative in the mid-1990s when big utilities started to place large orders for wind farms to benefit from government incentives. That firm, Gamesa Eólica, consisted of Gamesa Energía (51 percent), Vestas (40 percent), and the industrial holding company of Navarra's regional government, Sodena (9 percent).[32] By 1997 Gamesa Eólica controlled 70 percent of the Spanish wind turbine market. The joint venture with Vestas landed Gamesa "exclusive rights to manufacture, assemble and sell Vestas technology in Spain."[33] Gamesa made no attempts to hide the origins of its company's turbines in Vestas's technology. While other domestic turbine competitors emphasized the fact that their wind turbine technology was 100 percent Spanish, Gamesa frequently emphasized its ties to Vestas.[34] Gamesa also was able to build upon the Vestas technology with its own R&D activities. For example, in 1997 the company introduced the Ingecon variable speed system to its turbines, which it claimed increased nominal capacity by approximately 6 percent and was therefore an improvement on the Vestas design. The system was widely applied to its biggest seller, the 660 kW G47 turbine, based on the Vestas V47.[35]

In November 2001 Gamesa Eólica reached a deal with Vestas to buy out its share of the company and maintain the intellectual property rights to continue to utilize and build on Vestas's technology in the Spanish, and global, market. The split between the two companies resulted in Vestas relinquishing its 40 percent share in Gamesa Eólica to the Gamesa group for 287 million euros and began the phasing out of the technology transfer agreement between the two companies. As Gamesa continued to grow, it needed to be able to expand beyond the geographical confines of its technology transfer agreement with Vestas. The original agreement limited Gamesa Eólica's market mainly to the Iberian Peninsula and some Latin American countries, while Vestas remained free to supply the rest of the world. Following the separation deal, Gamesa Eólica acquired the right to sell anywhere in the world. It also retained its technology transfer agreement with Vestas for the two 850 kW models until the end of 2002 and for the G66–1.65 MW and V80–2 MW machines until the end of 2003.[36]

In early 2002 Gamesa Eólica also reached a deal with Germany's REpower Systems to manufacture and market its 1.5 MW turbine. In 2003 Gamesa acquired Spanish turbine manufacturer Made, which held 14 percent of

Spanish market share in 2003, further increasing its global reach.[37] Gamesa has since expanded globally, with manufacturing facilities based in Germany, India, China, the United States, Brazil, and eleven regions of Spain, and R&D facilities based in Spain, the United States, Denmark, Scotland, and China.[38] As of 2011 the company had received type certification on its new 4.5 MW turbine—the product of five years of work and 200 million euros of investment.[39]

Since its separation from Vestas, Gamesa has substantially expanded its operations in China and, among the foreign manufacturers, has done quite well there. Like Vestas, Gamesa established its first China-based manufacturing facility in Tianjin in 2006 through its wholly owned subsidiary Gamesa Wind Tianjin. Gamesa's Tianjin factory began by producing blades, nacelles, gearboxes, and control systems for its 850 kW turbine. In 2010 Gamesa introduced its 2 MW turbine model to the Chinese market by establishing a new manufacturing facility in Da'an city in Jilin province, and by 2011 it had six manufacturing plants across China.

Gamesa has been one of the most successful foreign turbine manufacturers in partnering with the largest state-owned power companies that are behind most wind power development in China. It has signed strategic agreements with Guangdong Nuclear Wind and Datang Renewable Power Co. Ltd. to develop and supply 1,315.3 MW of wind capacity between 2010 and 2013, and it is supporting Longyuan in its plans to develop wind farms outside China.[40] Gamesa and Guangdong Nuclear Wind began operating their first two wind farms in Taipingshan and Tangwangshan in 2010, and Gamesa is developing another 289.5 MW in Liaoning with Datang.[41]

A German Venture

Nordex's founders began manufacturing wind turbines out of a small boiler-making workshop in Denmark in 1985. By 1987 the company was producing the world's largest wind turbine at the time, with a rated capacity of 250 kW. By 1993 Nordex was already working on developing a 1 MW machine, and in 1995 it installed the world's first megawatt series turbine. While Nordex began operations in Germany in 1992, it was not until 1996 that it became a majority German-owned company. The company has merged with and acquired several wind technology companies over the years, including its 2003 merger with turbine manufacturer Sudwind Energy GmbH.

Nordex SE, headquartered in Rostock and Hamburg, includes Nordex Energy GmbH, Nordex Energy B.V., and further subsidiaries. All of Nordex's China operations (four subsidiaries or joint ventures as of 2011) are part of Nordex Energy GmbH. The company is currently expanding into larger and larger turbines, with a focus on offshore development. Current product models include two 1.5 MW machines (the N77 and N82) and three 2.5 MW machines (the N100, N90, and N80), plus the 2.4 MW N117 unveiled in December 2011. Nordex installed its first offshore wind turbine in 2003 (the N90/2300).[42]

In 2003 Nordex experienced major financial challenges. According to a letter to the shareholders, it "dipped deeply into the red in 2003," which came as a shock to many since up to then "everything pointed to growth."[43] The company believed that "old management had been mistaken regarding the further demand in our main markets . . . and had failed to adapt business processes to the new industrial standards."[44] In fiscal year 2002–03 Nordex recorded an annual loss of 154 million euros. As result, the CEO was suspended from duty by the company's Supervisory Board in February 2003; in August 2003 he resigned, citing personal reasons. That summer Nordex launched an operative restructuring program that streamlined the company and reduced losses. According to company reports, the financial troubles of the company had effects on operative business. Owing to the weak equity base, many banks rejected Nordex as a supplier of larger wind farms.

The company was able to work with investors to recapitalize and has since recovered from these more challenging times. Stock prices increased after 2005 through 2007 but then began to decline again in 2008. Despite being hit by a drop in year-on-year sales, Nordex increased its overall profitability in 2010 after reducing costs and generating greater income from realized projects.[45]

Although Nordex does not have the global reach of other leading wind turbine manufacturers, it has sales offices in eleven countries in Europe as well as in Japan, China, and the United States. In addition to its production facilities in Germany, it also has a nacelle production facility in the United States, and blade and turbine production facilities in China.[46]

Nordex was one of the early wind turbine manufacturers that came to China in the 1980s to evaluate the country's wind potential and demonstrate smaller turbines. The company began to import larger turbines from Germany to China in 1995, starting with a 250 kW model and then moving to a 600 kW model. Nordex established its Asia-Pacific headquarters in

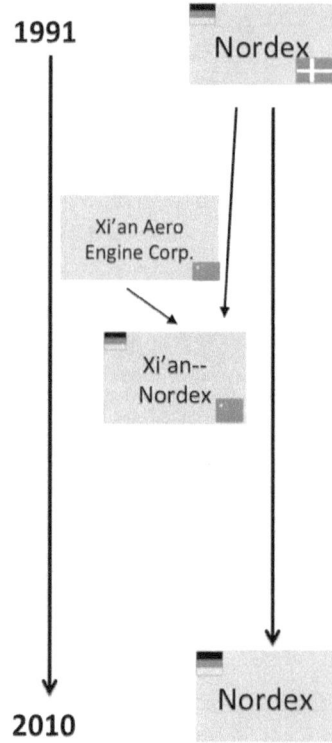

FIGURE 4.4 Timeline of Nordex Company Structure in China

Illustration by author.

Beijing in 1997 in an "optimistic but difficult market environment."[47] At this time it was still a second-tier wind manufacturer and virtually unknown to China. The former State Development Planning Commission's Ride the Wind Program, launched in January 1997, offered 400 MW of wind projects to attract foreign turbine manufacturers to team up with preselected local partners for technology transfer as part of its target for 1,000 MW of wind power capacity to be installed in China by 2000.[48] Nordex was selected to form a joint venture with Xi'an Aero Engine Corporation, and the two companies signed a contract forming the new company Xi'an-Nordex in May 1997, laying out their plans to manufacture turbines for 200 MW of new wind development.[49]

In July 1997 Nordex employee Carsten Pedersen stated that Nordex's intention was to meet this 200 MW target by the end of the year, and that the plan for the joint venture was to pursue an ambitious localization program that would increase the local content used in the turbines to 80 percent.[50] Neither of these goals was met. In fact, as of the end of 2009 Xi'an-Nordex had installed only forty-nine of its 600 kW turbines, totaling 29.4 MW. All these turbines were erected in China between 2000 and 2004.

Nordex's decision to set up a joint-venture enterprise in 1997 corresponded with the Chinese government's relaxation of policies that had previously restricted the investment of foreign capital in the electric power sector, though JVs were still required in many sectors. Nordex's joint-venture presence allowed for the China-based sales and technical support departments of Nordex Germany to be expanded. This expansion in China also included support for project development and implementation.

As part of the joint-venture agreement, Nordex transferred a license for production of its 600 kW turbine. Only some components, including the machine housing, blades, and masts, were manufactured by Xi'an-Nordex in China.[51] Even after transferring the technology to manufacture its 600 kW machine in China, however, Nordex continued to sell its larger, more advanced turbines, which were imported from Germany. In 2001 Nordex installed four 1.3 MW turbines in Liaoning province, which were the first MW-scale turbines to be installed in China. Some of the turbines used Chinese-made components, including blades from LM Glasfiber Tianjin and other components from the Balke-Dürr Zhangjiakou Installation Company.

While Xi'an-Nordex was initially responsible for the assembly of Nordex turbines in China, the development of local components, and the purchase of components on the local market, Nordex Germany had simultaneously established service centers in Shantou, Shanghai, Beijing, and Changchun in the 1990s to serve as Chinese outposts for Nordex Germany. In 1999 Nordex founded a company in Waigaoquai, Shanghai, to conduct service on wind turbines and the sales of spare parts; this company was 100 percent German owned. Spare-parts sales were also managed out of the Shanghai office. Nordex set up this separate company because the joint venture (JV) in Xi'an was not legally permitted to conduct service on products it had not produced, and Nordex wanted to continue to promote the sales of its more advanced turbines being manufactured in Germany.

According to Nordex executives, the company was aware of the fact that the joint-venture enterprise was unlikely to be as successful as the Chinese government promised. It was still willing to proceed with the technology transfer arrangement, though, because it believed it would bring value to Nordex in the form of "the reputation of a government approved and supported wind turbine manufacturer being in accordance with state policy." Nordex believed that the reason the Chinese government selected it for the Ride the Wind Program was that it was willing to "accept conditions required by the Chinese government that would have been unacceptable if considering only the joint venture itself," since it looked at the joint-venture plan as just part of its overall China market entrance strategy.[52] Examples of these otherwise unacceptable conditions included an equity majority share for the Chinese partner and the transfer of design technology in return for a technology transfer fee.

In the beginning "only manufacturing technology was transferred, with the design technology to be transferred and paid later" with the revenues earned by the joint venture. As the JV was being developed, Nordex was simultaneously installing sales and technical support departments in Beijing and a broad sales network around China. The company was not threatened by the JV since it knew that its Chinese partner had little experience in sales and marketing.[53] The JV turbines and the imported turbines would be marketed by the same people—the Nordex sales office in Beijing. As a result, Nordex was able to build up a strong sales network among customers, approval authorities, and power companies and, together with its rising reputation from the JV, increase its market share in China.

It is evident that Nordex decided that the costs and concessions associated with the technology transfer, including its taking a minority ownership share of the company and transferring the intellectual property rights of one of its best-selling turbine designs, were outweighed by the benefits of the arrangement. Perhaps Nordex knew that the transferred turbine model would quickly become outdated as Nordex Germany continued to develop larger and larger turbines. It may also have assumed that its participation in the joint venture would help its relationship with the Chinese government as well as increase its name recognition and reputation in a market that appeared to be on the verge of taking off, thereby helping the company to pursue its real goal of selling more imported turbines from Germany. Nordex was able minimize its upfront investment by renting a workshop from a Chinese company rather than building a new one. It was also given a say in the selection of the Chinese general manager of the joint venture and believed that despite its minority ownership it was able to "maintain control over all important decisions."[54]

Despite ties to the Chinese government, the JV company had a hard time competing with international companies with better reputations. Xi'an-Nordex moved ahead with developing its own prototype machine in the year 2000 (a 600 kW turbine installed in Xianrendao, Liaoning) even though sales were slow. Around the same time, SDPC started a subsidy program for developers using locally manufactured turbines, which resulted in an order for ten additional turbines to be delivered in 2001.

Nordex reportedly found that it could not achieve cost reductions through local production, at least not in the short term. No single component purchased in China was cheaper than a series product from Europe, and the transportation price from the coast to Xi'an and back resulted in an even higher cost than expected.[55] It also became increasingly apparent to the company that Xi'an was not a strategic location for a manufacturing plant since no wind farms were being developed there at that time. The company chose to locate in Xi'an purely because the government-selected Chinese partner was already located there.

Despite the setbacks of the Xi'an joint venture, Nordex Germany continued to sell its larger wind turbines in China, and the company continued to pursue joint ventures with other Chinese firms. In 2001 Nordex acquired a 66.67 percent stake in Qingdao Huawei Wind Power Co. Ltd.

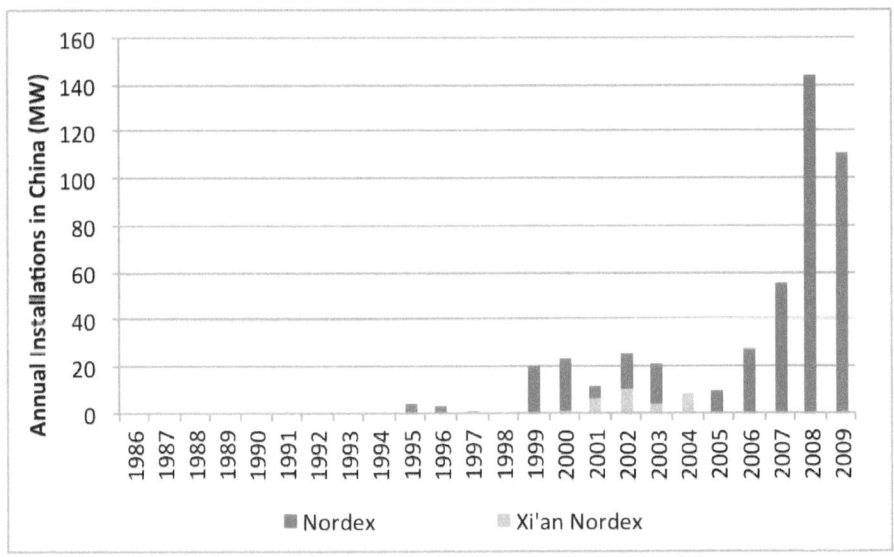

FIGURE 4.5 Wind Installations in China (Nordex and Xi'an Nordex)

Author's database.

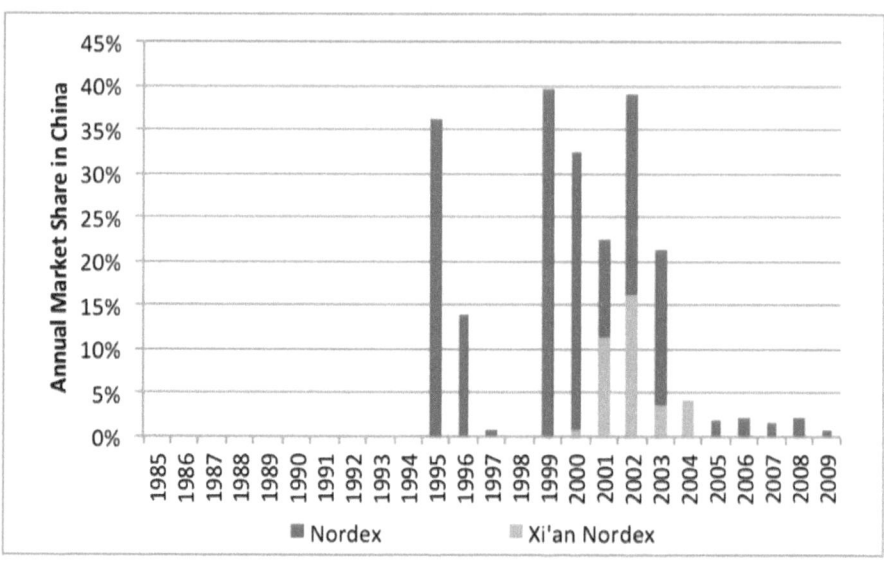

FIGURE 4.6 Annual Market Shares in China (Nordex and Xi'an Nordex)
Author's database.

of Qingdao in Shandong province in order to develop a large-scale wind farm there.[56] The project was completed in 2004 and financed with a 10 million euro loan from DEG, a German development finance institution that grants loans to private companies in developing and reforming countries. After many years of resisting local manufacturing in China, Nordex began moving toward manufacturing turbines with local content by cooperating with local partners other than its Xi'an joint venture. In December 2005 Nordex launched a joint venture with two companies in Yinchuan, Ningxia province, to produce 1.5 MW turbines.[57] Also in 2005 Nordex established a blade facility in Baoding; in 2006 it built a manufacturing facility for nacelles in Yinchaun; and in 2007 it created another blade factory in Dongying.

Despite constant changes to Nordex's manufacturing strategy in China, the company has yet to regain the market share, or annual sales numbers, that it had previously secured. Nordex's steep drop in performance in China is certainly due in part to the financial troubles described above but also to the fact that the company appears to have focused its sales efforts on markets outside China.

The American Newcomer

GE is a relative newcomer to the wind business in China compared with other foreign manufacturers, installing its first wind turbines there in 2004. General Electric, however, has been working in China for decades in a variety of industries, including electric power. A GE steam turbine installed near Shanghai in 1917 was only recently retired.[58] As China began to open up to foreign companies in 1980, Jack Welch, GE's former chairman and chief executive, reportedly went to China looking to sell aircraft engines, power turbines, medical equipment, and lightbulbs. Soon after, GE Power Systems, later renamed GE Energy, was selling coal- and natural gas-fired steam turbines to China's electric utilities.[59] By the early 1990s GE was assembling medical diagnostic machines in China, and more recently the company has opened a research center in Shanghai and expanded a medical manufacturing facility in Beijing.

GE sold its first wind turbines to China's Longyuan power company in 2004: ten 1.5 MW turbines to be installed in Huitengxile, Inner Mongolia.[60] A long line of U.S. turbine manufacturers, including U.S. Windpower, Kenetech, and Enron, all of which eventually were absorbed by GE or its predecessors, have had a much longer history of selling turbines to China. The first U.S.-made turbines installed in China consisted of five 100 kW U.S. Windpower turbines installed in Zurihe, Inner Mongolia, in 1989. Zond, which went bankrupt and was purchased by Enron Wind, is the U.S. manufacturer with the second largest installed capacity in China after GE: 16.5 MW installed in 1998 and 1999. Historically U.S. turbine manufacturers have played a smaller role in the Chinese market than the Danish and German manufacturers, although GE has been making a sizable contribution to annual installations in China since its market entry in 2004.

WIND POWER TECHNOLOGY DEVELOPMENT IN THE UNITED STATES

U.S. Windpower, founded in 1980, developed one of the only successful U.S.-designed wind turbines at that time and at one point was the largest wind company in the world. In 1992 its turbines made up about 30 percent of California's wind generation capacity.[61] Despite initial success, U.S. Windpower, which later changed its name to Kenetech Corporation, declared Chapter 11 bankruptcy on May 29, 1996. The unstable

U.S. market was blamed for the majority of the company's problems, but blade failures on several of its machines operating both in the United States and overseas placed additional financial burden on the already struggling company. It is reported that Kenetech had "compromised its turbine engineering to meet financial goals and had to declare bankruptcy after engineering problems emerged while the company was straining to meet ambitious development goals."[62]

In May 1996 the American Wind Energy Association (AWEA) declared that the bankruptcy was a clear sign of serious problems in the U.S. wind industry, and that the United States was in danger of losing out on a growing multibillion dollar global market, for which it blamed the U.S. government's inconsistent policies on wind development.[63] The U.S. market had been filled with ten-year power purchase agreements (PPAs) that dropped off after the eleventh year, causing many wind farm owners to find they were suddenly faced with a substantial decrease in revenue from electricity generated. In the quarter preceding its bankruptcy, Kenetech's average price for electricity received on its 178 MW of capacity dropped from $0.087 per kWh to $0.053 per kWh; its additional capacity would see a similar drop in price at the end of 1997.[64]

After Kenetech went bankrupt, its assets continued to play a role in the U.S. wind industry. Zond, often referred to as Kenetech's archrival, bought up several of Kenetech's undeveloped projects after Zond was acquired by Enron and became Enron Wind in 1997. Enron picked up Kenetech's U.S. patent on variable-speed wind power electronics, which, although not clear at the time, was probably one of Kenetech's most valuable assets.

Zond Systems was established in the mid-1980s and became a publicly traded company in 1994. Zond is credited with installing more than three thousand wind turbines and had a twenty-year history of wind project development. The purchase of America's largest remaining wind company by Enron, by then an established giant of independent power marketing, was an event without parallel in wind energy history.[65] However, it served as a harbinger of the even larger multinational companies that would eventually enter the industry, including GE and Siemens.

In 1997 Enron Corporation was one of the world's largest integrated natural gas and electricity companies, with $19 billion in assets and one of the largest natural gas transmission systems in the world. Prior to Enron, other large corporations, including Westinghouse, Boeing, MBB, and British Aerospace, had dabbled in the wind power industry, but their interest soon faded. For many the Enron acquisition of Zond Systems was proof that wind power had indeed come of age, since "even companies who made a great deal of money from oil and natural gas were now seeing the light when it comes to the bright future of renewable energy."[66] For others, the purchase served as a cynical attempt by a company spending $200 million in national advertising to cloak itself in the virtues of green energy while continuing to peddle more polluting forms of electricity to the majority of its customers.

In addition to purchasing Zond, in October 1997 Enron Corporation also acquired the former German wind turbine manufacturer Tacke Windtechnik, rescuing it from bankruptcy and bringing it under the Enron Wind umbrella. The Tacke assets were bought through a new German subsidiary, Enron Wind Holding GmbH. Enron kept the company operating under the name Tacke, and it continued wind turbine manufacturing and sales in Europe, India, the Middle East, and North Africa.

It is very likely that Enron's purchase of Tacke was spurred by an interest not only in Tacke's 1.5 MW turbine technology but also the gearbox technology that came from Tacke's link with its parent company, a former gearbox manufacturer, giving Tacke over one hundred years of experience in gearbox manufacturing.[67] The Zond Z-series turbines had faced high-profile problems with generator and gearbox design, necessitating major retrofit work that Enron was left to deal with.[68] The gearbox problem with the Zond turbines was analogous to the earlier problems experienced by Kenetech Windpower. The problems with the Z-series raised operational and maintenance costs on installed projects to levels higher than expected by Enron when it bought Zond in early 1997. Enron consequently made the decision to phase out the Z-series in the United States and instead assemble a version of the 1.5 MW Tacke machine under the direct supervision of Tacke's European technicians. Tacke was among the first companies to produce a commercially available 1.5 MW wind turbine, and at the end of 1996 it was the world's fifth largest wind turbine manufacturer. Some of GE Wind's turbine technology today is likely based on this original Tacke design.

By the 1990s Enron was the only remaining major U.S. turbine manufacturer. But Enron's career in the wind industry was to be relatively short-lived owing to the company's bankruptcy amidst scandal in December 2001. Enron left behind many valuable assets that had previously belonged to Zond, Tacke, and Kenetech, though the most valuable asset likely was the American patent on variable-speed wind turbine electronics. Many large companies explored purchasing what was left of Enron Wind, including

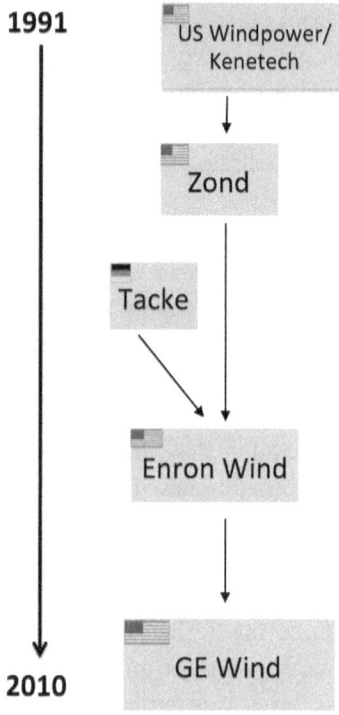

FIGURE 4.7 Timeline of U.S. Wind Companies in the Chinese Market

Illustration by author.

Danish manufacturer Vestas. However, in the end it was GE that succeeded, purchasing Enron Wind in May 2002.

With the purchase came the wind technology manufacturing knowledge of a long line of U.S. wind manufacturers, combining some of the best wind industry know-how from both the United States and Germany. Many saw GE's entry into the wind industry as an important turning point in the credibility of the industry. The decision of a large company like GE to enter the wind business meant that it acknowledged the industry's growth potential. It also may not have been fully coincidental that legislation extending the U.S. production tax credit (PTC) was passed only five weeks after GE had announced its market entry, since GE carried tremendous weight in Washington.

INTELLECTUAL PROPERTY AND PROTECTIONISM IN THE WIND INDUSTRY

GE Wind now holds a valuable card that has been passed along through the U.S. wind turbine manufacturing industry for decades—the patent for the variable-speed wind turbine. The American patent was awarded to U.S. Windpower in 1993. In 1995 U.S. Windpower (Kenetech) sued Germany wind company Enercon for patent infringement and won, thus preventing one of its major competitors from selling its variable-speed technology in the United States.[69] A year later the U.S. International Trade Commission banned Enercon technology from the United States, and Enercon lost a major wind plant order in Texas as a result. The patent changed hands four times among American companies before falling into the hands of GE as part of its purchase of Enron Wind. This much-sought-after intellectual property has been the subject of countless legal appeals, including an unsuccessful Enercon countersuit against Kenetech in 1997 for "losses sustained" and for abuse of the legal process.

The technology at the center of this dispute is the electronics necessary for the operation of variable-speed wind turbines, which—despite added complexity over stall or pitch-controlled models—have gained in dominance in the wind industry to the point that nearly all wind turbine suppliers today offer some form of variable-speed option. The economic advantages of variable-speed technology are project specific, but it generally offers greater versatility for meeting utility criteria for grid stability for large penetrations of wind power on electricity systems.[70]

The patent in question, U.S. patent 5,083,039 of the U.S. Patent Office, is not a patent of the variable-speed technology itself; rather it is its combination in a "unique and unusual" way that is patented.[71] German company Enercon has consistently argued that the patent granted to Kenetech was for technology already in the public domain, an argument made by citing a wind technology text book from 1989 and claiming that information that was publicly available cannot be patented.[72] However, American patent authorities and courts have consistently rejected Enercon's arguments. European manufacturers, including Vestas, have been forced to make special modifications to their turbine models for the American market to get around patent infringement. They have found this to be costly and inefficient, and they claim it has prevented technological progress in turbine development and the global dissemination of wind technology.[73]

The legal battles have not been limited to Enercon and Kenetech. Since the first lawsuit, there has been a long, drawn out legal battle between U.S. and European manufacturers. Since purchasing Enron Wind in

2002, GE has filed similar patent infringement suits in Europe and Canada against foreign manufacturers. Others in the industry have described GE's actions as "aggressive, negative, and restrictive," saying that "GE's move is diverting attention from the real business of building up a healthy wind industry" and that "it is a naked attempt to dominate the wind market with some dodgy old Yankee patent."[74] In rebuttal, Steve Zwolinski, former head of GE Wind Energy, said that GE believes it is important for the long-term growth of the industry that the intellectual property structure is upheld and clearly understood. "We are intending to invest quite a lot of money in this industry. We intend to make sure that the value of that investment is protected in the IP structure."[75] A partial settlement reached in June 2004 appeared to allow Enercon to sell turbines in the United States, but a legal battle remained in the European Patent Office. Enercon also filed a similar suit against GE to prevent it from selling turbines in Germany.[76]

Japanese manufacturer Mitsubishi was also the subject of a GE patent infringement claim, filed in 2008. In January 2010 the International Trade Commission ruled in Mitsubishi's favor, saying that it did not find Mitsubishi to have violated GE patents.[77] Not giving up that easily, GE filed a new complaint in February 2010 claiming that Mitsubishi infringed on 148 patents related to wind energy.[78] In May of that year Mitsubishi filed two lawsuits against GE, including an antitrust complaint alleging that GE deterred companies from purchasing Mitsubishi technology and a patent infringement case alleging that GE infringed on its patent for variable-speed wind turbines.[79] It remains to be seen whether GE's patent claims will continue to serve as a means of keeping foreign competition out of the U.S. market.

GE'S POWER SECTOR TECHNOLOGY TRANSFERS IN CHINA

As mentioned above, the wind industry is far from GE's first experience working in China. It has decades of experience working in China that are potentially valuable to its wind turbine business, specifically its extensive experience manufacturing and selling gas and hydropower turbines.[80] GE Energy, formerly GE Power Systems, has been in China for more than ninety years and has provided that country with 56 steam turbines, more than 160 gas turbines, and more than 100 hydro units.[81]

In 2003 GE was awarded a contract to supply thirteen units of 9FA gas turbine–based combined cycle systems to China in a contract worth $900 million. In conjunction with this contract, in February 2004 GE Energy signed an agreement to establish a joint venture with the Harbin Power Equipment Company (HPEC) of China. The $16 million joint venture, GE-HPEC Energy Service Company, was located in the development zone of Qinhuangdao city where HPEC already owned a manufacturing base.

This joint venture had GE owning the majority of the company (51 percent), even though majority foreign ownership had been prohibited until recently. The JV was to serve as a service company, providing repair and field services to the GE 9FA model heavy-duty gas turbines. In addition, the 9FA turbines were to be assembled at the facility by HPEC in a technology transfer agreement with GE. At the time the deal was made, GE's F-class gas turbine fleet included more than 900 units in operation worldwide—an installed fleet bigger than that of all other major competitors combined—giving the company a technological lead in advanced gas turbine technology experience.[82]

Before the deal with GE to transfer gas turbine technology to China was reached in 2003, Chinese manufacturers reportedly only had the technology required for making much less efficient steam-powered turbines. Chinese companies had acquired that technology in part through various previous joint-venture collaborations, including the power systems division of Westinghouse Electric Company (now owned by Siemens). At least two Chinese companies were already exporting gas turbines to Southeast Asia and the Middle East at that time. However, the Chinese government also wanted to acquire the technology behind the vastly more efficient gas-fired turbines.

As a result of the technology transfer from GE, Chinese officials reportedly hoped to get "drawings for the entire turbine, including the modeling and mathematics behind the shape of the turbine's blades, how the blades were cooled while rotating, the chemistry behind the blade's makeup, and the thermal protective coating on the first row of blades, where temperatures are the highest."[83] Negotiators for GE and the two Chinese companies worked at six different sites in China for more than three months to craft a deal. Under the final terms, GE agreed to have HPEC assemble GE's turbines at its factory in northeastern China and manufacture most of the less sophisticated components in the turbine. Additionally, GE formed a separate joint venture with Shenyang Liming Aero-Engine Group Corporation to transfer technology for the combustion systems in the turbines, allowing the Shenyang venture to manufacture the second and third rows of blades inside the turbine. This was technology that GE reportedly was not willing to give up initially. Included in the transfer were technical drawings of a key cooling system and the advanced metallurgy of the blades.

GE reportedly opted not to relinquish the "most secret elements of the turbine, including the design of the cooling system for the first row of blades and the technology behind a thermal protective coating for those blades," and Chinese negotiators eventually acquiesced.[84] Whether this was GE's own decision was unclear, as U.S. export rules may have prohibited the company from sharing this technology since it is used both in power-turbine production and in aircraft engines, which are theoretically protected under military security regulations. GE therefore opted to manufacture the first-stage blades at one of its own plants in South Carolina and then ship them to the Harbin factory for final assembly.

GE apparently believed that by maintaining majority ownership in the joint venture it could retain control over the most sensitive technology. The Chinese partner would not be able to fully exploit what it learned from the transfer until its engineers were able to replicate or build upon the advanced technology it obtained, which would likely take a while. By the time Chinese companies figured out how to replicate the technology, it would likely be outdated, and GE would have moved on to newer, more complex designs. GE executives "thought China would eventually get the technology one way or another."[85] Additionally, GE recognized that its potential for future market share in the United States was limited since the market for power generation equipment was growing much more slowly there than in China.

This example illustrates a relatively high-risk technology transfer model. To be considered for equipment contracts totaling several billion dollars, GE was required to form a joint venture with a state-owned company and to transfer technology and advanced manufacturing guidelines for its 9F turbine. Even though GE had spent more than half a billion dollars to develop the technology, the contract awarded was apparently worth the risk of transferring the technology to China.

Another GE technology agreement in China also took place in 2003 when GE Power Systems acquired majority ownership of Kvaerner Power Equipment Co., Ltd. (Kvaerner Hangfa) of Hangzhou, one of the leading suppliers of hydropower generation equipment in China, marking the largest acquisition by GE Power Systems in China up to that time.[86] The new company, GE Hydro Asia Co., Ltd., became integrated into the global operations of GE Hydro, a division of GE Power Systems, with GE holding a 90 percent ownership share and Hangzhou Industrial Asset Management Co. Ltd., Hangfa's parent company, holding only 10 percent.[87] This example is clearly quite different from the gas turbine example, in that GE was able to maintain majority control of the joint venture by establishing a new company under the GE name and integrating the Chinese partner into GE's operations.

GE'S WIND POWER BUSINESS IN CHINA

When GE entered the wind industry in China, it did not identify a local JV partner as it had done in other industries. Conditions were quite different in 2004 from what they had been in the early 1990s, when other leading foreign wind turbine manufacturers entered the Chinese market using imported technology. Now that the government had announced hundreds of megawatts of wind concession projects, mandating that a minimum of 70 percent local content be used in the chosen wind power technology, foreign turbine manufacturers were already exploring localization options. Although it was a relative newcomer to the wind industry, GE opted to build directly on its decades of experience in the power sector, which included a large supply chain in China and existing relationships with component suppliers. It also had name recognition in China that few foreign wind companies could compete with and therefore had a strong advantage in attracting new customers familiar with the GE brand.

GE believed its comparative advantage was in systems integration. While the company had the ability to meet China's then 70 percent local content requirement upon entering the Chinese wind market in 2004, GE was concerned that this was not going to be an acceptable business model to the Chinese government. Aware of the technology transfer requirements it had dealt with in the past, GE recognized that this model of using local components spread across many suppliers would result in far less technology transfer than many of its previous ventures in China.[88]

GE opted to enter the Chinese market with a locally manufactured 1.5 MW turbine, built with components sourced locally and certified and assembled by GE in existing GE factories.[89] GE relies on its Six Sigma certification system to guarantee the quality control of its products no matter where in the world they are manufactured, and it also offers a comprehensive technology warranty and extensive after-sales service.[90] As a result, GE

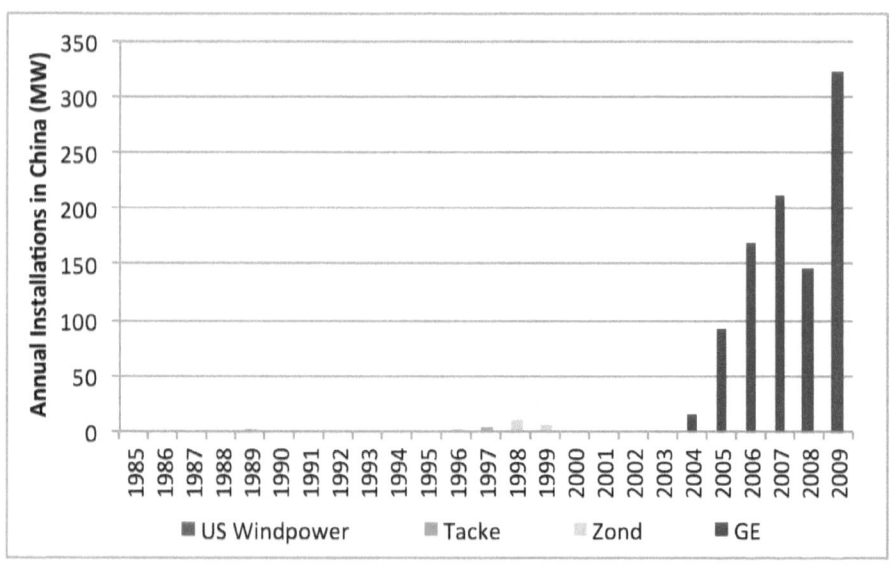

FIGURE 4.8 U.S. Wind Turbine Models Installed in China (U.S. Windpower, Tacke, Zond, and GE Wind)

Author's database.

was the first foreign wind turbine manufacturer to enter the Chinese market with a wind turbine that was predominantly made in China. The first turbines offered to the Chinese market not only could meet the 70 percent local content requirement but could utilize up to 90 percent local content, which was far higher than the level any other foreign wind turbine manufacturer could meet in 2004.[91] This model of localization also meant that GE did not have to transfer any technology to Chinese companies. Although GE likely trained Chinese companies that were supplying parts for its turbines, which may have resulted in some knowledge transfer, there was no direct transfer of IP rights in GE's initial foray into China's wind industry. This was in contrast to some of the models that GE has used in the past, including the transfer of gas turbine technology to Chinese partners.

While GE did not initially open a wind turbine–specific manufacturing facility in China, it utilized its China Technology Center in Shanghai as an initial base for wind technology R&D. GE's innovative capabilities have been attributed to the company's ability to harness global talent, which it accesses through its global innovation network. With major research facilities

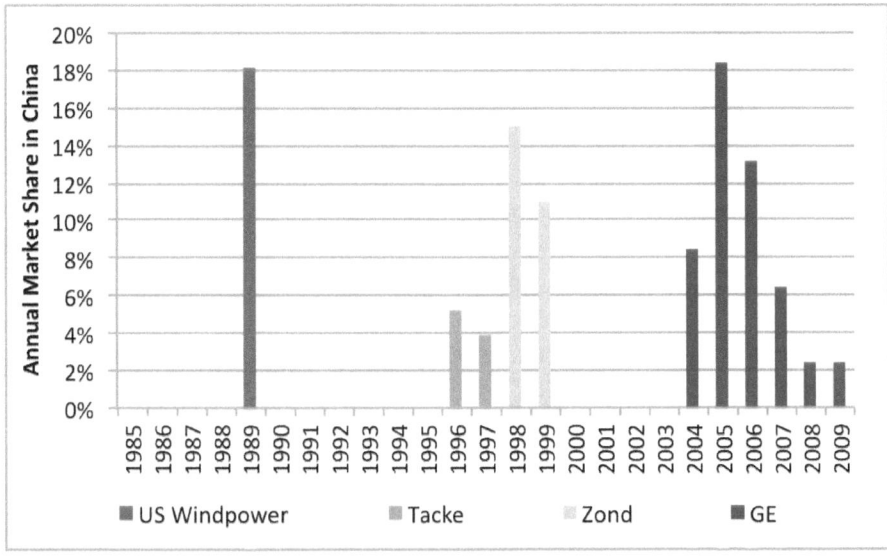

FIGURE 4.9 Chinese Market Shares of U.S. Turbine Manufacturers
Author's database.

located in Shanghai, Munich, Bangalore, New York State, and Rio de Janiero[92] that innovate in all areas of GE's businesses, and a diverse workforce that includes Americans as well as local workers, GE has been described as being more effective than other megacompanies with a global workforce in its ability to "mobilize experts from diverse disciplines and locales in pursuit of a common goal."[93] The GE China Technology Center in Shanghai includes many of the same features as GE's prominent training facilities in Crotonville, New York, including classrooms. Home to over one hundred staff focused on wind energy R&D, the Shanghai center has recently made increased investments in wind power technology; for example, in the first quarter of 2010, its R&D spending increased by 16 percent over the previous quarter, with the current focus being primarily on offshore wind technology.[94] In October 2009 the company completed the design of a 1.5 MW turbine customized for China's intertidal conditions on the coast. The wind R&D center in Shanghai has reportedly produced twenty GE patents for wind power technology in China.[95] In fact, GE holds the highest number of wind turbine patents in China, despite having a much smaller market share than the leading Chinese manufacturers.[96] Given the limitations of patents as a true measure of innovative activity in China (as described in chapter 3), it is likely that this is an illustration of GE's commitment to intellectual property rights protection (and its propensity to patent), described above.

GE opened its first Chinese facility exclusively dedicated to manufacturing wind turbines in 2006 in Shenyang, to manufacture its 1.5 MW turbines. Throughout GE's first few years in the Chinese market it was relatively successful, though like many other foreign manufacturers it has grown increasingly frustrated with the challenges of competing with emerging Chinese domestic manufacturers. While Chinese turbine manufacturer Goldwind was already selling its turbines in China when GE entered the market in 2004, other Chinese wind leaders Dongfang and Sinovel did not start large-scale production in China until 2005 and 2006, respectively.

In 2009 GE seemingly modified its model of wind power development in China by entering into a series of joint-venture arrangements. In January of that year the company announced that GE Drivetrain Technologies, a unit of GE Transportation, would partner with Chinese wind turbine manufacturer A-Power (Liaoning GaoKe Energy Group Co. Ltd), a relative newcomer to the Chinese wind industry. The agreement entailed GE supplying A-Power with gearboxes for its 2.7 MW turbines, a design that it had licensed from German company Fuhrlander, and A-Power and GE establishing a joint venture for a gearbox assembly plant. As a result, GE still would control the gearbox technology, and only the assembly would be performed at the joint-venture operation. In addition, the joint-venture gearbox assembly business was to be majority owned by GE Drivetrain Technologies, and GE planned to use it as a hub for gearbox assembly for the Southeast Asian region.[97] Since the core technology and the joint venture remained in GE's control, this partnership did not represent a major departure from GE's strategy in the wind business in China to date.

GE announced another joint venture in August 2009, however, which did represent a shift in approach. This time GE Drivetrain Technologies would be partnering with the Chongqing XinXing Fengneng Investment Company to produce large-diameter gears for the wind turbine industry at a new facility in Chongqing. Unlike the A-Power JV, GE was to be the minority owner in this operation. Initially the new business will manufacture gears for GE Drivetrain Technologies' own wind turbine gearbox production facility in Shenyang.[98]

On September 28, 2010, GE announced a new-joint venture arrangement to develop wind turbines for nearshore and offshore projects in China using direct-drive technology with Harbin Electric Machinery Company (HEC), a subsidiary of Harbin Power Equipment Company and the same company GE had partnered with in 2004 to provide gas turbines (described above).[99] The new joint venture is notable in that it focuses on direct-drive wind turbine technology and offshore projects in particular, while all of GE's wind turbine sales in China to date have been gearbox-based and onshore. GE owns a 49 percent share of this new company, and HEC 51 percent. In addition, HEC will acquire a 49 percent interest in GE's existing wind factory in Shenyang, which will continue to manufacture land-based wind turbines.[100]

GE's change in strategy may stem from new Chinese government regulations restricting majority foreign-owned firms from participating in offshore wind farm operation.[101] GE had been investing heavily in developing offshore wind technology for China and may have been looking for ways to participate in the extensive number of offshore projects that have been announced by the government (discussed in chapter 3). GE has been involved in several offshore technology demonstration programs in the United States, including a U.S. DOE–funded project to develop a 5–7 MW offshore turbine.[102] Another likely reason behind the partnership is GE's alleged interest in developing direct-drive wind turbine technology, which, while still used less frequently in the United States and Europe, is rapidly becoming the dominant wind power technology in China.[103]

Foreign Wind Companies in China

The Chinese wind power industry reflects the complex dynamics of any increasingly globalized industry, where lines of ownership across international borders are becoming more and more blurred. As companies become increasingly international or multinational, their country of origin becomes less and less significant. The players in China's wind power industry are typically grouped into three categories: Chinese firms, foreign firms, and Chinese-foreign joint ventures. However, a further investigation into the inner dynamics of these firms illustrates the limitations of these three categories.

TECHNOLOGY TRANSFER AND LOCALIZATION MODELS

Vestas, a Danish company from the beginning, has remained that way throughout almost two decades of operations in China. From about 1986 to 2006 it sold only imported turbines to China that were manufactured in Denmark and did not directly transfer any turbine technology to China. This model changed when the company realized that local manufacturing would be crucial to its survival in the Chinese market. While Vestas has remained a global market leader with 14 percent of world market share in 2010, it held only 5 percent of Chinese market share that year.

NEG Micon has its roots in Denmark and operated in China until 2003, both as a wholly owned Danish company and as a joint venture with Chinese company Goutou, where it maintained majority ownership share and the majority of the intellectual property rights associated with its turbine technology. All NEG Micon turbines sold to the Chinese market were manufactured overseas (primarily in Denmark, although at least one was made in the UK). Therefore, despite the joint venture with Goutou, the company remained primarily Danish, both in identity and in terms of technology ownership and production. One of the early leaders in the Chinese wind industry, NEG Micon was absorbed by Vestas before the local content requirements began to influence company strategies.

Gamesa has proven itself to be more than a Vestas spinoff and has also pursued a strategy of local manufacturing in China with the establishment of a production facility there in 2006. With 6 percent of global market share and 3 percent of Chinese market share in 2010, Gamesa has remained a steady global competitor with a strong foothold in the Chinese market.

Nordex, a German company with historical roots in Denmark, has operated in China both as a wholly owned German company and as a minority shareholder in the Sino-German joint venture Xi'an-Nordex. Even after transferring the design and licensing the rights to its 600 kW turbine to an ultimately unsuccessful government-arranged joint venture, Nordex opted to continue to pursue joint ventures with several other Chinese partners of its own choosing. These partnerships have allowed the company to transition to a locally sourced product line. In 2010 Nordex had 2 percent of global market share and less than 1 percent of the market in China.

GE's wind energy division is United States–based, although it is certainly the most multinational of all the turbine manufacturers mentioned in this chapter owing to its ties with General Electric Corporation. Its entry into the wind industry began with its purchase of Enron Wind, giving GE access to years of experience in the wind industry that can be traced back to both early U.S. and German wind turbine manufacturers. While some of the know-how associated with these acquisitions has reportedly been incorporated into the design of GE's turbines, the only tangible remnants of these now defunct companies are the patents held in GE's name.[104] GE's expertise in crossing national boundaries in its operations is illustrated in its wind turbine development and sales strategy for the Chinese market. The first foreign company to localize production in China, GE has used a variety of models to pursue Chinese market share, including several recent joint ventures. GE held a 9 percent global market share in 2010 while its Chinese market share had fallen to less than 1 percent.

In the global marketplace in 2010, the Danish, American, and German turbine manufacturers still lead in terms of market share that year, as well as in cumulative global installations, although Chinese manufacturers are right behind. In the Chinese market, however, market share held by foreign-owned wind turbine manufacturers has been steadily decreasing, as the Chinese manufacturers are increasingly successful. Despite the fact that the foreign firms were first movers in the Chinese market, the overarching Chinese industrial strategy to develop a homegrown wind industry eventually created successful Chinese competitors. Their ability to compete on

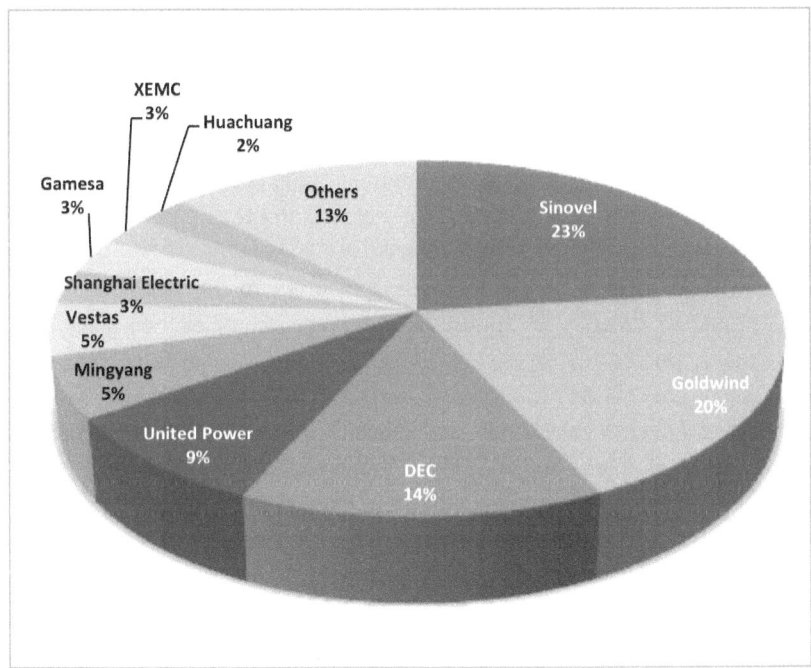

FIGURE 4.10   China Wind Turbine Market Shares

Shares are for 2010. BTM Consult (2011).

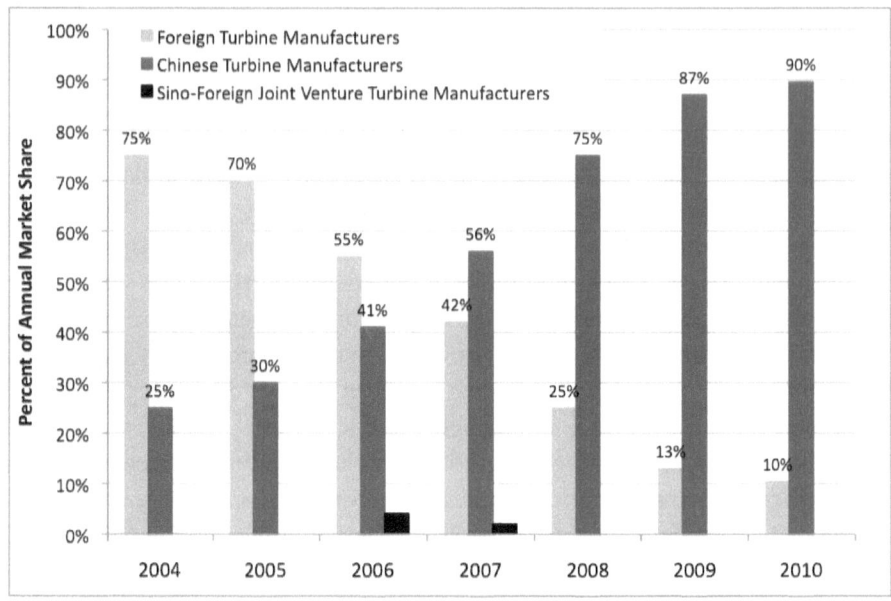

FIGURE 4.11 Chinese Market Share by Company Ownership Type

Li et al. (2010); BTM Consult (2011).

cost, in addition to the direct government support they have received (discussed in chapter 3), has proven to be detrimental to the performance of many foreign wind turbine manufacturers in China.

Each of the firm strategies for selling turbines to the Chinese market can be mapped along a continuum that ranks the strategies from those that fully retain control over intellectual property to those that fully transfer it to another firm. This is most clearly illustrated from the perspective of the foreign firms as they select their strategies to enter the Chinese market.

Foreign firms have been somewhat restricted in their actions in China in the past because of the regulations that govern foreign-owned firms, although such regulations have been relaxed in recent years in most sectors. Foreign companies can opt to enter the Chinese wind market by establishing a Chinese subsidiary company and are now able to maintain 100 percent ownership of the company if they wish. This allows the companies to maintain the highest degree of control over their foreign operations. Within this model of foreign ownership there are several options a wind energy company may pursue to produce its technology for the Chinese market: importing the entire turbine; importing key components and manufacturing or purchasing the rest locally; importing the minimum needed and manufacturing or purchasing the rest locally; or fully manufacturing turbines locally or using local materials.[105]

In the early stages of the Chinese wind industry, the Danish manufacturers NEG Micon and Vestas relied primarily on a model of selling turbines imported from Denmark. Nordex Germany continued to import turbines even after it had established a joint venture in China since it wanted to sell its more advanced, larger models of turbines, which were not being manufactured locally. In contrast, GE entered the wind industry in China by primarily using locally manufactured components for its wind turbines and importing the minimum necessary technology. After GE's demonstration of the benefits of local manufacturing, many foreign firms followed suit. Vestas, Nordex, and Gamesa all established factories in China, as did other foreign turbine manufacturers, including Suzlon and Siemens.[106]

There are many reasons a company may opt to establish a joint venture in China to manufacture wind turbines. The company may believe it can learn something from the Chinese partner and benefit from its involvement, perhaps through cost reductions or design expertise, or it may just hope to receive some sort of beneficial treatment from the Chinese government. The government has a history of encouraging foreign companies to establish joint ventures with Chinese enterprises to facilitate the transfer of advanced technology into China. In return it is often willing to offer special concessions to the foreign companies, such as a guaranteed market in the form of a large contract for goods produced in China.

The cost of establishing a joint venture in China may include a required technology transfer and the relinquishing of intellectual property rights to Chinese companies. A company transferring technology to China may fear that the Chinese partner will eventually be able to out-compete it, not only in the Chinese market but in the global market as well. Consequently there are often limitations written into technology transfer arrangements that restrict where a joint-venture enterprise is permitted to sell its technology globally.

NEG Micon, for example, established a joint-venture company with Beijing Goutou and transferred minimal and rather basic technology to this local enterprise. In contrast Nordex opted to be involved in a joint venture in which it held a minority share of the company and transferred a full turbine model to the joint venture, but it later pursued additional joint ventures with more limited technology transfers. GE opted to pursue multiple joint ventures with varying ownership structures but limited technology transfers.

The other option a foreign company can pursue to enter the Chinese market is to license the right to manufacture its technology to a Chinese company. In doing so the foreign company can elect to give up all involvement in the day-to-day operations of the company but may still profit both from the initial license sale and from any subsequent royalties agreed to under the licensing agreement for future turbine sales in China. REpower (formerly Jacobs Energie) began its involvement in China with the sale of a license to manufacture its turbines to then inexperienced Chinese turbine manufacturer Goldwind. It has since continued to be involved in the Chinese market, as well as other emerging markets, through additional licensing arrangements (discussed in chapter 6).

These different business models represent different degrees of protection of intellectual property rights. Maintaining independent ownership while operating in China allows a company to maintain control over both the day-to-day operations and the intellectual property rights associated with a proprietary technology design. If ownership is shared with another company, some degree of control over property rights likely also is relinquished. In the case of a full technology transfer, the transfer typically comes with some IPR in the context of a license arrangement for either components or a full turbine model to be manufactured locally.

It is difficult to draw conclusions about which model has been the most successful since the sample size is rather small and each company has pursued a rather different strategy. Given the local content constraint imposed by Chinese government policies, companies that had initially not chosen to locally manufacture their turbines opted to shift their production to China, but maintain control over IPR. Companies that initially pursued local manufacturing through a joint-venture arrangement that required them to relinquish control over IPR and company ownership experienced difficulty. Nordex, for example, found that partnership was difficult, particularly with a local partner that it did not choose itself, and that there was little incentive to make viable turbines when it was the minority shareholder and had less influence over the day-to-day operations of the company. Perhaps more important, since Nordex was able to continue to sell turbines directly through its German subsidiary, it had little incentive to see its joint venture produce successful turbines—or less expensive turbines—that could detract from the Chinese market share of the German subsidiary.

The other business model that appears to have proven successful is one in which both technology and IPR were transferred from a foreign company to a Chinese company and ownership was fully in the hands of the Chinese company. This eliminates conflicts of interest between the Chinese and the foreign companies either in daily operations but gives the foreign firm the incentive to see that the knowledge necessary to produce the turbine is wholly transferred so that it will be able to successfully reap the royalties from future turbine sales. This is what appears to have happened in the case of REpower's transfer of a license to produce its turbine to Goldwind. Since this licensing arrangement was made, many other Chinese companies have since pursued similar arrangements with other foreign firms.[107]

POLICY ENVIRONMENT FOR FOREIGN FIRMS

The current business models and ownership structures of wind turbine manufacturers operating in China have direct implications for the future development of the domestic Chinese wind turbine industry. The Chinese government has consistently conveyed the goal not just of promoting the increased utilization of wind turbines in China, but of increasing the use of domestically manufactured wind turbines and therefore fostering the establishment of a Chinese wind industry. This was perhaps best illustrated with the 1997 Ride the Wind Program, in which the government attempted to kick-start a domestic wind turbine industry with a series of Sino-foreign joint ventures. The program required technology transfers from the foreign partner to the Chinese partner and resulted in majority Chinese ownership of intellectual property and decision making. The hope was that these initial joint ventures would increase the technical capacity of Chinese firms in the area of wind turbine manufacturing and lead to a successful commercially produced Chinese wind turbine, and eventually many companies and many turbines.

The local content requirements that, until recently, were found in the conditions placed on wind concession project developers, as well as the developers of other wind farm projects in China, essentially caused every firm participating in China's wind industry to pursue local manufacturing. By about 2004 it was evident to most representatives of foreign wind companies operating in China that there was a need for them to move toward the local manufacturing of their wind turbines if they were to survive as competitors in the Chinese market.

For the Chinese companies, such as Goldwind, these policies had little impact on their overall business strategy since their goal all along has been to become technologically self-sufficient in the production of wind turbines and not have to import components from overseas. For the foreign firms operating in China's wind industry, however, these local content requirements have in many instances caused them to rethink their entire business strategy or, for new entrants to the industry, to shape their strategy to meet this requirement. Additionally, the later entrants to the Chinese market, such as GE, had the advantage of having observed the experiences of other foreign firms that were earlier entrants, including Nordex and Vestas, and could learn from their mistakes. There were several unsuccessful attempts at Sino-foreign joint ventures; Nordex was just one of the largest and therefore the one with the highest profile. The frustration that many of the foreign firms may have felt at being forced to take on a Chinese partner spread throughout the industry; this frustration was driven not so much by the partnership itself but rather by the government constraints it was forced to work around.

China's gradual transition to a market economy has been accompanied by a gradual loosening of the regulations that govern the operation of foreign firms in China, including policies that govern the investment, ownership, local content utilization, and technology transfer requirements of foreign firms. In the effort to develop indigenous high-tech industries, the Chinese government has become increasingly selective and restrictive in the type of imports and investments that it allows or officially encourages. In 1999 China amended its constitution to provide a legal basis for private-sector development, including foreign investment. Foreign firms may now create their own local subsidiaries in China under their complete ownership and control.

The increase in options for ownership structures of foreign firms in China is primarily due to China's 2001 accession to the World Trade Organization, upon which China committed to reduce the many restrictions on the private sector over time. Although China's own domestic regulations supposedly forbid geographical, price, or quantity restrictions on the marketing of a licensed product and foreign-invested enterprises retain the right to purchase equipment, parts, and raw materials from any source, many Chinese policies still encourage localization of production. Investment contracts often call for foreign investors to commit themselves to gradually increasing the percentage of local content utilized. In addition, officials carefully examine the sourcing of inputs at various stages in the approval process. Effective implementation of China's WTO commitments should eventually affect this bias, although clearly WTO restrictions have not changed Chinese policies in many sectors—including the wind power industry, where local content requirements were very widely used until 2009.

As regulations governing foreign companies in China were relaxed to permit wholly owned foreign subsidiaries, the unpleasant experiences of past firms, in combination with concerns over intellectual property protection, contributed to shaping the new generation of business models for wind turbine development in China. This new business model was one in which local manufacturing was achieved without the transfer of company ownership or the intellectual property rights associated with the wind turbine technology. Such a model allowed the foreign firms to meet the requirements of Chinese government policies without opening themselves up to risks associated with joint business ventures and shared technological know-how. This is the model that still persists today, but it is already facing challenges.

What was missing from this new business model was a complete technology transfer to China. The model was in fact an incomplete technology transfer—one in which modern wind turbines were being locally manufactured in China, but no blueprints or design documents were transferred to Chinese companies, and little to no local technological innovation occurred. The Chinese government may not have predicted that foreign companies would opt to localize their turbine manufacturing without formal technology transfers of the sort that would be facilitated through joint-venture arrangements. By choosing to maintain foreign ownership of companies while still localizing manufacturing, foreign companies were able to meet Chinese local content policy requirements while maintaining ownership over the IPR of their turbine designs and prevent or at least slow the rise of Chinese competitors. (In the meantime, however, smaller foreign firms were increasingly willing to transfer their wind technology to China, as will be discussed in chapters 5 and 6.)

The environment for foreign-owned wind turbine manufacturers in China is increasingly competitive as many new Chinese-owned manufacturers enter the market. Foreign companies face not only price competition but also a policy environment that is highly supportive for Chinese companies. The Chinese government is supporting research, development, and deployment in the wind industry in a manner not unlike that of Denmark and the United States in the 1980s. China's feed-in tariff policy to support wind electricity is quite similar to the one that led Germany to be the largest wind market in the world for many years. As government support for wind energy in the industrialized countries has waned in recent years, support in China has increased. This tension between local manufacturing and locally owned technology persists today and if anything has worsened. This increasing tension is illustrated in a series of trade disputes targeting China's renewable energy industries initiated in 2010 and 2011.

INTERNATIONAL TRADE TENSIONS

Regulations promulgated in 2001 related to China's WTO accession have generally improved the regulatory environment for foreign technology providers; however, in the wind industry there are myriad examples of the Chinese government requiring technology transfers from foreign firms wishing to do business there.[108] These practices have caught the attention of several foreign governments, including the United States.[109] In 1997 the United States Bureau of Export Administration commissioned a study to examine the prevalence of U.S. firms being forced to transfer technology in return for promises of market share in China and ultimately to examine the national security implications of such "forced" technology transfers.[110] The resulting report centers on role of the United States and of Western technology in aiding China's stated foreign investment and trade policy objectives of achieving the "modernization and self-sufficiency of China's industrial and military sectors." According to this study, although numerous complaints—both formal and informal—had been registered by U.S. companies with the U.S. government with regard to unfair trade practices in China, many companies were hesitant or unwilling to complain publicly or even privately about the numerous difficulties inherent in doing business in China.[111] In fact, the majority of industry representatives interviewed for the study clearly stated that "technology transfers are required to do business in China." In spite of this, most were still optimistic about their future business prospects in China, believing the technology transfer requirements were not too high a price to pay for access to Chinese markets. It is apparent from this study that China is able to leverage its enormous potential market and "play foreign competitors against one another in their bids for joint venture contracts and large-scale, government-funded infrastructure projects in China," often resulting in bidding wars over who is willing to transfer the most or the best technology.[112]

For example, GE executives have been quoted as saying that they are accustomed to negotiating an exchange of "technology for market" with the Chinese, or trading "short-term sales for long-term competition," and that China is always pushing for the "crown jewels of technology from companies that want access to China's exploding marketplace." It is also apparently understood that these demands "fall into a gray area of international trade law and economic development strategy." Blaming "the lure of mammoth Chinese markets," multinationals are hesitant to raise the issue; they are in a situation in which they want to use the newly negotiated WTO rules to their advantage in protecting their intellectual property and technology but are also eager to appease the Chinese government.[113]

In the wind power industry, local content requirements persisted long after China joined the WTO, and mandated technology transfers in return for market share are still widely used in China's electric power technology sector. In the context of the automobile industry, however, WTO violations appear to have received more attention. Prior to China's WTO accession foreign automobile firms had been required to use a certain percentage of Chinese-made parts because of local content requirements, but the Chinese government has since modified its position on this requirement at least in part.[114]

While China's local content and other potentially discriminatory requirements in its wind industry escaped international challenges for many years, a petition was filed with the U.S. Trade Representative to investigate these practices in September 2010. The petition notably came not from U.S. wind turbine manufacturers themselves, but from the major labor union for steelworkers in the United States, United Steelworkers (USW), steel being a key raw material input for wind turbines. The petition alleged that China was using many programs that violate its WTO obligations, including "discriminatory laws and regulations, technology transfer requirements, restrictions to access on critical materials, and massive subsidies that have caused serious prejudice to U.S. interests."[115] For discrimination, the petition singled out bidding preferences in the wind concessions and bases and the lack of projects going to foreign firms, the fact that foreign-owned projects cannot earn CDM credits without a majority Chinese partner, and several technology transfer arrangements in the wind industry where foreign companies were forced to transfer technology to Chinese partners as a condition for the partnership. The petition also highlighted the Ministry of Finance subsidies to Chinese wind producers, local content requirements for wind and solar (arguing that they persist even though China agreed to eliminate them in October 2009), and other EX-IM support and "trade distorting" domestic subsidies that went to Chinese green technology firms that came from stimulus and other government support.[116] The USW legal team researched the claims using Securities and Exchange Commission filings from various U.S. wind and solar companies. Most companies did not comment directly on the petition when it was released. In fact GE opted to release a statement praising China's clean energy policies and criticizing U.S. policies, perhaps out of concern for a possible backlash from this petition.[117] The primary industry association of U.S. wind companies, the American Wind Energy Association, also did not respond to the petition directly, instead using the opportunity to call for stronger policy support for wind energy in the United States.[118]

While the USW petition included a list of alleged WTO violations related to China's clean energy practices, the USTR initially targeted only one program—the Ministry of Finance Special Fund for Chinese wind producers. On December 22, 2010, U.S. trade officials announced that they had asked for WTO talks, beginning a sixty-day period for the two countries to resolve their disagreement through consultation. Prior to the formal WTO consultations with China held on February 16, 2011, according to United States Trade Representative Ron Kirk, Beijing agreed to remove requirements that foreign companies bidding for large-scale wind power projects in China have prior experience in China, recommitted to eliminating discriminatory local content requirements in wind manufacturing, and informed the United States that two other subsidy programs challenged by USW had been eliminated during the U.S.-China Joint Commission on Commerce and Trade (JCCT) talks in mid-December 2010.[119]

During the February WTO consultations, USTR "made clear its view that the subsidies provided to Chinese wind turbine manufacturers under the Special Fund program were prohibited because they were conditioned on the use of domestic over imported goods,"[120] as defined under article 3 of the WTO Subsidies and Countervailing Measures (SCM) Agreement.[121] Following those consultations, China "took action formally revoking the legal measure that had created the Special Fund program," as announced in a USTR press release on June 7, 2011, thus resolving the formal WTO dispute.[122]

USW responded to USTR's announcement of the dispute resolution by calling it "good news for our members, U.S. companies and American workers," while simultaneously calling for "continued action on our other complaints in our petition to ensure that China's protectionist and predatory practices in the clean tech energy sector are eliminated."[123] The response from Chinese wind industry stakeholders, however, was far more aloof. *China Daily* reported that "China's wind power companies expect to see little impact from the recent news that the country will end industry subsidies because of the investigation launched seven months ago following complaints from U.S. manufacturers," while Shi Pengfei, vice president of the China Wind Energy Association, said that it made sense for the Chinese government to end these subsidies now that Chinese manufacturers are strong enough to compete with international players.[124]

New trade disputes in the clean-energy arena are emerging outside of the disputes between the United States and China. For example, the feed-in tariff of the Canadian province of Ontario to support wind and solar power plants manufactured using locally produced content is the target of WTO disputes launched by Japan and the European Union.[125] The United States and China, however, remain at the center of most key debates over clean-energy manufacturing, subsidization, and deployment because of their key roles in all these areas.[126] While the two countries have been able to resolve many of their trade tensions through a combination of bilateral talks and official WTO consultations, trade tensions in the clean-energy sphere have not ebbed. This is clearly illustrated by increasing tensions over competition between U.S. and Chinese solar manufacturers in light of continuing economic struggles in the United States,[127] as well as a recent report from U.S. company American Superconductor Corporation (AMSC) of intellectual property theft by leading Chinese wind turbine manufacturer Sinovel.

AMSC expanded into the wind industry with its acquisition of Austrian firm Windtec in 2007. It entered the Chinese market through its partnership with Sinovel, engaging in joint development of several new wind turbine models.[128] What in the fall of 2009 AMSC characterized as a successful partnership[129] began to sour publicly by April 2011, when a company press release updating investors on fourth-quarter financial results stated that Sinovel had "refused to accept shipments of 1.5 MW and 3 MW wind turbine core electrical components and spare parts" that it had previously agreed to purchase, and that it has also failed "to pay AMSC for certain contracted shipments made in fiscal year 2010."[130] Sinovel was likely experiencing the overcapacity in turbine supply that was being widely reported by manufacturers throughout the Chinese market because of the rapid production expansions that had been made in response to aggressive government wind power targets.[131] This expansion triggered new Chinese government guidelines in 2010 and 2011 that aimed to consolidate the industry and targeted the most advanced Chinese wind turbine manufacturers for support at the expense of the smaller manufacturers.[132] Slower than expected grid expansion, along with the entry of numerous new wind technology suppliers in the Chinese market, resulted in an oversupply of turbines and a delay in the development of new projects.[133] These combined factors were likely driving down the prices that wind turbine manufacturers could obtain, causing them to look for opportunities to cut costs wherever possible.[134] When Sinovel sought to cut its own costs, it likely decided to cut back on imports from its American partner.

Following these initial allegations, concerns over IP theft emerged. In September 2011 AMSC filed a claim for arbitration against Sinovel with the Beijing Arbitration Commission, as well as a civil complaint, and announced plans to file a criminal complaint, all related to allegations of stolen intellectual property. AMSC claimed that Sinovel paid an AMSC systems integrator in Austria for source code and software that Sinovel used to upgrade hundreds of its wind turbines in order to meet proposed Chinese grid codes. In September the employee had pleaded guilty and was sentenced to one year in prison.[135]

Accusing Sinovel of stealing valuable trade secrets and copying protected software,[136] AMSC is likely regretting its decision to collaborate with Sinovel. Sinovel has been the leading wind turbine manufacturer in China since 2008 and has openly stated its goal of becoming the largest wind turbine manufacturer in the world by developing state-of-the-art wind power technology on numerous occasions.[137] The slogan streaming across the banner on the company Web site is "China No. 1 is just our starting point," referring to its number position in the Chinese market and goal to achieve this position in the global market as well.[138] AMSC's days in this partnership were likely numbered from the beginning since a continued reliance on foreign technology was never part of Sinovel's—or China's—plan.

It is clear that just as the environment for doing business in China has evolved dramatically over the past three decades, so have the strategies of foreign firms for business operations there. Early business operations in China were viewed as extensions of the parent firm, "passive implementers of headquarters' strategies, tactics, practices, procedures, and policies."[139] While many foreign firms were disillusioned with government-facilitated joint ventures, Chinese managers simultaneously desired foreign investment, managerial skills, and techniques, and the transfer of modern business technology.[140] Consequently there were often conflicts between foreign managers and local managers operating in joint-venture enterprises, contributing to mutual frustration, resentment, and sometimes failure. This was clearly illustrated in the early years of the Sino-foreign joint ventures in wind turbine manufacturing, as well as in many other industries that began with joint ventures—particularly those arranged by the government rather than by the firms themselves.

Frustrations with conducting business in China appear to be outweighed by the potential benefits China has brought foreign firms. Regulations now permit a foreign partner to hold a majority share of the joint-venture holdings or, in many cases, maintain full ownership over China-based operations, which was not the case when companies began selling wind turbines to China in the early 1990s. Although many joint ventures to produce wind turbines have failed, new ones are being established all the time.

There is a growing recognition that the role of Chinese firms in an increasingly globalizing world economy may not just be to serve as the factory floor for industrialized country firms, but also to serve as technological leaders and innovators in their own right. As Chinese firms are becoming serious global competitors in many industries, Western companies willing to enter the Chinese market and learn from Chinese business practices and knowledge bases are already acquiring new knowledge and transferring it to operations elsewhere in the world. Foreign firms increasingly are not only manufacturing in China but also investing in R&D activities, with foreign R&D already estimated to account for 25–30 percent of total business R&D in China.[141] In addition, China's move toward a more innovation-based economy can be expected to lead to improvements in IPR protection, particularly as Chinese enterprises seek to protect their own innovative activity. For example, Tsinghua University vigorously pursues instances of IPR infringement, and the Chinese Patent Office has conducted an active campaign to distribute information on IPR.[142]

The close proximity and small number of players in the Chinese wind industry encourage many informal learning networks that are influencing firm experience and strategy. Chinese firms entering the wind turbine industry have benefited from technology brought to China by foreign firms in the context of both official and unofficial—or unintended—technology transfers, and it is very likely that Chinese firms will be able to out-compete foreign firms in wind turbine technology sales in the near future if current trends continue.

While Chinese firms are only beginning to export wind power technology abroad, many have aggressive plans for international expansion. Even with these added protections, some foreign firms that previously believed it advantageous to domestically manufacture technology in China and take advantage of lower costs may now be more hesitant to do so in the face of potential competition from Chinese companies.[143] The opportunity provided by what is already the largest wind power market in the world continues to attract foreign firms, however, despite the apparent challenges of doing business in China.[144]

5
Goldwind and the Emergence of the Chinese Wind Industry

Goldwind was China's first leading wind turbine manufacturer. Now ranked among the top three domestic wind companies, Goldwind has benefited from a combination of sustained government support and an effective technology acquisition and development strategy. As the first Chinese-owned wind turbine manufacturer to produce a successful wind turbine design, it has developed a reputation for independent technology innovation.

Led by a CEO who is deeply committed to the company and to the industry, Goldwind's corporate culture encourages creativity in its employees and among its technology partners in a manner that is unusual among Chinese firms. An investigation of how Goldwind acquired its wind turbine technology provides a clear example of how China is obtaining advanced wind power technology through international technology transfers. While every firm in China has adopted a somewhat different strategy and established different technology partnerships, most firms have used licensing, merger and acquisition, and joint development strategies similar to those used by Goldwind. This chapter offers a case study of Goldwind in order to examine in detail one company's learning within the Chinese wind industry.

122 Goldwind and the Emergence of the Chinese Wind Industry

The Company

The origins of the Goldwind Science and Technology Company Limited were in the Xinjiang Wind Energy Company (XWEC), which was established in northwestern China in 1986, the same year that Danish industry leader Vestas installed the first utility-scale wind turbine in China.[1] Goldwind's chairman and CEO, Wu Gang, is a longtime scholar of wind power technology. He grew up as an engineer in China's Xinjiang Autonomous Region, a place of minimal opportunity but excellent wind resources. After learning about wind power, he helped to bring some of the first modern wind turbines to China, and he worked alongside the Danes as they constructed the first demonstration wind farms in Xinjiang.[2] Wu was awarded the World Wind Energy Award in 2006, and *Businessweek* named him one of China's most powerful people of 2009.[3]

Goldwind reached the top ten of global wind energy companies in 2006, rising to number nine in 2008, number five in 2009, and number four in 2010.[4] The Goldwind group of companies has expanded beyond wind turbine manufacturing to include the entire wind industry value chain, from R&D, to project development and financing, to wind farm construction and

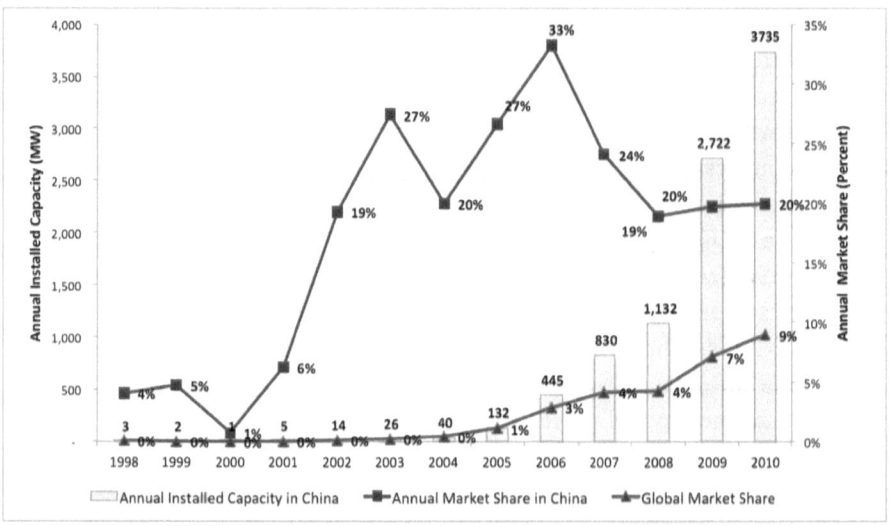

FIGURE 5.1 Goldwind's Installed Wind Capacity in China
Includes turbines installed by the Xinjing Wind Energy Company.

Author's database.

operation.⁵ It also owns mining operations for the rare earth materials that it uses in its permanent magnet generators.⁶ The company is beginning to expand its international presence, engaging in research, development, and demonstration projects on several continents.

Goldwind's sales approximately doubled every year from 2000 to 2008. In 2007 Wu took the company public, listing it on the Shenzhen stock exchange and raising nearly US $200 million. By 2009 Goldwind's total market capitalization had reached RMB 40 billion, and Morgan Stanley ranked its equity value second among global wind turbine manufacturers. The company's total revenue totaled RMB 10.7 billion in 2009, with over 99 percent of sales taking place in China. Net profit for the first nine months of 2010 surged 49.72 percent year on year to RMB 1.53 billion, and its revenue was RMB 10.89 billion during the period—61.78 percent more than in the same period of the previous year.⁷ In October 2010 Goldwind raised $917 million in its Hong Kong stock exchange IPO.⁸ Despite these public holdings, Goldwind still remains partially state owned.

TECHNOLOGY ACQUISITION

In 1989 XWEC helped to import and install 13 Bonus 150 kW turbines from Denmark at the Dabancheng wind farm in Xinjiang to form the largest wind farm in China at that time. In 1996 XWEC bought a license from Jacobs Energie, a small German wind turbine manufacturer, to manufacture 600 kW wind turbines in China.⁹ Since 1996 Goldwind has pursued a business model that allows it to implement modern foreign technologies while promoting its own technological advancement, with the goal of creating new ideas and eventually gaining benefit from its own products and the results of its research and development.¹⁰

The early turbines installed in China still relied on imported components; the five turbines installed in 1998 had about 33 percent local content. In 2001 Jacobs Energie merged with another company to form the REpower Systems Group.¹¹ That same year, Goldwind obtained a license from REpower for a 750 kW turbine.¹²

In 2003 Goldwind acquired a technology license from another German company, Vensys Energiesysteme GmbH, for the Vensys 62 1.2-MW turbine.¹³ Also in 2003 Vensys erected its first 1.2 MW prototype in Germany. Vensys originated as a wind turbine design company rather than a manufacturing company and therefore was complementary to a company like Goldwind, which had some manufacturing experience but little design experience. The Vensys direct-drive turbine technology was then (and is still) somewhat uncommon in wind turbine designs, but it is thought to have many advantages over the traditional gearbox design, as discussed below. When Vensys developed a low wind speed version of its turbine with a larger 64 meter diameter rotor that increased output to 1.5 MW, Goldwind acquired the license for that turbine as well.

In early 2008, when several other firms made a bid to purchase Vensys, Goldwind opted to purchase a 70 percent stake in the company outright so that it could continue its partnership. Becoming the controlling owner of the company gave Goldwind more power over the direction of Vensys's R&D activities, as well as fewer constraints over access to its intellectual property. It is also likely that Goldwind was concerned about relying exclusively on its licensing agreements with REpower since that company had been purchased by Indian wind turbine manufacturer Suzlon, one of Goldwind's competitors in the Chinese market, just the year before.[14] After its acquisition of Vensys, Goldwind began to jointly develop several new wind turbine designs in partnership with the company. Work began on the development of 2.5 MW and 3 MW turbines, as well as on 5 MW and 6 MW turbines with a view toward offshore applications. The components for Goldwind's wind turbines are now sourced almost entirely within China and constructed at one of several manufacturing and assembly facilities located throughout the country, although some core technologies are still imported.[15]

Goldwind, somewhat surprisingly, has opted to encourage rather than discourage Vensys's other partnerships, including its licensing arrangements with overseas companies. Current license holders of Vensys technology include Enerwind of Argentina, IMPSA of Brazil, ReGen Powertech of India, Eozen of Spain, and CKD NOVÉ Energo of the Czech Republic and Slovakia. Wu Gang, Goldwind's CEO, believes that it is important to give the designers at Vensys the creative freedom they need, and that by allowing them to engage directly in the manufacturing process (a type of learning by doing), they may improve the quality of their designs, which can then transfer to the Chinese part of the company.[16]

Goldwind appears to be expanding its technology partnerships beyond Vensys in order to pursue expertise in offshore wind turbine design. Reports have announced the company's plan to purchase an 85 percent stake in Golden Concord Wind Power of Jiangsu from Golden Concord Wind Equipment Holdings and a 15 percent stake from Shanghai Guoneng Investment,

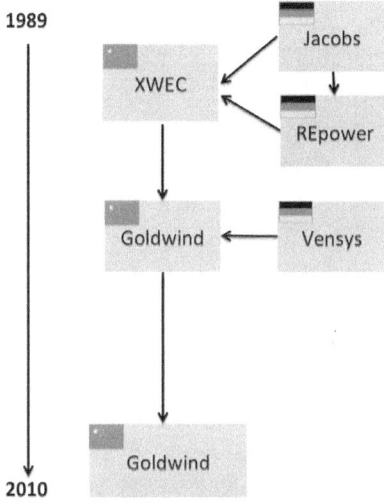

FIGURE 5.2 Timeline of Goldwind's Company Structure in China

Illustration by author.

reportedly costing Goldwind RMB 200 million and giving it full ownership of the company.[17] As Goldwind has recently established a wind equipment manufacturing facility in Jiangsu (discussed below), the know-how obtained from Golden Concord is expected to contribute to its offshore technology development, with a 6 MW design being the stated goal.[18]

RESEARCH, DEVELOPMENT, AND DEMONSTRATION

In 1997 the Chinese government recognized XWEC's early work as a national key project for technology development. The company received RMB 4 million from the Xinjiang government, most of which came from an R&D fund from the Xinjiang Science and Technology Commission, and Wu was able to use this government funding to leverage additional bank loans.[19] While the company's early attempts at wind turbine design proved challenging,[20] with foreign technology partnerships and additional government R&D support Goldwind was able to develop several successful wind turbine designs.

Goldwind has been increasing its total R&D investments annually. The company invested a reported $17 million in 2009 and an estimated $30 million in 2010, with total expenditures from 2007 to 2009 estimated at 4 percent of total sales revenue.[21] Goldwind reportedly planned to invest about $7 million in 2011 to produce 6 MW turbines.[22] In addition to being the recipient of government funding under China's 863 Program, Goldwind is also receiving funds from international investors such as the International Finance Corporation of the World Bank.[23] As the Chinese government has invested more money in Xinjiang in an attempt to encourage economic development and ease social unrest, some of these funds have been put toward emerging industries such as wind power technology.

Goldwind now has R&D facilities in Beijing, in Urumqi, and at the Vensys facility in Germany. The Urumqi center, designated a National Wind Power Technology Engineering Research Center, is supported by both the central and provincial governments, as well as by Goldwind's own R&D expenditures, and has been an important factor in the company's technology achievements.[24] As of 2010 the company had an estimated 500 employees conducting R&D out of a total staff of about 2,600. As the company has expanded, it has become increasingly able to compete for the most skilled workers in the wind turbine industry and has attracted former employees of leading wind companies such as GE, Gamesa, Vestas, Siemens, and Suzlon. This gives the company access not just to skilled wind technicians but to a broader global learning network, as many of these employees have worked for years in the wind industry outside of China.

Goldwind has also helped to established majors in wind power engineering at thirteen colleges across China to train potential future employees. Many of its staff members have been sent abroad to obtain advanced training by foreign companies or institutes, and several high-level managers have been sent abroad for MBA programs. Goldwind has also built up its R&D capacity through collaborations with universities and research institutes in both China and overseas, including the Xinjiang Agriculture University, Delft University of Technology in the Netherlands, Garrad Hassan, and Aerodyn.[25] In September 2011 Goldwind announced the founding of Goldwind University, the first corporate university within China's wind turbine manufacturing industry, which plans to "hold training courses and lectures for internal personnel by Goldwind's senior management," and will feature "notable professionals from within the wind power industry as well as academic advisors and professors from world-renowned educational institutions such as Stanford University, Peking University and Tsinghua University."[26]

As one of China's earliest wind turbine manufacturers, Goldwind has had a longer period of time over which to demonstrate its technology than most domestic manufacturers have had. Goldwind's backyard, the Xinjiang Autonomous Region, has some of the best wind resources of any province in China. Goldwind installed its first 1.2 MW turbines in 2005 and its first 1.5 MW turbines in 2008. Building on Vensys's original designs, Goldwind has developed variants of its 1.5 MW turbine for operation in low-temperature, high-altitude coastal areas as well as in high-humidity, high-temperature regions. Many of the designs for China that were codeveloped with Vensys were also demonstrated beforehand at the R&D center in Germany.

In 2007 Goldwind became the first Chinese turbine manufacturer to demonstrate an offshore wind turbine, with a project in the Bohai Gulf that it built in cooperation with the China National Offshore Oil Corporation. This project allowed Goldwind to gain valuable offshore demonstration experience that could inform its future offshore technology development strategy. The company's current R&D focus is on achieving mass production of the 2.5 MW turbine for both onshore and offshore use, finishing the design for the 5 MW turbine for offshore use, and developing a new 6 MW model. According to one report, the company plans to expand its coastal Jiangsu offshore technology facility in order to establish an offshore wind turbine test site.[27]

Other R&D being conducted by Goldwind is aimed at improving grid integration technology. The company has announced successful tests of low-voltage ride-through technology (LVRT) in its flagship 1.5 MW direct-drive turbine. LVRT allows turbines to maintain operation without disturbance during sudden voltage dips in the grid, and the company plans to proceed with LVRT for its larger turbine designs.

While Goldwind's early 600 kW and 750 MW turbine designs licensed from REpower were for conventional wind turbine technology, its newer turbine models that were developed with Vensys are permanent-magnet direct-drive machines. This technology eliminates the need for a gearbox, one of the most problematic wind turbine components when it comes to reliability. Referred to as "the Achilles heel of the wind turbine industry,"[28] several wind turbine manufacturers have experienced systemic gearbox failures, requiring expensive retrofits that in some cases resulted in bankruptcies.[29] The direct-drive design reportedly can reduce production costs since there are fewer components, lower maintenance costs due to increased reliability, and lower installation costs due to the fact that the turbines are lighter weight than conventional designs.[30] The design also increases turbine efficiency.[31] Few wind turbine manufacturers use the direct-drive technology design, however; Germany's Enercon was one of the first to manufacture the technology successfully. The venture into direct-drive technology therefore was a big gamble for Goldwind, but it appears to be paying off.

Experience in China

Hailing from Xinjiang, Goldwind originated in the part of China where the first utility-scale wind farms were constructed, and the company played an instrumental role in promoting China's wind power sector.

DEPLOYMENT

By the end of 2010 Goldwind had installed 9,086 MW of wind power capacity, which represented about one-fifth of all wind turbines installed in China at that time. Its factories are reportedly capable of producing 5,500 MW of capacity per year.

Over the past decade Goldwind has been selected by the government for several high-profile demonstration projects. At the end of 2003 the company received the order to supply the turbines for the Guangdong Huilai wind concession project, which was one of the first two government wind concessions.[32] At that time, the turbine prices being offered by Goldwind were about 25 percent less than the prices being offered by foreign-owned turbine manufacturers, making the Goldwind turbines particularly attractive to developers trying to win the projects by bidding the lowest price. Goldwind's selection attracted the attention of other developers in the Chinese market. As the large state-owned power companies have played an increasingly important role in developing wind power projects in China, they have developed partnerships with the leading Chinese wind turbine manufacturers.[33] Goldwind turbines since have been selected for several projects developed by Longyuan, Guohua, and China Guangdong Nuclear power companies.

In late 2006 Goldwind was awarded a contract to supply the turbines for a high-profile wind farm associated with China's hosting of the 2008

Olympic Games. Concerned with its global environmental reputation, China referred to the 2008 games as the "Green Olympics," using the event as an opportunity to demonstrate China's commitment to environmental protection and green-technology development to visitors from around the world. The wind farm, located near the popular tourist destination for viewing the Great Wall in Badaling, was the first utility-scale wind farm to be built in the Chinese capital. Electricity from the wind farm contributed to the Beijing government's goal of 20 percent of the Olympic venues to be powered by wind energy.[34]

In 2007 Goldwind partnered with the China National Offshore Oil Corporation (CNOOC) to build the first offshore wind turbine in China. The project featured just one of Goldwind's 1.5 MW turbines, but erecting the first offshore turbine represented a milestone in Chinese wind power development. CNOOC, one of the largest state-owned oil companies in China, contributed its expertise in building offshore oil platforms to construct the project in the Bohai Sea, the site of extensive offshore oil and gas drilling.[35] The Chinese government has extensive plans for offshore wind power development; the first 100 MW project off the coast of Shanghai went online in June 2010. The three largest wind turbine manufacturers in China—Goldwind, Sinovel, and Dongfang—have all established manufacturing facilities in the coastal provinces for the production of offshore wind turbine models.

As large-scale wind development continues in China, developers continue to select Goldwind's technology. As the Chinese government transitioned from megawatt-scale concession projects to gigawatt-scale wind bases, Goldwind earned contracts for these sites as well. For example, in 2008, the company announced contracts for seven wind farms in Gansu province totaling more than 800 MW of capacity and worth more than RMB 5.2 billion as part of the government's large wind base development.[36] In December 2010 the company announced that it had won a RMB 4.77 billion contract to sell 1.1 GW of wind turbine generators to five projects in Hebei province and three projects in Xinjiang.[37]

MANUFACTURING

Goldwind established its first large-scale wind turbine assembly plant in Xinjiang in 2002, with an annual production capacity of 200 units of wind turbine generators (WTG) ranging in size from 600 kW to 1 MW. Since then its production has expanded substantially, with factories now located in Beijing, Baotou (Inner Mongolia), Chengde (Hebei), Jiquan (Gansu), Nanjing (Jiangsu), Ningxia, and Xi'an, in addition to the original factory in Urumqi, Xinjiang.[38] An additional facility is being constructed in Suzhou (Jiangsu) for offshore development.

Goldwind sources its components from a variety of suppliers throughout China. Maintaining quality control across the entire supply chain, however, can prove to be a challenge. When Goldwind began manufacturing turbines, there was a major technology failure that was traced back to the blade bearings. The failure resulted in much negative publicity for the company. Goldwind spent a lot of time studying the problem and discovered that the failure was due to the steel used to make the bearing, which was apparently of low quality and lacked sufficient strength.[39] After a massive recall of the parts the problem was eventually fixed, but the experience illustrated how one small supply-chain error can result in a disastrous public relations problem for an emerging company in a highly competitive market.

Experience Overseas

By the end of 2010 Goldwind had exported only a handful of its own wind turbines from China to overseas markets, but it has announced plans for aggressive international expansion in the coming years. While many of its initial forays have been into established wind markets like the United States, it has plans for broad expansion into emerging markets as well, many of which are just beginning to utilize wind power.

Goldwind was widely reported to be the first Chinese turbine manufacturer to export wind turbines to the United States.[40] It exported three sets of its GW77/1500 turbines to the Uilk wind farm in Minnesota that began operation in January 2010. In late December 2010 Goldwind announced its first large-scale wind farm in the United States, a 106.5 MW project in Illinois called Shady Oaks. In January 2011 it announced two additional 10 MW wind farms in Montana, and by early 2012 it had expanded across seven states.[41]

Goldwind has also exported six turbines to Cuba and has contracts for projects at various stages of development located in Australia, Pakistan, Ethiopia, Chile, Ecuador, Cyprus, and Scotland. The company has plans to expand into the Canadian and South African markets, as well as to other countries in Europe, Asia, Africa, and South America, and it has established sales offices in Chicago and Sydney (in addition to its offices in China).[42] If the wind turbines deployed by its technology partner Vensys, and by Vensys's multiple licensees in Argentina, Brazil, India, Spain, and the Czech Republic, are included among Goldwind's exports, then the company's global experience looks quite different. Vensys and its licensees have deployed eight of the 1.5 MW machines in Canada, fifty-seven in Brazil, eight in Spain and Portugal, fourteen in Germany, three in Poland, three in Bulgaria, five in Pakistan, forty-five in India, and two in Russia, as of 2011.[43]

As the bulk of Goldwind's technology development shifts from Germany to China, it is looking to expand its global presence through its Chinese technology base rather than the Vensys base. Goldwind's strategy for the U.S. market, for example, is not just to export to a U.S. company, or even to partner with a U.S. company, but to be viewed as a U.S. company. This is not unlike the strategy pursued by Japanese auto manufacturers Honda and Toyota; while Americans may never see them as "American companies" in the same way that they see General Motors or Ford, they may care that these companies have factories based in the United States and that their cars are being made there.[44] While all companies are increasingly globalizing their supply chains, making nationality somewhat less meaningful, the national identification of companies remains optically important in an era in which discussions about job creation, and green jobs in particular, are increasingly political.[45] More than 60 percent of the content used in the wind turbines at Goldwind's Minnesota project was made in the United States, although the generators, nacelles and hubs were all shipped from China.[46]

Goldwind has opened a sales office in Chicago, Illinois, staffed with primarily American personnel and management. In May 2010 the company announced the appointment of its first U.S. CEO to head up Goldwind USA, Tim Rosenzweig. Rosenzweig had worked in the U.S. wind industry for many years, most recently as the senior vice president of finance and chief financial officer of First Wind, a small Massachusetts-based wind company with projects in Maine, New York, Utah, Vermont, and Hawaii.[47] The state-owned China Development Bank has reportedly given Goldwind a $6 billion credit line to fund its expansion into the United States.[48]

Having access to the financial backing of Chinese banks has given Chinese companies an edge, particularly in the U.S. market, where financial institutions are still recovering from the global financial crisis. Goldwind can therefore rely on Chinese bank financing until it can build a reputation in the United States, though it would eventually like to be backed in the United States by

U.S. banks.[49] While Goldwind has stated its intention to build a manufacturing facility in the United States, formal plans have yet to be announced. The company has already partnered with U.S. component suppliers, including a contract with Ohio-based manufacturer Timken to supply bearings.

Goldwind is beginning to recognize that while competing in established wind markets can be difficult, it has substantial advantages over its global competitors in emerging markets. The company has made forays into markets where few Western companies have a presence, such as Pakistan and Cuba. But its recent expansions into South Africa and Ecuador also illustrate the company's interest in moving into new wind power markets where products may be particularly well suited. For example, the permanent magnet direct drive systems need less maintenance, which is a huge benefit in a new market with minimal technical personnel or components on hand, and particularly in a developing country where projects may be remote and infrastructure may be limited. In addition, Goldwind offers products that are adapted to high-altitude wind sites which could be highly useful in much of South America.[50] Perhaps Goldwind's most important advantage, however, is that its turbines typically sell for substantially less money than those made by American and European companies.[51]

Goldwind is already on six continents. As it looks to sell its turbines in international markets, however, questions about quality control are bound to arise. While Goldwind has more years of operational experience with its turbines in China than many other Chinese manufacturers, its oldest models have been in operation for just over ten years, and its newer models for under ten years. Most investors expect wind turbines to operate for twenty to thirty years. Because of the uncertainties associated with the quality of relatively new technologies, international technology certifications could be important for a company like Goldwind to acquire. Back in 2000 Goldwind was the first wind turbine manufacturer in China to receive ISO 9001 certification, and its current turbine models have international certifications from TUV-Nord and Germanischer Lloyd, and Chinese certifications from the China General Certification Center and the China Classification Society Industrial Corporation.

Evidence of Learning

To assess the future ability of a Chinese wind company such as Goldwind to compete in the global marketplace, it is important to understand how past technology transfers and partnerships have influenced the company's ability to learn and its potential for innovative capacity.

TECHNOLOGY TRANSFER MODELS

Goldwind, like many other wind companies, began its entry into the wind industry by acquiring a technology license from a foreign company. As it became more experienced, it sought out a German design firm to work with rather than a manufacturer that could also be a competitor. Its relationship with Vensys started with a license, but once Goldwind acquired majority control of the company, the relationship evolved into more of an equal partnership, and one in which Goldwind has full access to Vensys's technological know-how. While Goldwind's initial contribution to the partnership was primarily its manufacturing expertise and Vensys's contribution was its design expertise, Goldwind CEO Wu Gang encouraged Goldwind staff to venture more into the design realm, and Vensys staff to venture more into the manufacturing realm, to create more opportunities for learning by doing.[52]

Now a much more sophisticated company than when it originated, Goldwind is increasingly able to expand its knowledge through additional mergers and acquisitions, as demonstrated though its acquisition of Golden Concord to build its offshore know-how. Its access to increasingly extensive lines of capital also makes it more able to acquire companies as a means of absorbing know-how rather than relying on licenses, which can be highly restrictive, particularly when it comes to technology exports. Many Chinese turbine manufacturers that have relied on licenses to build their turbine designs are prohibited from exporting the technology outside of the Chinese market.[53]

MANUFACTURING SCALE AND TECHNOLOGY
IMPROVEMENTS

As a late entrant into the global wind industry, Goldwind was able to begin its manufacturing experience with a 600 kW turbine. While this is a small model in today's terms, at the time it was still a rather popular model around the world. After moving up to a 750 kW design, Goldwind sought to increase its technological expertise and with it the size of its turbines.

Its flagship design is the 1.5 MW direct-drive model. Not content with just one megawatt-scale design, Goldwind has been moving rapidly toward the development of larger and more advanced wind turbines, with a 2.5 MW in demonstration, and 5 MW and 6 MW offshore designs under development. Goldwind's 2.5 MW turbine may be a good illustration of its technological achievements to date. According to the company it adds 1 MW of capacity to the 1.5 MW model and only 10 percent additional weight. Wu has called it the company's best machine yet.[54]

Goldwind's technology improvements have been driven by the demonstration opportunities granted to it by the Chinese market. China became the largest global wind power market measured by annual installations in 2009 owing to extensive government policy support for wind power development over the past decade. Goldwind has also been the beneficiary of direct government R&D support, as well as indirect financial support in the form of access to favorable loans to fuel its domestic, and ultimately its international, expansions.

ACCESS TO LEARNING NETWORKS

While China was barely on the map in terms of wind power development in the 1980s and 1990s, it was already becoming a global test bed for wind turbine technology. Early Chinese wind farms featured wind turbines from all over the world and wind power engineers of many nationalities working side by side as they tested their technology. By the year 2000, while only 344 MW of wind power capacity had been installed in China, these wind turbines had been manufactured by more than seventeen companies from six countries.[55] Even some of China's first wind farms in remote Xinjiang utilized turbines imported from Denmark and Germany. As a result, early Chinese wind turbine manufacturers like Goldwind were likely able to benefit from the learning networks created by foreign firms developing wind farms in localized geographic regions.

In recent years Goldwind has further expanded its access to global learning networks. Its acquisition of Vensys gave Goldwind access to a network of skilled engineers and a company with a different geographic focus, allowing it to better integrate European wind industry experience into its operations. Even more recently, as Goldwind looks to expand into the United States and Australian markets, it has hired American and Australian workers with extensive experience in their home markets to help it better understand how to operate within these domestic contexts. As

Goldwind deploys its technology abroad, it must make some changes in order to accommodate local conditions. For example, to install its turbines in Minnesota, Goldwind had to make software changes to facilitate grid compatibility, as well as minor hardware changes, primarily to meet U.S. safety standards.[56]

Goldwind has licensed its technology only through Vensys, not directly, and it is notable that the licensees have been primarily emerging economy firms. India was home to a thriving wind industry even before China, and Brazil is rapidly becoming a hotspot for wind development. There is also much interest in wind power development in eastern Europe. This challenges the traditional notion that technology is transferred exclusively from North to South. Since it is primarily companies in emerging economies and developing countries that are looking to acquire technology through transfers in the form of licensing agreements, it is likely that Chinese firms like Goldwind will increasingly become the source of such technology transfers, not just the recipients. Goldwind's plans for expansion into emerging markets further support this theory.

The many other Chinese wind turbine manufacturers have used technology transfer models similar to those used by Goldwind, including licenses, mergers and acquisitions, and partnerships with other firms to conduct joint technology development. Others have developed original designs, often in conjunction with research units at universities. The largest market share in China is held by Chinese firm Sinovel, a relative newcomer to the industry, with about a quarter of the market in 2009 and 2010. Sinovel obtained its 1.5 MW wind turbine technology through a licensing agreement with German firm Fuhrlander. It later partnered with American Superconductor and its wholly owned, Austrian-based subsidiary Windtec to jointly develop 3 MW and 5 MW turbines.[57] The other major Chinese manufacturer, Dongfang Electric Corporation (DEC), held a 14 percent market share in China in 2009 and 2010. Based in Sichuan province, DEC is part of a large, state-owned enterprise managed directly by the Chinese central government, and is one of the largest power plant construction firms in China.[58] It obtained its wind power technology through a licensing agreement with REpower for its 1.5 MW wind turbine. Another emerging Chinese firm, A-Power, obtained its technology through a licensing agreement with Fuhrlander for a 2.5 MW turbine, and with Danish firm Norwin for 225 kW and 750 kW wind turbine models, which also included the establishment of a joint venture company.[59] The sources and models of technology development used by Chinese wind companies are detailed in table 5.1.

TABLE 5.1 Technology Development Models in the Chinese Wind Industry

Chinese Company	Model of Technology Transfer	Source of Technology Transfer
A-Power (GaoKe)	License	Fuhrlander (Germany)
	License/joint development	Norwin (Denmark)
Beijing Beizhong	License	DeWind/ (Germany/UK/US/Korea)
Changzing	Self-developed	Developed with Shanxi Science and Technology University (China)
CSIC Haizhuang	License	Frisia (Germany)
	Joint development	Aerodyn (Germany)
CSR Zhuzhou	License	AMSC-Windtec (US/Austria)
DEC	License	REpower (Germany)
	Joint development	Aerodyn (Germany)
	Joint development	AMSC-Windtec (US/Austria)
Engga	Self-developed	Developed with the Tsinghua Industrial Academy (China)
Envision	Joint development	Supported by the European Clean Energy Fund (EU)
Goldwind	License	Jacobs/REpower (Germany)
	Joint development	Vensys (Germany)
Guodian United	License	Aerodyn (Germany)
Hadian	Self-developed	Developed with Harbin Power Planet Equipment Corporation (China)
Hafei	Joint venture	WinWind (Finland)
Harbin Steam Turbine Co.	License	DeWind/EU Energy (Germany/UK)
Hewind	Joint development	Aerodyn (Germany)
Huachuang	Self-developed	Developed with the Shenyang University of Technology (China)
Huide	License	Fuhrlander (Germany)
Jiuhe	License	Windrad Engineering (Germany)
Minyang	Joint development	Aerodyn (Germany)
New United	Self-developed	Developed with the Shenyang University of Technology (China)
REpower North	Joint venture	REpower (Germany)
SBW	Joint development	AMSC-Windtec (USA/Austria)
Sewind	License	DeWind/EU Energy (Germany/UK)
	Joint development	Aerodyn (Germany)

Sinovel	License	Fuhrlander (Germany)
	License	Windtec/AMSC (US/Austria)
	Joint development	Windtec/AMSC (US/Austria)
Tianwei	Joint development	Garrad Hassan (UK)
Windey	License	REpower (Germany)
	Self-developed	Developed with Zhejiang Institute of Mechanical and Electrical Engineering (China)
Wuhan Guoce Nordic New Energy	License	Deltawind/Nordic Windpower (Sweden)
XEMC	License	Zephyros/Lagerwey (Netherlands)
XJ Group	Joint development	AMSC-Windtec (USA/Austria)
Yinhe Avantis	Joint development	Avantis Energy (Germany)
Yinxing	License	Mitsubishi (Japan)

Sources: Author's own database and Paul Recknagel, *Mapping WTG Manufacturers in China* (GTZ Renewable Energy Program, 2010).

Note: There are reports of many smaller companies developing their own technology, but such companies have been omitted from this table.

ACCESS TO SKILLED WORKERS

Wu Gang has created a work environment at Goldwind that fosters creativity. Staff are encouraged to take music lessons and to participate in sporting events on site, with the company sometimes bringing in professional musicians and athletes to motivate workers.[60] Wu believes that if his staff have hobbies and enjoy life, they will be more productive employees. He describes the ideal corporation as being equal parts military, university, and business, in that it should be a place with strict rules but also a creative learning environment, oriented toward profit maximization.[61]

As Goldwind looks to international markets, it is hiring international wind industry talent as well, focusing on employees with experience in its target markets such as the United States and Australia. Goldwind has already been successful at hiring talent within China, including workers who had been trained by the many foreign-owned wind turbine manufacturers with operations in China.

Outlook for the Chinese Wind Industry

Since Goldwind is a product of the Chinese wind industry, its ultimate success as a company will depend on its ability to continue to thrive within the Chinese market, as well as its ability to expand into other markets.

PLANS FOR EXPANSION

Overall the outlook for the Chinese wind turbine industry is strong. An increasingly stable and favorable policy environment for wind in China will continue to make China one of the largest markets for wind power development in the world, and Chinese firms will continue to be awarded the majority of the domestic projects. In addition, Chinese firms will increasingly look to export markets and will likely be able to complete globally based on their ability to offer lower-priced products. While few Chinese turbines have been exported outside of China, this is likely to change in the coming years.

Chinese wind turbine manufacturers are well positioned to continue their expansion within China for years to come. In addition, it is very likely that several of the leading firms will begin to export their turbines for sale abroad. The entrance of new firms into the Chinese market has increased competition as turbine manufacturers vie for projects. More than thirty-five Chinese-owned companies have erected wind turbines in China, and eighty or more have technology at various stages of development. Among Chinese companies, firms attempt to differentiate themselves based on several factors. Some use name recognition, as many wind turbine manufacturers started in other industries in which they had already built a reputation. Others compete based on their relationship to large state-owned companies that may give them preferential financial support. As Chinese firms gain operating experience, however, the primary means of differentiation is likely to become product quality. To date, very few companies have operating experience beyond just a couple of years, and some do not even have that. Goldwind is somewhat unique in that some version of the company has been manufacturing wind turbines for over two decades. The other two leading Chinese wind turbine manufacturers, Sinovel and DEC, have been in the wind business for less than five years.

There is not much difference in price across Chinese manufacturers, but Chinese wind turbine manufacturing companies do have a competitive advantage over the majority of international wind turbine companies when it comes to price. This advantage currently exists in the Chinese marketplace even as foreign wind companies have shifted their manufacturing facilities to China and have been able to take advantage of comparably lower Chinese wage rates. This indicates that Chinese turbines may in fact be lower cost owing to factors aside from inexpensive labor—most likely due to less extensive warranties and after-sale service or a willingness to take power profits to gain early market share, though cost savings in the manufacturing process are also possible. If Chinese firms begin to export their turbines overseas, they will likely be extremely competitive in other markets as well. If, however, they shift production to the target markets, any labor-related advantages may be reduced.

Chinese wind turbine manufacturers have benefited, and will continue to benefit, from the large domestic market for wind power development in China, which gives China-based firms a huge opportunity to sell their products, develop their technology base, and gain operational experience. They have also benefited greatly from government support, particularly in recent years. While local content requirements instituted in 2003 gave local manufactures an early boost, 2008 Ministry of Finance programs to directly support the development of advanced Chinese wind turbine technology,[62] along with the awarding of concession and wind base projects to developers that utilize Chinese technology, have been extremely beneficial to emerging wind turbine manufacturing companies in China. These policies have given Chinese firms a strong advantage over their foreign competitors in the Chinese marketplace, and this may assist these firms as they expand into markets outside of China.

Despite an overall positive outlook for Chinese manufacturers, several key weaknesses exist as well. Very few firms in China have sufficient operating experience with their wind turbine technology. It is very common for companies in the early stages of developing a new product to experience technical challenges and setbacks. Goldwind, one of the few Chinese firms with several years of operating experience, experienced major failures in hundreds of the wind turbines it had installed across China, which was later traced to a material defect. Other Chinese wind companies have had their share of trouble as well—for example, Sinovel reportedly had several turbines collapse at different locations throughout China.[63] Unexpected technical failures can be extremely costly and can threaten the financial stability of a company. In addition, technical failures can be very harmful for a firm's reputation, particularly if the firm has not had years of successful performance to counter any difficulties that may arise.

Very few Chinese wind turbine manufacturers have built any sort of reputation outside of China, and this is one area in which the foreign turbine manufacturers that have been in the wind industry for decades have some advantage. As a result, the enhanced utilization of internationally established certification and testing standards would help Chinese wind turbine manufacturers improve their technology by identifying technical problems and would help companies differentiate their products in the marketplace with quality assurances. China would benefit greatly from a national renewable energy laboratory to conduct independent testing of wind power technologies, similar to the United States' National Renewable Energy Laboratory in Golden, Colorado.[64] While the State Grid Energy Research Institute has established a wind turbine testing facility in Zhangbei, some turbine manufacturers have reported difficulty in gaining access to the site.

Projections for wind power development in China for the next one to two decades vary dramatically. Recent projections published by Li Junfeng and Gao Hu of the Energy Research Institute of the NDRC and Shi Pengfei of the China Hydropower Engineering Consulting Group Corporation lay out an ambitious future for wind power in China. Their most conservative scenario projects 150 GW in 2020, 250 GW in 2030, and 450 GW in 2050, while their most ambitious scenario projects 230 GW by 2020, 380 GW by 2030 and 680 GW by 2050.[65] Now that national wind targets have been increased to 200 GW by 2020 and 1000 GW by 2050, however, even their more aggressive scenarios seem modest.

If such ambitious growth targets are to be achieved, China will have to make significant improvements to its transmission infrastructure. Improved transmission networks allow wind projects to be developed in areas with excellent resources but low electricity demand. The lack of an improved transmission infrastructure will be a crucial barrier to China's ability to fully exploit its wind power resources. There are widespread reports of significant delays in connecting wind farms to the power grid; if wind developers cannot connect their projects, they cannot meet their contracts to supply electricity. As a result, additional studies need to be done to look at high-penetration wind scenarios on China's power grids, particularly in parts of China where a smaller, regional grid is not interconnected to a larger network.

There are political as well as technical reasons that wind projects in China have experienced challenges in obtaining access to the power grid.

While some grid operators in China fear that wind energy will threaten the stability of their entire power system, others are reluctant to support the use of renewable energy because of the additional regulatory and economic burden it may bring. While advanced and "smarter" grid technologies may be able to help facilitate the integration of much larger amounts of wind power, political barriers to wind integration will likely require a different set of solutions.

Other challenges to integrating wind into regional power grids in China, and in wind-rich northern China in particular, stem from the use of combined heat and power (CHP) systems. The North China power grid, which includes the Northeast grid, the North grid, and part of the Inner Mongolian grid, has some high-quality wind resource locations but also many coal-fired CHP plants that provide both heat and power. This means that they must be kept running in order to produce heat in the cold winter months. While this region used to have many smaller coal plants in the 50 to 100 MW range, many of these have been targeted for shut downs as a result of the national energy-efficiency programs. These coal plants, while dirty and inefficient, provided one of the few more flexible sources of power in the region that could be used to balance intermittent wind resources. As a result, wind power is frequently curtailed, which is highly inefficient in that it essentially results in the wind-generated electricity being wasted. The inflexibility of this system is further exacerbated as a result of insufficient capacity for cross-provincial transmission of wind power. For all these reasons, a crucial topic for further study is how to improve wind integration in northern China.

OBSTACLES ABROAD

By the end of 2011 few Chinese turbine manufacturers had exported turbines outside of China. In addition to Goldwind's plans for expansion discussed previously, several other Chinese manufacturers have either exported turbines or announced plans for forthcoming exports, including Sinovel (to India, the United States, Greece, and Ireland), DEC and Sewind (to India), Shanghai Electric (to Thailand and the United Kingdom), XEMC (to the Maldives and Ireland), Hewind (to Chile), New United (to Thailand, Turkey, and the United States), Huachuang (to Pakistan), Huide (to the United States), Minyang (to Sweden and the United States), and Guodian United, Haizhuang, and Zhuzhou CSR (both to the United

States). China's Minyang had announced plans to export 900 MW of its turbines to the United States through a partnership with GreenHunter Energy of Texas, but the deal reportedly fell through when Minyang fell behind on production. Other Chinese wind farm deals have faced major political barriers as well. For example, Shenyang Power Group announced that it would be supplying 2.5 MW A-Power turbines made in China for a wind farm in west Texas, but the deal raised many concerns, particularly from members of the U.S. Congress, that China was trying to compete with the United States in its own domestic market in an industry that the government had specifically been trying to promote with tax credits and other green jobs initiatives.[66]

Protectionism is proving to be as much of a challenge for Chinese turbine suppliers to the U.S. market as U.S. and other foreign suppliers have long claimed it has been in the Chinese market.[67] Protectionism over green-technology manufacturing is not just a U.S.-China battle, however. Local content requirements have been used widely to encourage local wind turbine manufacturing around the world, including in Quebec and Ontario, Canada, as well as in Spain, Brazil, India, Australia, and Portugal.[68] In September 2009 Japan filed a complaint against Canada with the World Trade Organization, claiming that the local content requirement in Ontario's green energy plan constituted a prohibited subsidy since it discriminated against imported Japanese technology.[69] As other markets look to promote wind power industries at home, there is increasing hesitance to rely on the same local content requirements that China and other counties have utilized to promote their domestic industry because of concerns about becoming the subject of international trade disputes.[70]

Other obstacles faced by Chinese manufacturers abroad include concerns among investors and developers about the quality of their products. As one U.S.-based energy consultant put it, "the issue is not that there is a quality problem. It's that we don't know there isn't."[71] The challenges to exporting to the U.S. market are some of the very same reasons the market is a desired destination for Chinese companies. Because of its reputation for quality products, the U.S. market could boost the global reputation of Chinese turbine manufacturers if they can successfully develop a project there.

As one of China's earliest wind turbine manufacturers, Goldwind provides the longest history over which to assess the technological progress of a single Chinese wind company. Goldwind, like hundreds of other companies in China, started with a single license purchased from a foreign company. Over about two decades, however, it has successfully transitioned to become an innovator in its own right, making it one of China's leading wind turbine manufacturers and a relatively rare case for the Chinese wind industry. Goldwind is also a company of many firsts: installing some of the first Chinese-made turbines in China, supplying the first large wind farm in the nation's capital for China's first Olympics, manufacturing the first offshore wind turbine demonstrated in China, exporting one of the first Chinese wind turbines abroad, establishing the first U.S. subsidiary of a Chinese wind company, and hiring the first American CEO to lead a Chinese wind company overseas.

Since Goldwind is a product of the Chinese wind industry, its ultimate success as a company will depend on its ability to continue to thrive within the Chinese market, as well as its ability to expand into other markets. A consistently stable and favorable policy environment for wind power will continue to make China one of the largest markets for wind power development in the world, and Chinese firms likely will continue to be awarded the majority of the domestic projects. The Chinese market will also become more competitive, however, owing to an increasingly crowded market with more and more new entrants. Continued expansion of wind power development in China also faces numerous challenges, particularly related to transmission and integration. Goldwind reportedly has already experienced financial losses tied to wind power curtailment in China's northwestern and northeastern provinces.[72] If the government's ambitious growth targets are to be achieved, policy makers and power and grid companies will have to make significant improvements to China's transmission infrastructure and address increasingly complex technical and political problems related to integrating wind into the existing power mix.

Chinese wind turbine manufacturers, including Goldwind, have benefited greatly from government support in recent years. While this support is currently the subject of emerging trade disputes, it has likely already served its purpose. Even if the trade disputes result in the removal of some of the programs that have benefited Chinese wind manufacturers as they developed and demonstrated their technologies, three Chinese companies have already reached the top ten of wind turbine manufacturers worldwide and therefore have likely already received the early support they needed from their home market to enable their expansion.[73]

The ultimate success of the Chinese manufacturers, however, will be tested as they look to export to overseas markets. Current events seem to signal an increase in protectionism in green-technology industries around the world. But protectionism is not new to the wind industry, and China is not the only target. As a result it is unclear to what extent Chinese companies will face barriers to exporting their technology any more than European companies have faced barriers in the United States, or Japanese manufacturers have faced barriers in Canada.

Perhaps the biggest challenges lie in the fact that very few Chinese wind turbine manufacturers have built any sort of reputation outside of China. Most of these companies, including Goldwind, are still relatively young and are selling products with relatively few years of operating experience. As a result, they still face real risks of potentially disastrous technology failures just as their global reputations are being earned and solidified. Most of the concerns with Chinese wind power technology today stem not from poor designs but rather from poor quality control, both in the manufacturing process and in how the technology is operated and maintained over time. Even Goldwind CEO Wu believes that the Chinese wind industry lags behind the rest of the world in terms of R&D, management, and internationalization.[74] Goldwind's ability to address these issues successfully will in many ways set the tone for Chinese expansion in this sector since for years the company has served as a harbinger for the rest of the Chinese wind industry.

6
Wind Energy Leapfrogging in Emerging Economies

While modern wind power technology originated in Europe and the United States, the emerging economies are quickly becoming the hub of the global wind power industry.[1] While many emerging economies are now beginning to pursue wind power development, China and India are the only ones already among the top wind power utilizers in the world. India was the early emerging economy leader in wind power development, but it was surpassed by China in 2009. China and India are also the only two emerging economies with top-tier wind turbine manufacturers. South Korea is still a relative newcomer to the wind industry, but the recent entry of many large Korean industrial firms makes it well positioned for future growth.

While there are many potential benefits to local wind manufacturing, there are also significant barriers to entry into an industry containing companies that have been manufacturing wind turbines for more than twenty years. In emerging economies, limited indigenous technical capacity and quality control can make entry even more difficult. International technology transfers can be a solution, although leading companies in this industry are unlikely to transfer proprietary information to companies that could become competitors. This is even riskier for technology transferred from

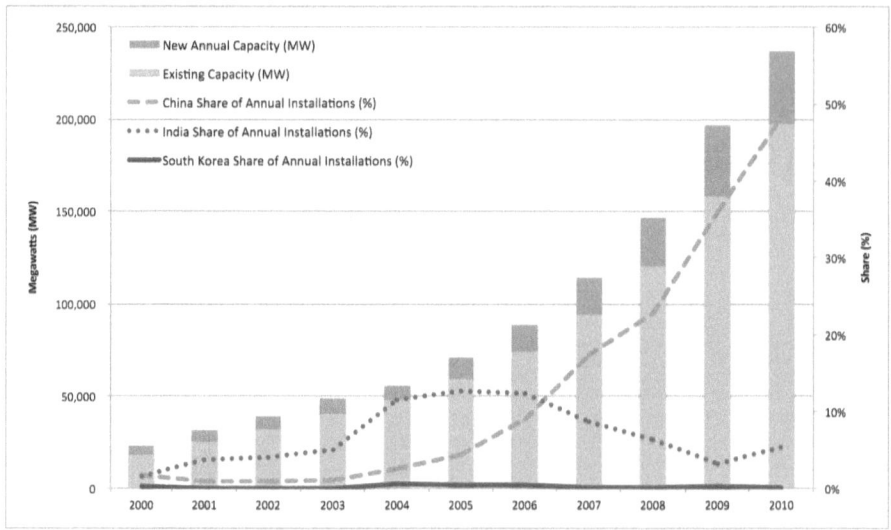

FIGURE 6.1 Role of China, India, and South Korea in Global Wind Power Development

Author's database.

developed to developing countries, where an identical but cheaper turbine potentially could be manufactured.

This chapter compares the strategies used by Chinese firms to develop wind power technology, as detailed in earlier chapters, with the strategies used by the domestic wind power technology companies in India and South Korea, arguably the two other most successful "latecomers" to the wind industry. It begins by reviewing the evolution and current status of wind power development in India and South Korea. Turning to the current industry leaders in each country, it examines how these companies acquired the technological know-how and intellectual property rights associated with their respective wind turbine designs, how the domestic and international contexts in which these companies operate shaped their technology development strategies, and whether differences in their respective technology development strategies contributed to differences in the performance of the companies in the marketplace. It concludes with an assessment of the outlook for the future development of the Chinese, Indian, and Korean wind power industries and a discussion of the policy lessons for facilitating low-carbon technology transfer that can be gleaned from these three cases.

Industrial Strategy in the Newly Industrializing Countries

Late entrants to the global wind power industry have used different strategies to foster the development of their own domestic large wind turbine manufacturing companies.[2] A common strategy has been to obtain a technology from a company that has already developed advanced wind turbine technology. Technology transfers can occur through different models, including a licensing agreement that gives the licensing firm access to a certain wind turbine model, often with some restrictions on where it can be sold, or the establishment of a joint-venture partnership either to share a license or for collaborative R&D. Firms also can opt to collaborate to jointly develop a new technology design and then share the associated intellectual property. If a firm has the capacity and means, it can also obtain access to technology through the purchase of ownership rights in a company with the desired technology or other forms of mergers and acquisitions.

All these models of technology transfer have been used within the Chinese wind industry. Goldwind, China's first successful wind turbine manufacturer, began by licensing technology, then acquired majority ownership of Vensys, a company with which it had previously had a licensing arrangement. It eventually began to jointly develop new technology designs with the same partner.

The larger domestic context in which the innovative activity is taking place, often referred to as the national innovation system, is likely an important determinant of the ultimate success of a technology transfer, particularly concerning a country's ability to adopt an externally sourced technology and apply it internally. The organization and distribution of innovation-related activities often differ among countries and regions, with some similarities among the technological catch-up models employed by the Asian "late-industrializing countries."[3] China, India, and South Korea have all been involved in technological catch up, frequently through international technology transfers, although the national innovation systems in these three countries differ from one another.

Regional and global learning networks have likely played a large role in the development of wind turbine technology over time, as discussed in previous chapters, and emerging Asia has become one such regional learning network in the wind industry. The wind industry—characterized by its small number of firms, highly specialized technology, and geographically specific hubs of innovation (often near wind development locations)—is likely to exhibit many of the characteristics of the regional learning networks that have been observed in other industries and locales.[4] Just as the early wind development in Denmark and the United States provided a crucial learning ground in the 1970s and 1980s, the emerging wind markets of India, China, and South Korea are serving as valuable regional learning networks for new firms. The increasingly global reach even of new firms, facilitated by technology transfer partnerships with overseas firms, has also provided a valuable resource to global learning networks of knowledge and innovation.

As the two largest developing country wind power markets in the world, China and India are now pertinent places to examine models of technology transfer that have facilitated the emergence of several leading wind power technology firms. South Korea, several years behind China and India in terms of its own domestic wind power development, has a very strong industrial base on which to build a wind power industry. The similarity in technology transfer models being used by South Korean, Indian, and Chinese firms, as well as the impending global competition among these countries in the wind industry, provides a rich basis for comparative analysis.

India's Wind Industry

As of 2010 India ranked fifth in the world after the China, the United States, Germany, and Spain in cumulative wind power installations, with 13,066 MW. The Indian government had set a target for 10,500 MW by 2012 as part of its Eleventh Five-Year Plan, which it has already surpassed. It also has a target for renewable energy to contribute 10 percent of total power generation by 2012. Wind power alone, however, constitutes less than 1 percent of India's total electricity generation.

The potential for wind power in India is estimated at 45,000 MW, though the lack of detailed national resource assessment means that the actual number may be far higher.[5] The best wind resources are in the eastern and southern parts of the country, particularly near the coasts. The highest wind energy potential is believed to be located in the states of Kamataka (11.5 GW), Gujarat (10.6 GW), and Andhra Pradesh (8.9 GW), followed by Tamil Nadu (5.5 GW), Rajasthan (4.8 GW), and Maharashtra (4.5 GW).[6]

India's wind power industry began to take off in the early 1990s, though it has experienced periods of boom and bust over the past two decades. In the late 1990s in particular the industry experienced a slowdown, reportedly owing to the reduction in government tax benefits, delays in processing land approval, and technical problems related to poor installation practices in the preceding years. Growth started to take off again with the 2003 Electricity Act. In recent years both the government has succeeded in injecting greater stability into the Indian market by encouraging larger private- and public-sector enterprises to invest in wind power.

Tax exemptions and accelerated depreciation for up to 80 percent of project costs in the first year, in addition to a generation-based incentive (GBI) scheme, have served as key incentives for wind power development. In June 2008 the Ministry of New and Renewable Energy (MNRE) announced a national GBI scheme for grid-connected wind power projects less than 49 MW, providing an incentive of 0.5 rupees per KWh. In early 2009 this was expanded to all projects to offer this incentive to investors for a period of ten years, provided they do not claim the depreciation benefit. This expanded tariff incentive was meant to provide an incentive for wind development that was attractive to a broader range investors that may have not been interested in the depreciation-based benefit because of limited tax liability.

In addition to the MNRE incentives, many Indian states have set feed-in tariffs to support wind power development. Tariff rates range from 3.14 rupees per KWh in Kerala to 4.5 rupees per KWh in Rajasthan. The 2003 Electricity Act required each state to fix its own minimum percentage for purchase of renewable energy, taking into account the availability of such resources in the region and the impact on retail tariffs. As a result, most states have established mandatory renewable energy shares. One of the more aggressive quotas for wind is found in Tamil Nadu, where the standard was set at 10 percent between 2008 and 2009, increasing to 13 percent between 2009 and 2010, and 14 percent between 2010 and 2011.[7]

India has taken some direct steps to encourage local wind turbine manufacturing. For example, it has structured customs duties in favor of importing wind turbine components over importing complete machines.[8] It has also developed a national certification program for wind turbines, administered by the Ministry on Non-Conventional Energy Sources (MNES) and based in large part on international testing and certification standards. The Indian government has been supporting R&D in wind power technology since the 1980s.[9]

India now has a rather concentrated local wind power industry of relatively few but powerful turbine manufacturers and developers. The country has a solid domestic manufacturing base, led by Suzlon, which held 55 percent of Indian market share in 2009. While Suzlon originated in India, it now sells turbines all over the world. Other key players in the Indian market include Germany's Enercon, with 16 percent market share in 2009; Vestas, with 7 percent; and RRB, with 9 percent. RRB was formed through a 1987 joint venture with Vestas that dissolved in 2006.[10] There are also some smaller manufacturers, including Pioneer Wincon, SWL, Inox Wind, and Ghodawat Energy. Several other international turbine manufacturers have established production facilities in India, including GE, Gamesa, Siemens, ReGen Power Tech, LM Glasfiber, WinWinD, Kenersys and Global Wind Power. Overall a dozen international companies now manufacture wind turbines in India, either through joint ventures under licensed production, as subsidiaries of foreign companies, or as Indian companies with their own technology.[11] The current annual production capacity of wind turbines manufactured in India is about 3,000–3,500 MW, projected to rise to 5,000 MW by 2015.[12]

SUZLON

Indian wind turbine company Suzlon is now well established in the international wind market beyond India, operating in twenty countries and supplying turbines to projects in Asia, North and South America, and Europe. Suzlon is owned by Indian entrepreneur Tulsi Tanti and his siblings. Tanti started in the textile industry and turned to wind turbines to power his business when faced with soaring power costs and the infrequent availability of power. This led him to establish Suzlon, India's first homegrown wind power company. Within five years Suzlon had made the list of top ten wind companies, and the company has remained there since. Co-investors include two major American investment funds, City Group and Chryscapital, each of which injected $25 million into the company. In 2004 Suzlon established its international headquarters in Aarhus, Denmark, strategically selecting Denmark because of its base of wind energy expertise and extensive network of components suppliers.[13] Suzlon since has expanded its operations to India, Australia, China, Brazil, Spain, Portugal, Italy and the United States. It has thirteen manufacturing facilities across China, India, and the United States, and R&D centers in Denmark, Germany, the Netherlands, and India.[14]

Suzlon first obtained its wind turbine technology in a 1995 technical collaboration agreement with a German company, Sudwind, in which Sudwind shared technical information relating to the manufacturing of its

270, 300, 350, 600, and 750 kW wind turbine models in return for royalty payments. Then, in 2001, Suzlon obtained a license to manufacture rotor blades from Aerpac B.V. and entered into an agreement with Enron Wind Rotor Production B.V. in which Suzlon made a one-time payment to acquire the necessary molds, production line, and technical support to produce another model of rotor blades in India.[15]

In 2005 the firm began manufacturing generators through a subsidiary, Suzlon Generators. Suzlon owned 74.9 percent of the company, a joint venture with Elin EBG Motoren GmbH of Austria. In 2006 Suzlon purchased Belgian company Hansen, the second largest gearbox manufacturer in the world, expanding its access to gearbox technology and marking the second largest foreign corporate takeover by an Indian company in any industry at that time.[16] Suzlon also has an arrangement with Winergy AG, the leading gearbox supplier in India, which allows for the use of domestically manufactured gearboxes while it continues to work to advance its own technology. In May 2007 Suzlon acquired 33.85 percent of REpower's shares, and by December 2009 it had acquired 92 percent. As of early 2012 the two companies were still operating somewhat independently of each other, however, with little technology cooperation or knowledge exchange, although this is likely a temporary situation.[17]

Suzlon currently offers wind turbines that range in size from 600 kW to 2.1 MW. The company's manufacturing strategy has been to build on the licensing and joint-venture agreements described above with its own research and development, and to manufacture as many wind turbine components as possible in-house. The firm believes that increasing its in-house manufacturing capabilities will help to lower costs by giving it greater control over the supply chain and enable quicker, more efficient assembly for faster delivery times to customers.[18] This strategy of developing integrated manufacturing capability is particularly aimed at supporting high-growth regions, including India, China, and the United States. Like several other leading global wind turbine manufacturers, Suzlon established a large production facility in Tianjin, China, in response to the local content requirements promulgated by the Chinese government.[19]

A technological setback in 2007, when instances of blade cracks were discovered during the operation of some of Suzlon's wind turbines in the United States, required the company to retrofit its total fleet of 1,251 blades.[20] After this incident there were reports of order cancellations.[21]

Suzlon has over 550 staff engaged in technology development and R&D activities, split between India and Western Europe.[22] Its investments in

R&D, including design changes and technological upgrades as well as certification, product development and quality assurance, have increased substantially in the past few years.[23] One research center based in the Netherlands benefits from local Dutch expertise in turbine blade development, while another center located in Germany benefits from local gearbox expertise.

South Korea's Wind Industry

South Korea has been installing a stable but relatively small annual wind power capacity in recent years. It ranks thirtieth worldwide in terms of total installed national wind capacity, with 364 MW installed by the end of 2010.[24]

South Korea's promise lies more in its domestic manufacturing base than its domestic wind development potential. Wind resources in Korea are adequate, but land area is limited. As a result, most wind development to date has been focused in the coastal areas, including on Jeju Island, and there is a lot of interest in pursuing offshore development. The Korean Wind Industry Association estimates South Korea's theoretical onshore wind resource potential at about 369 GW, with 18.5 GW of technical potential. Its offshore potential is estimated at 309 GW, with 31.4 GW of technical potential at an average depth of 20 meters. There are currently 8 GW of offshore wind projects either under development or in the planning stages.[25]

South Korea's national energy plan set a target for the share of new and renewable energy in total primary energy consumption to be 3 percent in 2006 and 5 percent in 2011.[26] Wind generation was expected to provide the largest contribution (up to 25 percent or 5.2 TWh) of the total generation by new and renewable sources in 2011, but it fell short. However, new pledges by the government to reduce GHG emissions by 30 percent below projected emissions by 2030 have led to a reinvigorated "green growth" strategy and an announcement of a government-supported 2.5 GW wind project to be built between 2014 and 2019.[27] In addition, the government's Energy Vision 2030 plan targets a 9 percent share of renewables by 2030.

To achieve these targets, the government is providing attractive incentive programs, such as a fifteen-year guaranteed feed-in tariff, tax incentives, and subsidies for the local wind market.[28] Wind generation is eligible for a fifteen-year feed-in tariff of 107.29 won per kWh that is to be reduced by 2 percent every year after October 2009.[29] There is also a cost reduction of one-tenth from income or corporate taxes for the installation of a new renewable energy facility, and import duties on grid-connected wind generators and blades have been reduced.

Other programs that support wind power development include compensation by the government for losses to commercial banks when long-term project financing to renewable energy construction is offered at lower than commercial rates. In addition, a renewable construction facility can make a proposal to Korea Energy Management Corporation (KEMCO) for a maximum $20 million loan that is payable over ten years following an initial five-year grace period.[30] South Korea's January 2009 Green New Deal Stimulus Package included additional funding for renewable energy development, and the Comprehensive R&D Plan on Green Technology called for a twofold increase of R&D spending on green technology by 2012 in twenty-seven key technology areas.

The autonomous government on windy Jeju Island has promised to support the installation of wind power plants on the island, having set a target for 500 megawatts by 2020, including 300 megawatts from maritime wind power.[31] Electricity from wind currently accounts for just 3.4 percent of power demand for the island's population of 560,000, but the Jeju government aims to increase the figure to 20 percent by 2020 and 50 percent by 2050. Jeju has also become the site of Korea's first electric smart grid, allowing for real-time monitoring of electricity demand and output with digital technology that enables communication between consumers and utility firms.

LEADING MANUFACTURERS

Traditionally an importer of equipment for wind power projects, South Korea does not have a long history of manufacturing wind turbines. Since 2006, however, many Korean firms have entered the wind industry, including some of the country's largest industrial conglomerates. While several Korean firms are poised to succeed in the industry, no one firm has emerged as an industry leader. Today Korean firms undertaking wind turbine technology development at various stages include Daewoo, Doosan, Hyosung, Samsung, Hyundai, Hanjin, STX, Rotem, and Unison.

Daewoo Shipbuilding and Marine has been developing a 2 MW onshore turbine since about 2005 that was expected to enter serial production around 2011. Daewoo acquired ownership of manufacturer DeWind from its U.S. owner Composite Technology Corporation, giving Daewoo immediate access to a product program consisting of a 1.25 MW model and two 2 MW models, along with R&D facilities and production lines in Germany and the United States and the development rights for the 629 MW Little Pringle project in Texas.[32] DeWind's origins were in Germany, though it has had British and American owners at various times.

Doosan Heavy Industries, South Korea's top power-plant builder and the world's biggest seawater desalination plant provider, has partnered with AMSC affiliate Windtec to develop a 3 MW direct-drive offshore wind turbine. The first prototype was installed onshore in October 2009 on Jeju Island, with serial production expected in late 2010 for onshore applications and in 2012 for offshore applications. Hyosung is another company that has been involved in the wind industry for several years, though primarily through R&D activities, including the development of 750 kW and 2 MW onshore turbines as well as a 3 MW offshore turbine.

Samsung Heavy Industries began developing a 2.5 MW onshore turbine with UK design firms Romax and Garrad Hassan in 2008, with its first turbines installed in 2010. It includes the use of blades from LM Glasfiber in conjunction with a five-year supply contract. It has also begun work on the development of a 5 MW offshore turbine, with production targeted for 2013. Samsung announced plans for continued overseas expansion including in the United States and Germany, and in 2011 it stated that it was building 2 GW of wind power projects in Ontario, Canada.[33]

Hyundai Heavy Industries Co Ltd., the world's largest shipbuilding company, announced its entry into the wind industry in 2008 when it signed a deal with AMSC-Windtec to license technology for 1.65 and 2 MW wind turbines. Hyundai's marketing and sales rights for both turbines under the license extend to dozens of countries around the world, including the United States.[34] Hyundai is also working with Avantis Energy (Germany) to develop a 2.5 MW turbine under license. Rotem, a subsidiary of the Hyundai Kia motor group, has received support from the Ministry of Knowledge Economy to develop a 2.0 MW low wind speed direct-drive wind turbine, along with cooperation from several other research institutes and companies. In June 2010 AMSC announced plans to jointly develop a 5 MW wind turbine with Hyundai for offshore use.[35]

Hanjin, a relatively small company within the plastic and synthetic fiber industry, began developing a 1.5 MW turbine in 2003 in cooperation with German firm Idaswind that has been in serial production since 2008. STX

Corporation, the world's fourth largest shipbuilder, has been involved in the wind industry since 1999 when it developed projects in South Korea using Vestas turbines. In 2009 STX signed a deal to completely acquire Harakosan Europe BV, a Dutch manufacturer of gearless wind turbines.

The Unison Corporation began developing its 750 kW wind turbine in 2001 under a Korean government-sponsored consortium that included Bokuk Electric, Hankuk Glass Fiber, and the Pohang University of Science and Technology (POSTECH). In 2005 it began development of a 2 MW model. It also owns the first two commercial wind projects in South Korea, which used Vestas 2.0 MW and Suzlon 2.1 MW turbines.

While Korean manufacturers entered the wind industry late in the game, their technology acquisition and transfer strategies have focused on advanced wind power technology and offshore wind technology in particular. This is likely due to the fact that there is very little potential for domestic market sales in onshore turbines within Korea, but there is potential for offshore development. Korean firms are also primarily targeting export markets with offshore potential. Samsung, Daewoo (via DeWind), and Hyundai have already announced orders for the U.S. market, while Hyundai and Unison have received orders for onshore projects in the Chinese market.[36]

Through either acquisitions or partnerships with smaller wind turbine manufacturers or design firms, Korean firms are attempting to leapfrog directly to advanced wind turbine technology. The fact that most of the Korean firms are not small companies but huge conglomerates with significant industry experience and a worldwide client base ensures good

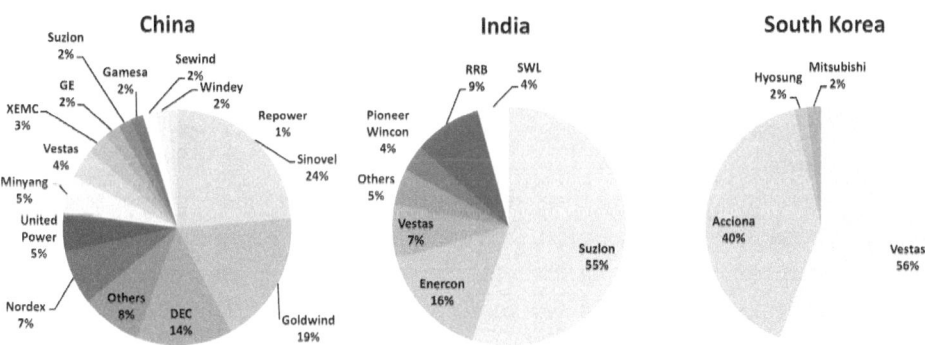

FIGURE 6.2 Wind Turbine Market Shares in China, India, and South Korea
Shares are for 2009.

Li et al. (2010); MAKE (2010); GWEC (2009); BTM Consult (2010).

financial backing and resources for mergers and acquisitions. Most South Korean firms are heavily reliant on foreign engineering and design firms, even if they are developing in-house technology and intellectual property. There are still many key components being imported rather than locally manufactured for the Korean market.

Contrasting Wind Industry Development in China, India, and South Korea

Firms in China, Korea, and India have used different strategies to acquire or develop wind power technology, with varied results. The origins of the wind power technology acquired by firms in the three countries, however, reveal many common sources of such knowledge. Companies have different advantages and face different obstacles to their continued success based on the characteristics of their respective domestic environments and the competition they face in other markets.

TECHNOLOGY TRANSFERS AND ACQUISITION STRATEGIES

Although there are several technology transfer models available to a company looking to enter the wind industry, wind power technology firms from China, India, and South Korea share many similarities in the models they have adopted. Three primary models have emerged: licensing, mergers and acquisitions, and joint development. In addition, there are several common sources of technology information that have worked with firms across these three countries.

Licensing
Several companies began their ventures into the wind industry by setting up licensing agreements, most commonly with small European wind turbine companies. The acquisition of technology from overseas companies is one of the easiest ways for a new wind company to quickly obtain advanced technology and begin manufacturing turbines that may already have been field tested or even have substantial operating experience.

There is a disincentive, however, for leading wind turbine manufacturers to license proprietary information to companies that could become competitors, particularly when technology is transferred from developed to developing countries, where a similar technology potentially could be manufactured with less expensive labor and materials. Consequently, developing-country manufacturers often obtain technology from smaller wind power companies that have less to lose in terms of international competition and more to gain in license fees. The technology obtained from these smaller suppliers may not necessarily be inferior to that provided by the larger manufacturing companies, but it likely has been utilized less and therefore has less operational experience. Alternatively companies may be willing to license outdated models of their technology (often smaller turbine sizes) or to license technology that comes with restrictions on any turbine exports outside the market in which the home manufacturer is based.

Suzlon began its wind turbine manufacturing with a license from German company Sudwind. Goldwind similarly began its operations based on licenses from German firms Jacobs and REpower. Numerous Chinese wind turbine manufacturers have relied on licenses as well. Sinovel and Beijing Beizhong have benefited from licenses acquired from German firms Fuhrlander and DeWind. Dongfang Electric Corporation (DEC) and Windey, like Goldwind, have both licensed turbine designs from REpower. China Shipbuilding Industry Corporation Haizhuang Windpower Equipment Co., Ltd (CSIC) and Guodian United both obtained licenses from German firm Aerodyn. AMSC-Windtec, an American owned firm with roots in Austria, has licensed wind turbine technology to Chinese firms Shenyang Blower Works (Group) Co., Ltd. (SBW), XJ Group, Sinovel, and CSR Zhushou, as well as Korea's Hyundai and several smaller Indian wind turbine manufacturers.

Mergers and Acquisitions

As wind companies become more established, or if they have sufficient financial resources, mergers and acquisitions provide another strategy for technology transfers. M&A gives more authority and flexibility to the acquiring company in how it decides to use the technology, unlike a licensing agreement, which typically has strings attached. Technology acquisitions through M&A can be successful only if the acquiring company has the ability to integrate the new business knowledge into its current business. In addition, there can be a significant financial investment involved.

While Suzlon began its operations based on licenses, it later acquired majority ownership of REpower. Goldwind similarly began its operations based on licenses and later acquired majority ownership of Vensys.

In contrast, the large industrial Korean conglomerates Daewoo and STX used M&A to obtain wind turbine technology early on, purchasing American-owned firm DeWind and Dutch firm Harakosan Europe BV, respectively. While Goldwind's acquisition of Vensys seems to have resulted in the sharing of knowledge, as witnessed through the joint development of new turbine designs, Suzlon's acquisition of REpower has been somewhat restricted by M&A regulations, and the operations of the two companies are still somewhat separate.

Joint Development

As firms develop their own design and manufacturing expertise, they may be more interested in codeveloping wind turbine technology with firms that bring a different set of experiences to the partnership. An advantage of joint development is that there is no initial concern about market competition, and when multiple manufacturers are involved, arrangements for the sharing of any resulting IPR are almost always made prior to the start of the joint work. These arrangements can be more straightforward when joint development involves a firm that primarily focuses on design working with a firm that primarily focuses on manufacturing. The risk with this model, however, is that if the design firm has no manufacturing experience and if manufacturers have no design experience, the resulting product may look great on paper but fail in the factory or in the field.

Several Korean firms are pursing the joint development of wind turbine designs: Hyundai with Avantis, Doosan with AMSC-Windtec, Samsung with Romax and Garrad Hassan, and Hanjin with Idaswind. This form of technology acquisition is also becoming increasingly common in China, particularly among the larger firms. Examples include Sinovel's joint development with AMSC-Windtec, DEC's with AMSC-Windtec and with Aerodyn, Goldwind's with Vensys, A-Power's with Norwin, and Hewind and Sewind's with Aerodyn.

Several firms, particularly in China and Korea, have relied on government support for R&D to design wind turbines, often in conjunction with a consortium of research institutes or universities. While this is a less common model, it is being used by Hyundai Rotem and Unison in Korea, as well as several smaller Chinese manufacturers such as Windey, which originated at China's Zhejiang Institute of Mechanical and Electrical Engineering.

Global Learning Networks

The global reach of a firm's innovative activities can also play an important role in its technology development strategy. Of particular note is the difference in strategy pursued by Suzlon and Goldwind in this regard. Many of the differences between the original technology development strategies of the two companies are related to how they opted to position themselves with respect to domestic and global learning networks.[37]

Suzlon established many overseas operations in order to build on the knowledge gained through its technology licenses even before it had established a substantial market share in its home market of India. This combination of licensing arrangements with foreign firms and internationally based R&D and manufacturing facilities, complemented by the hiring of skilled personnel from around the world, created a global learning network that Suzlon customized to fill in the gaps in its technical knowledge base. Suzlon has been able to draw on this self-designed learning network to take advantage of regional expertise located around the world, such as in the early wind turbine technology development centers of Denmark and the Netherlands. This is in contrast to Goldwind's early years of technology development, in which it remained almost exclusively focused on the Chinese market and conducted very little R&D or manufacturing outside of China.

In recent years, however, Goldwind too has expanded its access to global learning networks, most notably through its acquisition of Vensys. This gave Goldwind access to a network of skilled engineers and a company with a different geographic focus, allowing it to better integrate European wind industry experience into its operations. Even more recently, as Goldwind looks to expand into the U.S. and Australian markets, it has hired American and Australian workers with extensive experience in their home markets to help it better understand how to operate within these domestic contexts.

While they are in an earlier stage of technology development, South Korea's new wind industry entrants are also looking globally for their technology partnerships. Since the Korean companies are already looking to export markets outside of Korea and need to be positioned to compete with global industry leaders, they are not restricting their technology development activities to Korea itself.

As firms enhance their presence around the world by expanding manufacturing bases or R&D facilities, they are also increasingly able to tap into an expanded global knowledge base. Just a few years ago Goldwind was principally a Chinese wind turbine company, operating its manufacturing and R&D facilities primarily in China. This domestic focus changed with the acquisition of Vensys in 2008, when it began to increase its R&D activities in Germany. In contrast, Suzlon has been a company with a global presence for much longer than Goldwind has, beginning with European partners early on in its technology development process. It established many overseas operations to build on the knowledge gained through its technology licenses, even before it had established a substantial market share in its home market of India. Suzlon also has overseas operations across five continents, including subsidiaries, research centers, and sales offices. Although it conducts R&D abroad, Suzlon still relies primarily on components made in India, most of which are made in-house based on experience gained through its overseas research efforts.

COMMON KNOWLEDGE SOURCES

An investigation of the origins of the wind power technology being acquired by firms in India, China, and South Korea reveals the many common sources of such knowledge. When a firm shares licenses with multiple firms, or engages in joint development with multiple firms,

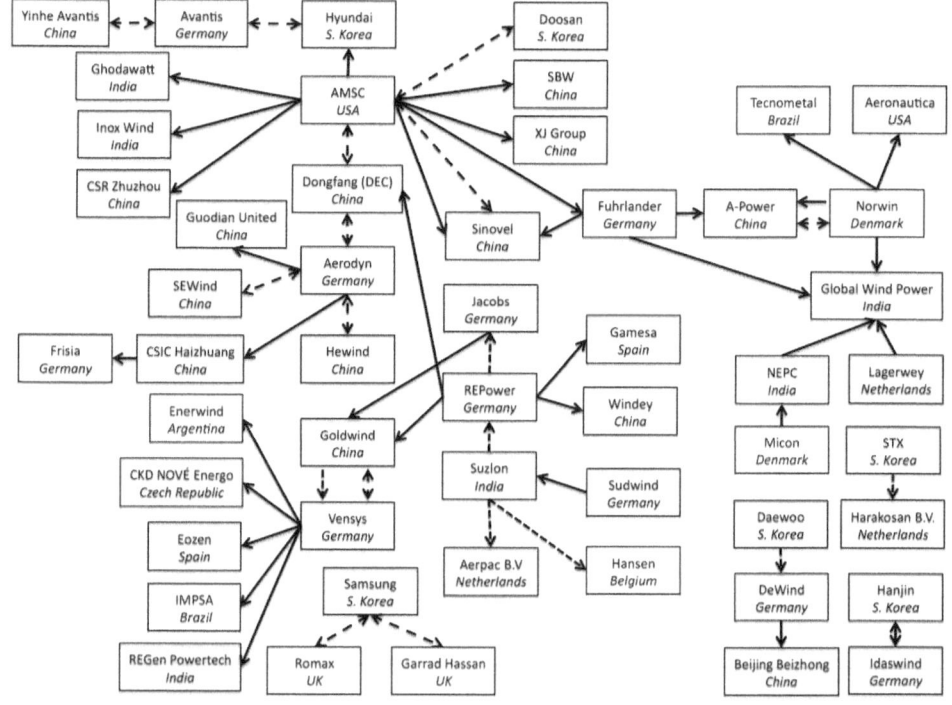

FIGURE 6.3 Wind Power Technology Transfer Networks in China, India, South Korea, and Beyond

Illustration by author.

this creates a network of firms through which knowledge can be shared. While such sharing of information is often restricted through contractual agreements, in other cases it is encouraged. This can have both positive and negative consequences for firms. Such networks increase access to global learning and experience worldwide, which is likely beneficial. However, networks that facilitate information sharing in this way can also create competitors and make it harder to safeguard valuable or sensitive information.

Key companies that have served as the source of wind power technology transfer for many of the manufacturers located in China, India, and South Korea (and beyond) are Avantis, Windtec, REpower, Aerodyn, Fuhrlander, Norwin, and Vensys. It is notable that these companies are either small manufacturers that are not competing with the companies they have licensed to in the Chinese, Indian, or Korean markets or are primarily engineering design firms with little to no manufacturing experience. One exception is REpower, which has become a top-tier global manufacturer in recent years and is now selling directly to many overseas markets.

Avantis, based in Germany, is working with both Hyundai in Korea and Yinhe in China to develop wind turbines. Windtec, now a subsidiary of AMSC, not only has transferred wind turbine technology to China (Sinovel, DEC, CSR Zhuzhou, SBW, and XJ Group), South Korea (Hyundai and Doosan), and India (Inox Wind and Ghodawat) but has also partnered with companies to produce wind turbines in Germany, Japan, Turkey, and Taiwan.[38] China's DEC has also benefited from a license from German firm REpower, the same firm that is now owned by India's Suzlon and provided licenses to China's Goldwind. DEC is also conducting joint development with German firm Aerodyn, the same firm that works with Chinese firms Guodian United, CSIC Haizhuang, Hewind, and Sewind. German firm Fuhrlander has licensed wind turbine technology to China's Sinovel and A-Power as well as India's Global Wind Power; Fuhrlander originally obtained this technology from Windtec. A-Power is also working with Danish firm Norwin, which has licensed and jointly developed wind technology with A-Power and with Global Wind Power of India, Tecnometal in Brazil, and Aeronautica in the United States. Global Wind Power also licensed technology from Dutch firm Lagerwey and NEPC of India; the latter originally obtained its technology from Micon of Denmark. German firm Vensys, now owned by Goldwind, has licensed wind technology to several firms around the world, including Enerwind of Argentina (primarily selling to the Brazilian market), IMPSA in Brazil, ReGen Powertech in

India, Eozen in Spain, a Canadian subsidiary of Vensys, and most recently CKD NOVÉ Energo in the Czech Republic and Slovakia.

DOMESTIC ENVIRONMENTS

While the leading wind turbine manufacturers in China, India, and South Korea have to some extent used similar models of technology acquisition, they have different advantages and face different obstacles to their continued success based on the characteristics of their domestic environment. This includes their home country's wind resource regimes and the domestic policy environments in which they originated.

Only a handful of the leading global turbine manufacturing companies did not primarily rely on their home market in the early stages of their technology development. One of them, Mitsubishi (Japan), may be the model for the new Korean manufacturers. Korea and Japan are pursuing little domestic wind power development, primarily due to lack of land availability and wind resource constraints.[39] While China and India have proven to have sufficient wind resources to support a domestic market, South Korea would need to rely primarily on offshore sites to develop a domestic market for wind power. This is a key reason that Korean manufacturers have had to leapfrog directly to larger, offshore wind turbine technology rather than starting with smaller, onshore models as the Indian and Chinese companies have. Since the Korean manufacturers are well-established firms with sophisticated manufacturing and innovation capabilities, this also reflected a strategy to compete with more established manufacturers from other countries.

China, India, and now South Korea have all benefited from aggressive government subsidy policies to support wind power development, although China's support has arguably been the most stable in recent years. Feed-in tariffs, which provide long-term price support for wind energy, have been particularly effective in all three markets. Despite the importance of national policy support, there are clear limits to understanding the success of these firms based exclusively on the national innovation systems in which they operate. The presence of these companies in different international markets, the frequency with which the firms look globally to pursue forms of technology development or acquisition outside their national borders, and the clear linkages between the origins of technological know-how among companies in different countries point to the need for a more global approach to examining innovation systems.

COMPARATIVE ADVANTAGES

A continually stable and favorable policy environment for wind will continue to make China one of the largest markets for wind power development in the world, and Chinese firms likely will continue to benefit from a domestic policy and business environment that awards them the majority of domestic projects. As the domestic market becomes saturated and as Chinese technology becomes more advanced, however, Chinese firms will increasingly look to export markets and have to compete globally. They may be able to compete in markets where they can offer lower-priced products, though there are already some concerns about quality. Few firms in China have sufficient operating experience to fully assess the quality of their technology, and it is very common for companies in the early stages of developing a new product to experience technical challenges and setbacks.[40] Recent data on wind farm performance in China have in fact raised concerns about quality control in Chinese wind turbine manufacturing, with performance further impeded by the suboptimal siting of wind farms and integration challenges with the Chinese electricity grid.

Goldwind's success in assimilating the technology it acquired is reflected in its increased knowledge and sophistication in the industry, its financial status, and its global reputation. Its increased technical knowledge is reflected in its increasing turbine sizes: the company first developed a 600 kW turbine in 1999—with a purchased license—and in 2010 was developing a 5 MW turbine prototype—through joint development with a design firm that it owns. Goldwind has added financial security owing to its ties with the Chinese government and support from domestic banks. While it still has a very limited global reputation and has exported only a handful of turbines outside of China directly, its acquisition of Vensys has expanded its global reach.

The outlook for India's wind power industry is somewhat uncertain. While the policy environment for wind power in India has improved in recent years, the industry is still heavily dependent on tax incentives that tend to attract a narrow range of investors. In addition, the Indian power sector is plagued with inefficiencies and severe reliability problems that create a difficult environment for wind power growth. Indian wind turbine manufacturers, though still relatively new entrants into the wind power industry compared with some early European and American firms, have many more years of operating experience than most Chinese and Korean firms. But while the Indian manufacturers possess advanced technology and solid operating experience, their global reputation is still somewhat uneven.

Within the Indian market there only a handful of firms competing for market share, unlike in China, where the total number of wind power firms may exceed 80. Indian firms like Suzlon have developed a global reach, allowing them to sell their product in leading markets around the world, while most Chinese firms have yet to leave China's borders. With this global reach comes market flexibility—when the Indian market slows, Indian firms can continue to sell their products to other markets. This reduces their vulnerability to one particular policy system. With this global reach, however, also comes increased risk. The larger and more distributed a firm's operations, the more financially extended it may become.

Suzlon has been able to successfully absorb the knowledge gained by its technology licenses and acquisitions and, through its own research and development, to manufacture increasingly sophisticated technology. Beginning with a 270 kW turbine in 1995, it is now selling primarily 2.1 MW machines. While Suzlon's acquisition of REpower illustrated its financial power and technical foresight, it now seems that assimilation of the know-how and experience that REpower holds will be more difficult than Suzlon initially anticipated.

South Korean firms, unlike Chinese and Indian firms, do not have a sizable domestic market or favorable domestic policy environment. This means that most Korean firms are externally focused, targeting export markets around the world. As a result, they will be competing with manufacturers already active in those markets, including Chinese and Indian manufacturers. A clear advantage of Korean wind firms is their focus on advanced turbine designs, including offshore turbines. The ability of Korean firms new to the wind industry to leapfrog directly to some of the most advanced technology available is a result of their strong existing industrial base and ample resources to support technology partnerships with leading international firms. Without a large domestic market to demonstrate their technology and build market share, however, Korean firms are embarking on a high-risk venture into a competitive marketplace full of firms with many more years of experience.

It is worth noting that Chinese, Indian, and South Korean wind turbine manufacturers all have plans for expansion into North American wind markets. For example, both Samsung and Sinovel are developing projects in Ontario, Canada. In Minnesota, where Goldwind already has a small wind farm, Unison and Sinovel are also looking at projects. Samsung and

A-Power are developing projects in Texas. Other projects using Hyundai, Minyang, Guodian United, and Windey turbines, among others, are also in the early planning stages (see chapter 5). Suzlon already has multiple projects in the United States and plans for further expansion, as does its affiliate REpower. It will be interesting to see how the emerging Asian wind turbine manufacturers compete against each other in the Canadian and American markets in the coming years.

POLICY IMPLICATIONS

As national governments consider policies and regulations to promote a wind power industry, these cases show that companies in China, India, and South Korea have benefited not only from policy support for wind power deployment but also from direct support for local manufacturers. This has particularly been the case in China, where local content requirements and preferential project selection for Chinese manufacturers have been instrumental in helping them build an industry. Such policies have not necessarily resulted in technology transfers, however, since only companies that do not complete in the Chinese market, or design firms that do not manufacture, have been the primary sources of technological know-how for Chinese firms. Other programs may better facilitate technology and learning transfer—for example, China's recently established low-carbon industry clusters, which provide incentives for wind turbine manufacturers to develop their technology in specific regions of the country.[41]

Since the wind turbines coming out of China, India, and South Korea have little operating experience, there is still room for government regulation to ensure quality control in products now being manufactured in all three countries. This could include establishing independent research and testing centers or requiring companies to meet international standards for technology certification, particularly as domestic companies look to export markets. For wind power ultimately to be successful in these countries, a focus on the operation and maintenance of the wind farms, and not just the manufacturing of the wind turbine technology, will also be crucial. This is an area in which all these markets can benefit from experience in countries that have been operating wind farms for decades.

As governments negotiate an international climate change treaty, at the forefront of discussions is how to best facilitate low-carbon technology transfer. This book illustrates how payments for licensing the intellectual property for commercially available technology have not necessarily served as barriers to technology transfer, and that technology transfers are occurring between private companies, via commercial channels, with little government interference. There may be a role for the government in facilitating technology transfers in the private sector, however, through large-scale procurements that aggregate demand and can make a particular market look more appealing to the transferring company. In addition, with advanced or with precommercial technology, leaders may not be as willing to give up IP to competitors. As a result, the government may be able to facilitate collaborative R&D and joint design for precommercial technologies that are deemed of great need but for which private companies are less willing to collaborate. Examples may include carbon capture and sequestration technology, as well as advanced solar power technologies.

This chapter has provided a look at how three emerging economies have acquired and assimilated advanced technologies in relatively short amounts of time. Such insights are crucial to facilitating international technology transfers, which will be an important component of any technological leapfrogging strategy to achieve lower greenhouse gas emissions in the developing world.

The primary technology transfer and acquisition strategies utilized by firms in China, India, and South Korea include licensing arrangements, mergers and acquisitions that resulted in the transfer of technology ownership, and the joint development of new technology. All these strategies were conducted within the constraints of national and international intellectual property law. The origins of the wind power technology developed in these three countries stemmed from many common companies, highlighting a vast network of knowledge transfer spanning industrialized and emerging economies. As technology development becomes increasingly global, firms in emerging economies can take advantage of their growing access to technological know-how that was previously developed primarily by and for the developed world.

It took firms in China, India, and South Korea less than ten years to go from having no wind turbine manufacturing experience to having the ability to manufacture complete, state-of-the-art wind turbine systems that are either already available or soon to be available on the global market. Each country already had an existing industrial base, however, and similar technology development or manufacturing capabilities may not be easily replicable in countries without such a base, particularly less developed countries. It is clear that licensing is a relatively inexpensive way to acquire knowledge and has allowed many of the early entrants in latecomer countries to enter the wind industry. The future potential of licenses is limited, however, particularly if the structure of the license includes little technological know-how, which limits future design modifications or innovations, or comes with market restrictions that limit global expansion. Access to global learning networks can be highly valuable for assimilating technological expertise, which can be done through international research, development or demonstration partnerships, or even something as simple as hiring specialized workers from abroad. For firms with access to substantial financial resources, mergers and acquisitions are a way to acquire technical knowledge and assimilate expertise among firms with similar corporate cultures and baseline expertise, as long as M&A regulations can be successfully navigated. Partnerships to jointly develop new models of wind turbine technology often work well between design firms and manufacturing firms with little conflict of interest, though IP sharing and market access arrangements must still be navigated.

7
Engaging China on Clean Energy Cooperation

Climate change is one of the most pressing international challenges of our time, and thus appropriately a great deal of research scrutinizes the international negotiations that seek to reduce global carbon emissions. But outside these efforts, an expanding but distinct area of international coordination with the potential for climate change mitigation has emerged: bilateral clean energy cooperation.[1] Cross-border collaboration on clean energy research, development and deployment promises to speed the rate at which innovation occurs. Policy makers also clearly believe in the potential of such models of collaboration. In 2009 the Obama administration launched several major new initiatives in which the United States and China agreed to undertake research collaborations on a range of clean energy topics. Other countries have also launched clean energy cooperation programs with China, including Japan, several EU countries, and most recently South Africa and Brazil. These arrangements vary widely in scope, however, targeting different stages of the technology development process and encompassing different actors across the public and private sectors.

As the largest energy consuming nation and largest greenhouse gas emitter in the world, China plays an increasingly important role in the geopolitics of energy as well as climate change mitigation. A longtime proponent of multilateralism, China has been increasing its engagement and its seniority in various multilateral forums, including the United Nations.[2] It has, however, been viewed by many industrialized nations as an obstructionist player in the ongoing international negotiations under the United Nations Framework Convention on Climate Change.[3] China, for many reasons, plays a very different role in the multilateral context from the role it plays in a bilateral one. The international climate change negotiations are full of political posturing and colorful displays of diplomatic rhetoric camouflaging fundamental disagreements on the state of the world.[4] In these negotiations, China time and again has served the role of spokesperson for the developing world, defending its right to emit in the name of economic development. In a direct bilateral discussion with a leading industrialized nation like the United States, however, China wants to be seen as an equal, and as the global superpower that it has become. For this reason direct bilateral discussions can lead to more effective platforms for cooperation on topics that become politicized in a larger negotiation.[5]

Bilateral cooperation allows for the identification of clear technical areas of mutual interest between the countries involved. For example, while China and the United States frequently disagree over core issues in the UN climate negotiations, the technical challenges behind their mitigation options are distinctly similar. Both countries have abundant domestic coal resources that provide energy security benefits but are a significant source of emissions. While both China and the United States also have excellent renewable resources, including wind and solar, the best resources and locations for renewable power development tend to be located far from population centers and electricity demand and thus will require expanded and modernized transmission infrastructures. Both countries have realized the potential energy efficiency gains they can achieve but still lag behind Europe, Japan, and others in developing a more efficient energy system.

China and the United States have a somewhat unique positioning in the international community on energy and climate change issues, as the two largest economies, the two largest energy consumers, and the two largest greenhouse gas emitters on the planet. Owing to the similarities in energy systems shared by the two countries, there are many areas where both the United States and China could benefit from cooperation on climate change and clean-energy development. This chapter therefore examines the status of and the opportunities emanating from the unique and decisive U.S.–China relationship on energy and climate. It also offers recommendations

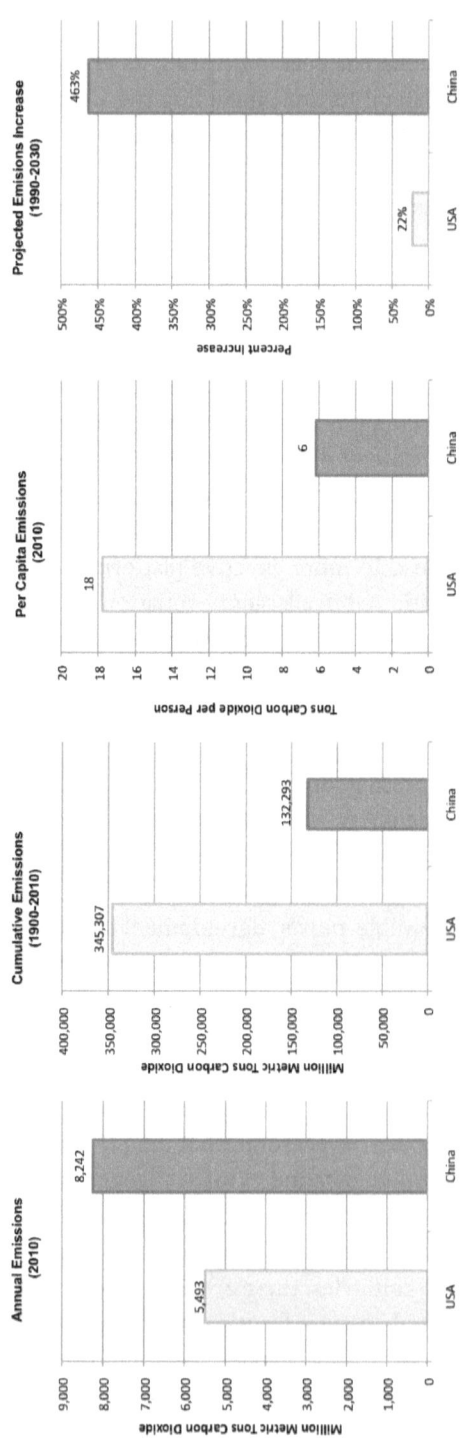

FIGURE 7.1 Carbon Dioxide Emissions in the United States and China
CDIAC (2011); World Bank (2011); EIA (2011).

for expanded U.S.–China cooperation on clean energy in order to advance the global transition to a low-carbon economy.

U.S.–China Energy and Environmental Collaboration

The United States and China have a long history of bilateral energy and environmental cooperation, both through official governmental channels and between universities and nongovernmental organizations. Bilateral talks on climate and energy issues between the United States and China are critically important because they can focus on advancing clean-energy technology developments that are of direct interest to both countries. The United States has increased its bilateral activities with China on clean energy and climate change just as multilateral climate change negotiations have stalled. As a result they may also be crucial to improving mutual understanding and ultimately facilitating a multilateral climate agreement that involves both countries.

China and the United States together comprise almost half of global emissions. As a result, any global solution to climate change must include the participation of these two countries. Fundamental differences exist, however, in how they each view their bilateral relationship and their roles in the multilateral system. These must be carefully navigated.

OFFICIAL BILATERAL COOPERATION

The U.S.–China Agreement on Cooperation in Science and Technology (S&T Agreement) was signed in 1979, soon after the normalization of diplomatic relations between the two countries. This agreement established a framework for many of the subsequent agreements on energy-related cooperation between the United States and China signed over the next thirty years. Many of the agreements signed during the 1970s and 1980s focused on promoting collaboration and understanding in basic research related to many core energy technologies of this period, including the 1970 Agreement on High Energy Physics, the 1983 Protocol on Nuclear Physics and Magnetic Fusion, and the 1985 Protocol on Cooperation in the Field of Fossil Energy Research and Development.

Cooperation on basic research topics continued during the 1990s, but a new set of cooperation agreements were signed focusing more on energy policy discussions. These included policy discussions on fossil and nuclear energy, energy efficiency, and renewable energy. In 1995 U.S. Secretary of Energy Hazel O'Leary signed several bilateral agreements with China on high-level energy policy consultations, nuclear energy, renewable energy, energy efficiency, coal-bed methane, and climate research. That year the initial Protocol for Cooperation in the Fields of Energy Efficiency and Renewable Energy Technology Development and Utilization was signed; it was subsequently amended with new programs of cooperation (through annexes to the agreement) in the years following.

In 1997 Vice President Al Gore and Premier Li Peng cochaired the first session of the U.S.–China Forum on Environment and Development in Beijing. The purpose of the forum was to expand cooperation and intensify dialogue between the United States and China on issues related to sustainable development, particularly protection of the global environment. Also in 1997 President Jiang Zemin visited the United States, and Secretary of Energy Federico Pena and State Planning Commission Vice Chairman Zeng Peiyan signed the Energy and Environment Cooperation Initiative (EECI). The EECI was an outgrowth of the U.S.–China Forum on Environment and Development designed to focus cooperative efforts on the intersection of energy and environmental science, technology, and trade. The initiative targeted urban air quality, rural electrification and energy sources, and clean-energy sources and energy efficiency and involved multiple agencies as well as participants from business sectors, linking energy development and environmental protection. The second meeting of the U.S.–China Forum on Environment and Development was held in April 1999 in Washington, cochaired by Vice President Gore and Premier Zhu Rongji.[6]

The agreements signed from 2000 to 2010 built upon the agreements of the previous two decades, which allowed for a further broadening of the energy topics being discussed bilaterally, as well as further linkages to other bilateral discussions taking place between the United States and China on economic and security issues. In 2006 the U.S.–China Strategic Economic Dialogue (SED) was founded by Vice Premier Wu Yi and U.S. Treasury Secretary Henry Paulson. The biannual, cabinet-level dialogue involved several agencies, including the DOE, the U.S. Environmental Protection Agency (EPA), the NDRC, and China's Ministry of Science and Technology, and included a specific track for energy and environment discussion. In 2008 the fourth SED led to the establishment of the U.S.–China Ten-Year Framework for Cooperation on Energy and Environment (TYF). On the U.S. side the TYF included participation by DOE, Treasury, State, Commerce, and

EPA; on the Chinese side it included participation by the NDRC, State Forestry Administration, NEA, MOF, MEP, MOST, and MFA. The TYF initially established five joint task forces on the five functional areas of the framework: (1) clean, efficient, and secure electricity production and transmission; (2) clean water; (3) clean air; (4) clean and efficient transportation; and (5) conservation of forest and wetland ecosystems.[7] These five areas were further elaborated in seven specific action plans for implementation[8] and were later expanded on in the July 2009 Memorandum of Understanding to Enhance Cooperation on Climate Change, Energy, and Environment, initialed by Secretary of State Clinton, Secretary of Energy Steven Chu, and Chinese State Counselor Dai Bingguo.[9] In April 2009 the Obama administration rebranded the SED as the U.S.–China Strategic and Economic Dialogue (S&ED), with the State Department and Treasury Department now cochairing the dialogue for the United States.[10] An additional Joint Working Group Meeting for the TYF was held in Washington, D.C., in May 2010, producing the U.S.–China Joint Statement on Energy Security Cooperation, as well as twenty-six specific outcomes on energy security and climate change.[11]

Between 2009 and 2010 the Obama administration placed a renewed focus on energy and climate change issues in the U.S.–China bilateral relationship in establishing a new set of agreements on clean-energy topics. The first Obama administration announcement on U.S.–China energy cooperation came in July 2009 in conjunction with Secretary of Energy Chu's first trip to China.[12] Minister of Science and Technology Wan Gang and National Energy Administrator Zhang Guobao, along with Chu, signed a protocol announcing plans to develop a U.S.–China Clean Energy Research Center (CERC) that would facilitate joint R&D on clean energy by teams of scientists and engineers from the United States and China, as well as serve as a clearinghouse to help researchers in each country. Priority research topics for the center were to include building energy efficiency, clean coal (including carbon capture and storage), and clean vehicles. At the July meeting the United States and China together pledged $15 million to support initial activities, with each government pledging equal amounts, while the details of the center were elaborated in subsequent visits.

The U.S.–China Presidential Summit in Beijing in November 2009 resulted in a significant set of new agreements on joint energy and climate cooperation between the two countries.[13] First, the details surrounding the aforementioned CERC were formally announced, including the fact that the center would be supported by public and private funding of at least

$150 million over five years, split evenly between the two countries.[14] As elaborated in the Protocol for Cooperation on a Clean Energy Research Center between the U.S. DOE, and China's MOST and NEA, each side is as to fund only the research activities of scientists from its own country. Any intellectual property rights created under this protocol through cooperative activities are to be jointly owned by both parties, with respective contributions pre-agreed by both sides under technology management plans for each project.[15] In addition, the DOE and MOST/NEA were to jointly establish the U.S.–China Steering Committee on Clean Energy Science and Technology Cooperation to provide high-level guidance for research activities and secretariats based in each country to coordinate the joint activities.[16] In March 2010 Secretary Chu announced the availability of $37.5 million in U.S. funding, as well as matching funds from U.S. grantees, over the next five years to support the CERC for a total of $75 million, in addition to $75 million in Chinese funding.[17]

During the 2009 summit the presidents also announced the launch of the U.S.–China Electric Vehicles Initiative.[18] Building on the U.S.–China Electric Vehicle Forum held in Beijing in September 2009, this initiative included joint standards development, demonstration projects in more than a dozen cities, and technical roadmaps and public education projects.[19] Also announced was a new U.S.–China Energy Efficiency Action Plan targeting buildings and industrial and residential sectors through the development of energy-efficient building codes and rating systems, energy-efficiency benchmarking of industrial facilities, training of building inspectors and energy-efficiency auditors for industrial facilities, harmonizing of test procedures and performance metrics for energy-efficient consumer products, exchange of best practices in energy-efficiency labeling systems, and convening of a new U.S.–China Energy Efficiency Forum to be held annually, rotating between the two countries.[20] The summit also produced the announcement of a new U.S.–China Renewable Energy Partnership. According to the U.S. Department of Energy, "both Presidents embraced a vision of wide-scale deployment of renewable energy including wind, solar and advanced bio-fuels, with a modern electric grid, and agreed to work together to make that vision possible."[21]

Other agreements announced at the November 2009 Presidential Summit include the "21st Century Coal" pledge to promote cooperation on cleaner uses of coal, including large-scale carbon capture and storage demonstration projects;[22] the Shale Gas Resource Initiative;[23] and the U.S.–China Energy Cooperation Program (ECP) to leverage private-sector

7.1 TIMELINE OF MAJOR EVENTS IN U.S.–CHINA CLEAN ENERGY COOPERATION

Year	Event
1979	U.S.–China Agreement on Cooperation in Science and Technology
	Memorandum of Understanding on Bilateral Energy Consultations
	U.S.–China Protocol for Cooperation in Atmospheric Science and Technology
	U.S.–China Agreement on High Energy Physics
1983	U.S.–China Joint Commission on Commerce and Trade
	U.S.–China Protocol on Nuclear Physics and Magnetic Fusion
1985	U.S.–China Protocol on Cooperation in the Field of Fossil Energy Research and Development (Fossil Energy Protocol)
1988	Sino-American Conference on Energy Demand, Markets, and Policy
1993	Establishment of the Beijing Energy Efficiency Center (BECon)
1994	China releases Agenda 21 report
1995	U.S.–China Protocol for Cooperation in the Fields of Energy Efficiency and Renewable Energy Technology Development and Utilization
1997	U.S.–China Forum on Environment and Development
	U.S.–China Energy and Environment Cooperation Initiative
	Establishment of the U.S.–China Energy and Environmental Technology Center
1998	U.S.–China Agreement of Intent on Cooperation Concerning Peaceful Uses of Nuclear Technology (PUNT)
	U.S.–China Joint Statement on Military Environmental Protection
2002	U.S.–China Statement of Intent on Cleaner Air and Cleaner Energy Technology Cooperation
	U.S.–China Joint Working Group for the Green Olympics Protocol
2003	FutureGEN Near Zero Emissions Coal Project
	Carbon Sequestration Leadership Forum
2005	First U.S. DOE office opened in China
	U.S.–China Energy Policy Dialogue
2006	Asia-Pacific Partnership on Clean Development and Climate
	U.S.–China Strategic Economic Dialogue
2007	U.S.–China Memorandum of Understanding on Cooperation on the Development of Biofuels
	U.S.–China Bilateral Civil Nuclear Energy Cooperative Action Plan
	U.S.–China Westinghouse Nuclear Reactor Agreement
2008	U.S.–China Ten-Year Framework for Cooperation on Energy and Environment

(continued on next page)

TIMELINE OF MAJOR EVENTS IN U.S.–CHINA CLEAN ENERGY COOPERATION *(continued)*

Year	Event
2009	U.S.–China Strategic and Economic Dialogue
	U.S.–China Memorandum of Understanding to Enhance Cooperation on Climate Change, Energy, and the Environment
	U.S.–China Climate Change Policy Dialogue
	U.S.–China Memorandum of Cooperation to Build Capacity to Address Climate Change
	Protocol for Cooperation on a U.S.–China Clean Energy Research Center
	U.S.–China Electric Vehicles Initiative
	U.S.–China Energy Efficiency Action Plan
	U.S.–China Renewable Energy Partnership
	U.S.–China Cooperation on 21st Century Coal
	U.S.–China Shale Gas Resource Initiative
	U.S.–China Energy Cooperation Program

resources for project development work in China across a broad array of clean-energy projects. The ECP is a nongovernmental organization that coordinates directly with the DOE and includes more than twenty-two companies as founding members, encompassing collaborative projects on renewable energy, smart grid, clean transportation, green building, clean coal, combined heat and power, and energy efficiency.

During the May 2010 meeting of the S&ED in Beijing, three clean-energy forums established by the above agreements were held, including the U.S.–China Renewable Energy Industry Forum, the U.S.–China Advanced Biofuel Forum, and the U.S.–China Energy Efficiency Forum. All forums included representatives from both government and industry and were accompanied by the announcement of many new public- and private-sector partnerships.

At the 2010 Biofuel Forum, eight new memorandums of understanding (MOUs) were signed covering topics such as aviation biofuel and cellulosic ethanol.[24] Many private-sector partnerships were also announced, including one between Boeing and PetroChina to evaluate developing a sustainable aviation biofuels industry in China; an expanded research collaboration between Boeing Research & Technology and the Chinese Academy of Science's Qingdao Institute of Bioenergy and Bioprocess

Technology on algae-based aviation biofuel development; and an inaugural flight using sustainable biofuel derived from biomass grown and processed in China conducted by Air China, PetroChina, Boeing, and Honeywell.[25] At the Renewable Energy Forum, Applied Materials and China Energy Conservation and Environmental Protection Group signed a MOU to explore projects to accelerate the development and deployment of solar energy including through a 5 MW thin-film photovoltaics (PV) project in Inner Mongolia.[26]

In 2011 the various programs of cooperation initiated during the November 2009 Presidential Summit continued to move ahead. For example, additional renewable energy forums took place in Washington in September 2011, and the three CERC consortia continued to make progress on their work plans.

NONGOVERNMENTAL COOPERATION

In addition to official government cooperation, there are many forms of U.S.–China energy cooperation between academic institutions, nongovernmental organizations, foundations, and the private sector. Examples of such programs include several nongovernmental partnerships focused specifically on engaging the private sectors in the United States and China and encouraging business partnerships, such as the U.S.–China Green Energy Council, the American Council on Renewable Energy's U.S.–China Program, the Clean Air Task Force's Asia Clean Energy project, Joint U.S.–China Collaboration on Clean Energy (JUCCCE), and the U.S.–China Green Tech Summit.

Other organizations have convened groups of stakeholders to provide high-level recommendations to the U.S. and Chinese governments on energy and climate cooperation, such as the Asia Society's Initiative for U.S.–China Cooperation on Energy and Climate and the U.S.–China Clean Energy Forum. Track II U.S.–China dialogues such as those convened by the Brookings Institution and the Carnegie Endowment for International Peace comprise leading thinkers outside the government or former government officials and provide opportunities for high-level exchanges in a nonofficial environment.

In addition, many U.S.–based nongovernmental environmental organizations now have sizable offices in China and engage in cooperative activities with Chinese partners, including the Natural Resources Defense Council, the Environmental Defense Fund, and the World Resources Institute.

Many universities have official research collaborations on energy and climate issues, for example, the Tsinghua-MIT Low Carbon Energy Research Center. One of the largest nongovernmental organizations engaged in U.S.–China cooperation is the China Sustainable Energy Program (CSEP), established in Beijing by the San Francisco-based Energy Foundation in 1999. Staffed by Chinese nationals and supported by international experts, CSEP supports China's energy-efficiency and renewable-energy policy efforts by funding studies conducted by Chinese agencies, experts, and entrepreneurs and linking them with expertise from around the world.[27]

The United States has also played a role in establishing leading clean energy research organizations in China. For example, the Beijing Energy Efficiency Center (BECon) was founded in 1993 with support from the Pacific Northwest Laboratory and Lawrence Berkeley Laboratory of the United States, along with the World Wildlife Fund for Nature. BECon is administratively part of the NDRC's Energy Research Institute but operates as a nonprofit, independent nongovernmental energy conservation and promotion organization.[28]

Barriers to Cooperation

Looking at the list of past and ongoing clean-energy cooperation efforts between the governmental, nongovernmental, and private sector in China and the United States, it is clear that there has been quite a bit of activity. While the official governmental track is certainly not the only means of bilateral cooperation, nor always the most effective, it is clearly important for cooperation to occur through official as well as unofficial channels. Despite the long list of official bilateral agreements signed between the United States and China in the area of clean energy and climate change, there have been many challenges to following through on the successful implementation of the activities agreed upon. Official bilateral cooperation has suffered in the past from a lack of consistent funding as well as from insufficient high-level political support and commitment. Cooperation is also hampered by the increasingly competitive relationship between the United States and China in the global economic marketplace.

While the list of agreements signed has been well documented by both governments, less attention has been paid to the results of these programs. The level of funding support provided to each initiative is generally also quite difficult to track, in many cases because the MOUs or initiatives signed were not backed by secure funding commitments. As a result, there has been some skepticism surrounding government agreements for bilateral cooperation that are not accompanied by both high-level political support and dedicated funding appropriations. This skepticism has played a role in U.S.–China bilateral relations and has contributed to some mistrust and some reluctance to pursue future cooperation initiatives.

The cancellation or downscaling by the United States of several key clean-energy projects has led to an understandable skepticism in China on the prospects for stronger long-term cooperation. Recent examples include the two-plus-year expiration of the U.S.–China Protocol on Energy Efficiency and Renewable Energy (followed by its eventual renewal) and the postponement and significant restructuring of the FutureGen project to build, in partnership with China, a commercial-scale advanced-generation coal plant with carbon capture and storage.[29]

It is particularly notable that more bilateral clean-energy and climate-change agreements were signed in 2009 than in any prior year. The fact that the president of each country signed the majority of these agreements illustrates political support at the highest level on both sides. The true success of these agreements, however, can only be measured by the results achieved by their implementation, which will in turn be contingent on the resources devoted to ensure that they are implemented successfully. Many of the details are yet to be worked out and real challenges remain, particularly regarding stable funding resources. The agreements outlining the new Clean Energy Research Center and the U.S.–China Renewable Energy Partnership, for example, both point to existing funding sources for implementing domestic actions in both countries, with minimal additional funding sources for collaborative projects. Recent congressional announcements have tried to restrict international cooperation activities, which may have direct implications for the agreements.[30] While it is clearly important that both sides bring resources to the table, if nothing new is allocated for these agreements it is unclear how they will result in any deviation from current practices. In addition, if both sides are paying their own way and there is no financial incentive for cooperation, activities must be in the clear interest of both or there is little reason for either to come to the table.

Cooperation and Competition

Cooperation is increasingly common between the United States and China in areas ranging from basic research to joint business ventures. At the same time, the two countries are competitors for resources, talent, and economic markets. While competition can be an engine for innovation, and clean-energy development in particular is an area where innovation will be vital, it is hard for any country to put long-term global interests ahead of near-term domestic interests—particularly in the fast-moving clean-energy sector.

Fears that U.S. climate regulations would help Chinese companies outcompete American companies have led to the inclusion of trade measures aimed at large developing countries—primarily China—in several draft proposals for climate-change legislation in the U.S. Congress.[31] Trade measures were included in several legislative proposals of the 110th Congress (2007–08), including S.1766, the Low Carbon Economy Act introduced by Senators Jeff Bingaman and Arlen Specter, and S. 2191, America's Climate Security Act, introduced by Senators Joseph Lieberman and John Warner. In the 111th Congress, similar provisions were contained in the American Power Act introduced by Senators John Kerry and Joseph Lieberman. With a stated purpose of protecting against foreign countries' undermining of a U.S. objective of reducing greenhouse gas emissions, the legislation stipulates that U.S. importers must buy international reserve allowances to offset lower energy costs of manufacturing certain goods coming from certain countries.[32] While some least-developed countries are excluded from these requirements, most developing countries must comply unless they have taken policy action at home deemed to be of comparable stringency to U.S. action. The impact of such measures on "leveling the carbon playing field" between the United States and China has been questioned,[33] but it is widely believed that any eventual U.S. legislation will contain some form of carbon-leakage provision (or a "China provision") aimed at appeasing labor interests, which have widely supported and helped shape the provision.

Trade measures are not the only means of addressing the competitiveness issue between the United States and China. Fashioned carefully, closer collaboration on clean energy could enhance the economic prospects of both nations while avoiding establishment of an unfair competitive advantage for either country. Recent events, however, have illustrated ongoing tensions in both countries surrounding access to clean-energy markets.[34] For example, announcements in October 2009 that Chinese wind turbine manufacturer Shenyang Power Group was supplying 2.5 MW turbines made in China for a wind farm in western Texas raised many concerns, particularly from members of Congress, that China was trying to compete with the United States in its own domestic market in an industry that the government had specifically been trying to promote with tax credits and other green-jobs initiatives.[35] The discussion over the Texas wind farm occurred close to the time that Commerce Secretary Gary Locke traveled to China to ask for the removal of a seven-plus-year policy requirement that wind turbines installed in China must be locally manufactured, essentially restricting any imported turbines. China agreed, and just a few weeks later in mid-November 2009, Shenyang's parent company, A-Power Energy Generation Systems Ltd., announced that it had partnered with the U.S. Renewable Energy Group to build a wind turbine production factory in the United States.[36]

The U.S. solar industry has also pointed fingers at Chinese solar manufacturers in the wake of the Solyndra bankruptcy,[37] claiming it is impossible for U.S. solar firms to compete with Chinese firms supported by lucrative Chinese government policy support—including subsides that may become the target of countervailing measures.[38] While Chinese solar companies now dominate global solar PV markets, U.S. companies are still playing an important role in the industry. A significant portion of the revenue from solar projects comes not from manufacturing the panels themselves, but from site preparation and system installation, which must be done locally with local workers.

Many of the most successful U.S. solar companies import the panels they sell from China but can still compete because of the locally based service they provide. Despite importation of solar modules from China, the United States was a net exporter overall of solar technology in 2010, exporting almost 2 billion dollars in solar-energy products.[39] The largest U.S. export is polysilicon, the feedstock for making crystalline silicon PV panels, and second largest is the capital equipment used to manufacturer solar PV products. While the United States had a 278.3 billion dollar trade deficit with China in 2010, it had a positive trade balance with China in the solar industry.[40] This is due to the fact that one of the lowest-value segments of the solar PV supply chain is made in China, not to mention one of the most energy-intensive.

China is paying a significant environmental cost for its competitive advantage in the solar industry. The boom in PV manufacturing directly conflicts with China's national policies to decrease the energy and carbon intensity of its economy.[41] Amidst reports of increasing numbers of environmental protests in China, in the fall of 2011 there were reports of protesters in eastern China camped outside a solar panel manufacturer in response to toxic waste from the factory being dumped into a nearby river, killing fish.[42]

Since most solar panels made in China are exported, China is not benefiting from the emission-free electricity the technology provides. However, new Chinese policies are trying to change this, namely, the national feed-in tariff policy for solar photovoltaics that was announced by NDRC in July 2011.[43] The program provides a subsidy to encourage the deployment of solar energy within China, which will likely be necessary to meet the recently augmented solar power capacity targets of 10 GW installed by 2015 and 50 GW by 2020.[44] While the previous target was for 20 GW by 2020, as recently as 2009 the national target for solar PV installations by 2020 was just 1.6 GW.[45] The recent price drop in global PV module costs, reportedly down from $3.50 per watt in 2008 to $1.75 in 2010 and $1.15–$1.20 in August 2011,[46] is likely one reason for the change of heart among Chinese policy makers, in addition to reports of overcapacity in the Chinese solar market, and their wanting to see China reap the environmental benefits of the solar technology and not just the costs.

Crystalline solar PV technology dominates the global solar market. Since crystalline module costs have fallen sharply in recent years, it is harder for other solar technologies to compete—particularly ones that have less demonstration experience and less manufacturing scale. But U.S. companies can still compete in advanced solar technologies. These technologies have high risks but potentially high rewards.[47] It is comparatively easy for countries like China to import the polysilicon feedstock and the capital equipment (production line) for solar module manufacturing, but harder to import the technological know-how to enable innovation in this sector.

While the United States and China may argue over where to build wind turbines and solar panels, the world would stand to benefit from the best, lowest cost wind and solar power technology available, and healthy competition should encourage both countries to try to produce it. Many Chinese wind and solar companies have benefited from cooperation with U.S. companies, and several Chinese firms, including wind company Goldwind and solar company Suntech, have indicated their plans to build factories in the United States employing American workers. But continued debates over market access in both countries, as well as trade-distorting subsidies, create an increasingly tense environment for U.S.–China cooperation on clean energy. The increasing number of trade disputes illustrates that there is clearly still a long way to go toward building the trust that will be crucial to scaling up clean-energy cooperation between the United States and China.

Areas for Expanded Cooperation

China has three principal options to reduce greenhouse gas emissions in the power-generation sector: using cleaner coal-based power, developing renewable and other low-carbon sources of energy, and promoting energy efficiency.[48] While China can benefit from expanded cooperation in all these areas, renewable energy presents opportunities for collaboration in basic research in fields that can contribute to future breakthroughs in renewable-energy technologies, in joint R&D in advanced renewable-energy technologies, in the joint demonstration of precommercial technologies, as well as in the sharing of best practices in policy making and in the development of joint strategic studies advising policy makers.

While other countries have led much of the earlier development in renewable energy, the United States and China are poised to become the largest markets for renewable-energy deployment in the coming years. In 2008 the United States and China became the two largest wind power markets in the world, and they are expected to continue lead the market for years to come. U.S.–China cooperation is crucial in two specific instances: where cooperation would trigger the potential for a scale of deployment within the two countries that would result in substantial technology cost reductions; and where deployment of a technology is crucial to climate mitigation but is unlikely to happen without both shared resources and a shared commitment to fronting the cost of a technology commercialization that would benefit the world.[49]

The United States and China share many common interests in low-carbon energy technologies, having similar energy resources and similar domestic technology priorities. In the area of renewable energy, expanded cooperation in solar PV, concentrating solar power (CSP), and wind power technologies should all be considered, as elaborated below.

SOLAR PHOTOVOLTAICS

While China lags behind the United States in solar PV deployment, it leads in solar technology production, and recent government signals show a commitment to expanded domestic utilization. Chinese solar R&D priorities include the development of world-class crystalline silicon purification technology, grid-integration technology for PV (including high-power, high-efficiency inverters); automatic-control technology for large-scale PV arrays; and sun-tracking technology.[50] China's PV manufacturers may be ahead of the world in terms of total production capacity, but they lag in advanced technologies as well as automation of manufacturing. As such, a priority for commercial manufacturers is to further refine and automate their crystalline solar manufacturing processes. An announcement of U.S.–China commercial cooperation in this area has been recently announced between Amtech and Yingli, using a technology designed at the Energy Research Center of the Netherlands.[51]

Concerns about quality control of PV modules made in China, along with the lack of clear technical standards, have led to an interest in building a national China PV Test Center to conduct quality control and enforce technical standards for PV systems. Such a center should be based on leading global standards, and the United States could play an important capacity-building role in establishing it.

In addition, China is interested in expanding the development of thin-film solar technology, particularly in light of the silicon shortages in recent years that may have slowed Chinese production growth.[52] Thin film represents only about one-fifth of global solar PV production, although annual growth rates for thin film PV have been higher than for crystalline PV. While cooperation on commercial technologies can pose competitiveness concerns, the United States and China could greatly expand collaborative R&D on precommercial PV and PV-supporting technologies.

CONCENTRATING SOLAR POWER

Solar thermal electric power has excellent theoretic potential as a renewable energy technology.[53] One study by the Chinese Academy of Sciences and McKinsey sees CSP as playing a pivotal role in CO_2 abatement in the United States and China. It estimates that the deployment of 75 GW in the United States and 100 GW in China by 2030 could account for 22 percent of the CO_2 abatement needed in both countries combined by 2030 for a 550 ppm stabilization scenario.[54] An even more optimistic study by Greenpeace estimates that CSP could supply 7 percent of the world's power needs by 2030 and 25 percent by 2050.[55] Compared with solar PV, CSP has the significant benefit of being both storable and dispatchable, as well as more easily integrated into current electric grids.[56] Current capital costs are high, however, and a lack of demonstration experience makes investors wary. The U.S. and Chinese governments could therefore play an important role in helping to further demonstrate this technology in the United States and China.

Currently the cost of energy from a CSP plant is estimated to be about three times the cost of energy of power from a coal plant. There are different system designs for CSP plants that include parabolic trough power plants and power tower designs, among others. For parabolic trough technology, the cost in the United States is estimated at $0.165 per kWh (with electricity storage), or $6,000 per kW installed.[57] For tower technology, the cost is more uncertain owing to lack of demonstrations but is estimated to range from $0.165 to $0.30 and $8,000 per kW (with storage). Projections, while ever uncertain, estimate that by 2020 the levelized cost of energy will be close to that of coal.[58]

Based on a rough estimate of solar resources and potential land available, there may be 7,800 GW of CSP potential in the United States and 6,000 GW in China.[59] While no commercial-scale CSP plants have been built in China to date, there were 2.5 GW of projects in the pipeline as of mid-2011.[60] A small 70 kW demonstration solar power tower plant was built in Nanjing in 2005, and a 1 MW solar power tower plant is currently being built in Beijing by the Chinese Academy of Sciences. There are feasibility studies under way for parabolic trough systems in Guangdong and Inner Mongolia. Solar thermal power technology is noted as a government priority under the Eleventh Five-Year Plan (2006–10) of National High-Tech Program of Research and Development, and in the National Program for Medium- to Long-term Scientific and Technological Development (2006–20). The United States also has minimal experience with deploying CSP technology but currently leads the world in terms of projects in the development pipeline with 8.67 GW. As of mid-2011 the United States had installed 507 MW of CSP capacity.[61] The U.S. DOE is also implementing several CSP demonstration projects in the United States.[62]

U.S. companies are increasingly recognizing the benefits of demonstrating precommercial technology in China where projects can be built faster and cheaper. CSP, along with fossil power plants with carbon capture and sequestration technology, is one such technology that could benefit from increased demonstration in China. Jointly supported U.S.–China demonstration projects using a variety of CSP technologies would help future commercial investors to better understand the cost structure of these technologies in both countries.

WIND POWER

Despite great strides in wind power development, there are still some major problems plaguing wind development in China that threaten the future viability of the wind industry. For example, recent studies have highlighted the fact that capacity factors for wind farms in China are much lower than in the United States.[63] This is problematic because it is making wind energy less competitive than it should be; if less power is being generated, less electricity is sold, and the project is more expensive overall.[64] The relatively low performance of wind farms in China has been attributed to the suboptimal siting of wind farms, limitations of the Chinese electricity grid in integrating wind power, and wind turbine technology performance. As a result, China could benefit from increased international guidance on wind resource assessment, project siting, operation and maintenance, integration experience, and technology standards and certification practices. Chinese government officials are particularly interested in expanded U.S.–China cooperation on wind power forecasting tools and technologies, and on the use of expanded smart grid technologies to facilitate wind power integration.[65] As China establishes technology standards and grid codes for wind turbines, government leaders have discussed the establishment of a national wind energy technology center to certify and test turbine performance. Such a technology center might be based on the National Wind Technology Center located at the U.S. National Renewable Energy Laboratory, and therefore could serve as another platform for U.S.–China cooperation.

Wind will play an important role in providing low-carbon electricity to the United States and China for years to come, and it is in both countries' interest to ensure that strong growth in this sector continues. China is currently the only country in the world building GW-scale wind power plants. As a result it is experiencing many challenges associated with integrating such large amounts of wind power into the power system that could provide valuable lessons for other countries looking to do the same. In this way China's current wind bases can serve as a laboratory for global learning, with many countries standing to benefit from the lessons of China's rapid wind power expansion.

An Outlook for the Future

The conversation between the United States and China on clean energy and climate change is in many ways just beginning. While bilateral activities have been in place for decades, and both countries are playing an increasingly central role in the multilateral climate negotiations, the role that both will play in the global climate change solution is just starting to be defined. Both countries have taken positive steps at home to promote low-carbon energy sources and increase energy efficiency. Neither country, however, has adopted economy-transforming, mandatory restrictions on carbon emissions.

While China must do much more to cut its rapidly increasing emissions if the world is to avoid dangerous climate impacts, China's leaders know that such reductions will not come easily or cheaply. A real decoupling of energy growth from greenhouse gas emissions will require a calculated move toward a low-carbon economy. The beneficial impacts of China's policies to deploy renewable-energy technology domestically and develop leading clean-manufacturing sectors are finally coming to fruition, but progress is still inadequate.

Of course, the United States faces similar challenges in its own low-carbon transition. The longer the United States waits to act on climate change, the more it is sacrificing its international credibility on the issue. U.S. commentators tend to criticize the ambition of China's climate actions while simultaneously fearing its superiority in clean-energy development. As tensions escalate, cooperation to accelerate the transition to a low-carbon global economy is increasingly important. If the United States fails to follow through on its clean-energy agreements with China, it will not only miss a huge opportunity to increase cooperation on key technologies to help address climate change but also undermine China's trust.

As the United States and China enter a challenging period in their broader political and economic relationship, bilateral platforms for cooperation are increasingly important. Although bilateral cooperation alone cannot solve the global climate challenge, it is essential to working out key differences, facilitating dialogue among business and policy leaders, and implementing workable solutions to climate change in incremental but concrete steps.

There clearly can be no solution to global climate change without the United States and China, and such a solution will depend on the ability of these two countries to see eye to eye. It will take many years to build the trust needed to overcome their differences on this issue, to develop and adopt low-carbon technologies, and to transform their economies. In the meantime, as the entire world looks to the United States and China to move forward, the fate of the global climate system remains in their hands.

Notes

Preface

1. The early research for this book (prior to 2005) was collected for my doctoral dissertation in Energy and Resources at the University of California Berkeley, "From Technology Transfer to Local Manufacturing: China's Emergence in the Global Wind Power Industry," filed in August 2005.

2. I conducted many interviews with Chinese officials at the central, provincial, and local government levels. Central government officials interviewed were primarily located within the National Development and Reform Commission (NDRC), the former State Development Planning Commission, and the State Economic and Trade Commission. I also collected information from researchers in the Center for Renewable Energy Development (CRED) of the Energy Research Institute (ERI) and officials in the former Energy Bureau (both affiliated with NDRC), as well as officials in the National Energy Administration (NEA) and the Ministry of Foreign Affairs (MFA). Provincial government representatives interviewed were primarily from the provincial Development and Reform Commissions; meetings included representatives from the provincial and local governments of Inner Mongolia, Xinjiang, Sichuan, Fujian, Guangdong, Shanghai, Beijing, and Baoding.

3. I conducted interviews with employees of NEG Micon, Nordex, Vestas, GE, Gamesa, LM, Suzlon, Goldwind, AMSC, and Guodian United and made informal inquiries of many other companies (for example, at wind energy conferences and

exhibitions). In many cases I interviewed former employees of these companies as well. When possible I made site visits to company factories, R&D centers, and wind farms where turbines were in operation.

4. Conducting research in a foreign language, and particularly in China, can pose certain challenges, such as issues of reliability surrounding language barriers. For example, a nonnative speaker who cannot identify hidden meanings embedded in interview responses may misunderstand the social contexts of language use. Additional biases involved in this study surround "the conspiracy of courtesy" related to the researcher's status as a "foreigner." (See Gerard J. Gill, *O.K., the Data's Lousy, but It's All We've Got (Being a Critique of Conventional Methods)*, Gatekeeper Series [London: Sustainable Agriculture Programme of the International Institute for Environment and Development, 1993].) An added complexity to conducting interviews with Chinese government officials is the possibility that respondents will fail to be completely forthcoming in providing information in order to uphold various government positions or will withhold information they deem to be politically sensitive.

5. See Valerie J. Gilchrist and Robert L. Williams, "Key Informant Interviews," in *Doing Qualitative Research*, ed. Benjamin F. Crabtree and William L. Miller, 2nd ed., 71–88. (Thousand Oaks, Calif.: Sage, 1999).

6. See, e.g., Jonathan E. Sinton, "Accuracy and Reliability of China's Energy Statistics," *China Economic Review* 12 (2001): 373–83.

1. Green Innovation in China

1. There is some disagreement over how to count the total installed capacity in China in 2010 owing to reported delays in connecting wind turbines to the electric grid. The Global Wind Energy Council (GWEC), which collects its figures from Chinese wind industry sources, reports that China installed 44,733 MW in 2010, and BTM Consult reports 44,781 MW, while *Windpower Monthly* reports a lower figure of 38,280 MW. According to researchers from China's State Grid Energy Research Institute, the GWEC and BTM figures include a considerable amount of capacity that was installed but had not yet begun to deliver electricity to the grid by the end of 2010. Bloomberg New Energy Finance estimates that only 78 percent of wind capacity was connected to the grid by August 2011. GWEC, *Annual Market Update 2010* (Brussels: GWEC, 2011); BTM Consult, *International Wind Energy Development: World Market Update 2010* (Denmark: BTM Consult ApS and Navigant Consulting, 2011); "Windicator," *Windpower Monthly* (January 2012): 64.

2. Stabilizing greenhouse gas emissions at a concentration of 500 parts per million carbon dioxide (a level that would likely avoid the worst of the impacts of climate change) requires a reduction of 175 gigatons of carbon (GtC) below projected emissions globally between now and 2050. If wind power were to generate about 20 percent of global electricity production in 2050, global cumulative carbon emissions

reductions from wind could exceed 35 GtC. GWEC, *Global Wind Energy Outlook 2008* (Brussels: GWEC, 2008).

3. CNA, *National Security and the Threat of Climate Change* (Alexandria: CNA Corporation, 2007); Daniel Moran, ed., *Climate Change and National Security: A Country-Level Analysis* (Washington, D.C.: Georgetown University Press, 2011); IPCC, "Summary for Policymakers," in *Managing the Risks of Extreme Events and Disasters to Advance Climate Change Adaptation*, ed. C. B. Field et al., 1–19. Special Report of Working Groups I and II of the Intergovernmental Panel on Climate Change (Cambridge: Cambridge University Press, 2012).

4. As noted in China's November 2011 Climate Change White Paper, the Twelfth Five-Year Plan for National Economic and Social Development has "established the policy orientation of promoting green and low-carbon development, and expressly set out the objectives and tasks of addressing climate change for the next five years." People's Republic of China, Information Office of the State Council, *Zhongguo yingdui qihou bianhua de zhengce yu xingdong 2011 bai pi shu* [China's Policies and Actions for Addressing Climate Change, 2011 White Paper] (Beijing: State Council, 2011).

5. Jose Goldemberg, "Leapfrog Energy Technologies," *Energy Policy* 26, no. 10 (1998): 729–41; Alice H. Amsden, *The Rise of the Rest: Challenges to the West from Late-industrializing Economies* (New York: Oxford University Press, 2001); Sanjaya Lall, "Technological Capabilities in Emerging Asia," *Oxford Development Studies* 26, no. 2 (1998): 213–43; Keun Lee and Chaisung Kim, "Technological Regimes, Catching Up, and Leapfrogging: Findings from the Korean Industries," *Research Policy* 30, no. 3 (March 2001): 459–83; Richard R. Nelson, "The Changing Institutional Requirements for Technological and Institutional Catch Up," *International Journal of Technological Learning, Innovation and Development* 1, no. 1 (2007): 4–12; United States Congress, Office of Technology Assessment (U.S. OTA), *Technology Transfer to China* (Washington, D.C.: U.S. OTA, 1987); Denis Goulet, *The Uncertain Promise: Value Conflicts in Technology Transfer* (New York: New Horizons, 1989); E. Mansfield, *Intellectual Property Protection, Foreign Direct Investment, and Technology Transfer*, Discussion Paper (International Finance Corporation, 1994); Kelly Gallagher, *China Shifts Gears: Automakers, Oil, Pollution, and Development* (Cambridge: MIT Press, 2006); Ogunlade Davidson, Bert Metz, and Sascha van Rooijen, eds., *Methodological and Technological Issues in Technology Transfer: A Special Report of the Intergovernmental Panel on Climate Change* (New York: Cambridge University Press, 2000).

6. Pluvia Zyuga and Dominique Guellec, *Who Licenses out Patents and Why? Lessons from a Business Survey* (Paris: OECD, 2009); John Barton, *New Trends in Technology Transfer. Implications for National and International Policy* (Geneva: International Center for Trade and Sustainable Development (ICTSD), 2007); South Centre and Center for International Environmental Law, *The Technology Transfer Debate in the UNFCCC: Politics, Patents and Confusion*, IP Quarterly Update (2008); Rachel Diamant, Helen Davison, and Meir P. Pugatch, *Promoting Technology Transfer in*

Developing Countries: Lessons from Public-Private Partnerships in the Field of Pharmaceuticals (Stockholm: Stockholm Network, 2007).

7. AnnaLee Saxenian, *Regional Advantage: Culture and Competition in Silicon Valley and Route 128* (Cambridge: Harvard University Press, 1996); Leonard Dudley, "Learning and the Interregional Transfer of Technology," *Southern Economic Journal* 40, no. 4 (April 1974): 563–70; Linda M. Kamp, Ruud E.H.M. Smits, and Cornelis D. Andriesse, "Notions on Learning Applied to Wind Turbine Development in the Netherlands and Denmark," *Energy Policy* 32, no. 14 (September 2004): 1625–37; Dieter Ernst, "A New Geography of Knowledge in the Electronics Industry? Asia's Role in Global Innovation Networks," *Policy Studies* 54 (2009); Philip Cook and Olga Memedovic, *Strategies for Regional Innovation Systems: Learning Transfer and Applications*, Policy Papers (Vienna: United Nations Industrial Development Organization, 2003).

8. Susan L. Shirk, *The Political Logic of Economic Reform in China* (Berkeley: University of California Press, 1993); Kenneth Lieberthal, *Governing China: From Revolution Through Reform* (New York: Norton, 1995).

9. David C. Mowery and Joanne E. Oxley, "Inward Technology Transfer and Competitiveness: The Role of National Innovation Systems," *Cambridge Journal of Economics* 19 (1995): 67–93; Bo Carlsson and Rikard Stankiewicz, "On the Nature, Function and Composition of Technological Systems," *Journal of Evolutionary Economics* 1, no. 2 (1991): 93–118; Bengt-Åke Lundvall, *National Systems of Innovation: Toward a Theory of Innovation and Interactive learning* (London: Anthem Press, 2010); Lundvall et al., "National Systems of Production, Innovation and Competence Building," *Research Policy* 31, no. 2 (February 2002): 213–31.

10. Dan Breznitz, *Innovation and the State: Political Choice and Strategies for Growth in Israel, Taiwan, and Ireland* (New Haven: Yale University Press, 2007); Yu Zhou, *The Inside Story of China's High-Tech Industry: Making Silicon Valley in Beijing* (Lanham: Rowman and Littlefield, 2008).

11. On China's S&T capability, see Richard P. Suttmeier and Cong Cao, "China Faces the New Industrial Revolution: Achievement and Uncertainty in the Search for Research and Innovation Strategies," *Asian Perspectives* 23, no. 3 (1999); Denis Fred Simon and Cong Cao, *China's Emerging Technological Edge* (Cambridge: Cambridge University Press, 2009). On China's role in the global economy, see Daniele Archibugi and Jonathan Michie, "The Globalization of Technology: A New Taxonomy," *Cambridge Journal of Economics* 19, no. 1 (February 1995): 121–40; Dan Breznitz and Michael Murphree, *Run of the Red Queen: Government, Innovation, Globalization, and Economic Growth in China* (New Haven: Yale University Press, 2011); Edward S. Steinfeld, *Playing Our Game: Why China's Economic Rise Doesn't Threaten the West* (Oxford: Oxford University Press, 2010); Adam Segal, *Advantage: How American Innovation Can Overcome the Asian Challenge* (New York: Norton, 2011); Dieter Ernst, "China's Innovation Policy is a Wake-up Call for America," *AsiaPacific Issues* 100 (May 2011): 12.

2. China's Energy and Climate Challenge

1. Portions of this chapter are adapted from Joanna I. Lewis, "China's Strategic Priorities in International Climate Negotiations," *Washington Quarterly* 31, no. 1 (2007): 155–74; Lewis, "China," in *Climate Change and National Security: A Country-Level Analysis* (Washington, D.C.: Georgetown University Press, 2011); Lewis, "Environmental Challenges: From the Local to the Global," in *China Today, China Tomorrow*, ed. Joseph Fewsmith (Lantham: Roman and Littlefield, 2010); Lewis, "The State of U.S.-China Relations on Climate Change: Examining the Bilateral and Multilateral Relationship," *China Environment Series*, no. 11 (December 2010): 7–39; Lewis, *Decoding China's Climate and Energy Policy Post-Copenhagen*, Policy Brief (German Marshall Fund of the United States, June 2010); Joanna I. Lewis, Jeffrey Logan, and Michael B. Cummings, "Understanding the Climate Challenge in China," in *Climate Change Science in Policy*, ed. Stephen H. Schneider et al. (Washington, D.C.: Island Press, 2009).

2. S. Solomon et al., eds., "Summary for Policymakers," in *Climate Change 2007: The Physical Science Basis. Contribution of Working Group I to the Fourth Assessment Report of the Intergovernmental Panel on Climate Change* (Cambridge: Cambridge University Press, 2007), 1–17; Thomas Conway and Pieter Tans, "Trends in Atmospheric Carbon Dioxide," NOAA/ESRL, November 2011. http://www.esrl.noaa.gov/gmd/ccgg/trends/global.html.

3. Solomon et al., "Summary for Policymakers." This estimate represents warming through 2005, though recent years have had similar global average temperatures so this number is still reasonably accurate.

4. Solomon et al., "Summary for Policymakers."

5. Erda Lin et al., "Synopsis of China National Climate Change Assessment Report (II): Climate Change Impacts and Adaptation," *Advances in Climate Change Research* 3 (2007): 6–11; Erda Lin et al., "Climate Change Impacts on Crop Yield and Quality with CO_2 Fertilization in China," *Philosophical Transactions of Biological Sciences* 360, no. 1463 (November 29, 2005): 2149–54; "Baogao chen qing cang gaoyuan bingchuan meinian tuisuo mi tu" [Tibetan Plateau Glaciers Retreat 7.8 m Each Year], Chinanews.com, November 17, 2011 (media report citing the newly released *Second National Assessment Report on Climate Change* of the China Meteorological Administration, Ministry of Science and Technology, and Chinese Academy of Sciences). Note that land surfaces have warmed more than ocean surfaces, so the global average given in the IPCC report is lower than a global land average. The China number reported is a land average, and since higher latitudes and higher altitudes have warmed more than lower ones, China's warming is likely higher than the global average.

6. Lin et al., "Synopsis of China National Climate Change," 1.

7. Solomon et al., "Summary for Policymakers."

8. Qingtai Yu, "Special Representative for Climate Change Negotiations of the Ministry of Foreign Affairs Yu Qingtai Receives Interview of the Media," September 22, 2007, http://www.chinaembassy.org.in/.

9. Elizabeth Economy, "China vs. Earth," *Nation*, 2007, http://www.thenation.com/doc/20070507/economy.

10. Lin et al., "Synopsis of China National Climate Change," 7.

11. Eun-shik Kim et al., "Sustainable Management of Grassland Ecosystems for Controlling Asian Dusts and Desertification in Asian Continent and a Suggestion of Eco-village Study in China," *Ecological Research* 21, no. 6 (November 2006): 907–11.

12. "Qihoubianhua guojia pinggu baogao weilai woguo hai pingmian jiang jixu shangsheng" [National Assessment Report on Climate Change: China's Sea Level Will Continue to Rise in Future], *Xinhuanet.com*, November 16, 2011. http://news.xinhuanet.com/tech/2011-11/16/c_122286961.htm.

13. "Baogao chen qing cang gaoyuan bingchuan meinian tuisuo mi tu."

14. Daidu Fan and Congxian Li, "Complexities of China's Coast in Response to Climate Change," in *Advances in Climate Change Research* 2, Supplement 1 (2006): 54–58; Congxian Li et al., "Some Problems of Vulnerability Assessment in the Coastal Zone of China," in *Global Change and Asian Pacific Coasts: Proceedings of the APN/SURVAS/LOICZ Joint Conference on Coastal Impacts of Climate Change and Adaptation in the Asia-Pacific Region*, Kobe, Japan, November 14–16, 2000, Asia Pacific Network for Global Change Research, 49–56; "Qihoubianhua guojia pinggu baogao weilai woguo hai pingmian jiang jixu shangsheng."

15. Economy, "China vs. Earth"; J. G. Titus, "Greenhouse Effect, Sea Level Rise, and Land Use," *Land Use Policy* 7, no. 2 (April 1990): 138–53.

16. Estimated using 2009 data from the National Bureau of Statistics, *2010 Zhongguo tongji nianjian* [2010 China Statistical Yearbook] (Beijing: China Statistics Press, 2010).

17. "Baogao chen qing cang gaoyuan bingchuan meinian tuisuo mi tu."

18. World Health Organization, *Climate Change and Human Health: Risks and Responses* (Geneva: World Health Organization, 2003).

19. Energy Information Administration, U.S. Department of Energy, *International Energy Outlook 2004* (Washington, D.C.: Energy Information Administration, 2004). Includes only carbon dioxide emissions from fossil fuel combustion. Note that almost all projections of China's emissions, including by Chinese researchers, were at least as far off as the U.S. DOE projections.

20. Energy Information Administration, U.S. Department of Energy, *International Energy Outlook 2006* (Washington, D.C.: Energy Information Administration, 2006).

21. Netherlands Environmental Assessment Agency, "China Now No. 1 in CO_2 Emissions; USA in Second Position," 2007, http://www.pbl.nl/en/dossiers/Climatechange/moreinfo/Chinanowno1inCO2emissionsUSAinsecondposition.html; BP, *BP Statistical Review of World Energy*, (London: BP, 2007).

22. The 2011 International Energy Outlook's reference case estimates China's CO_2 emissions in 2030 will be 463 percent above 1990 levels, while the high oil price case and low oil price cases estimate 567 percent and 396 percent above 1990 levels, respectively. Energy Information Administration, U.S. Department of Energy, *International Energy Outlook 2011* (Washington, D.C.: Energy Information Administration, 2011).

23. All estimates are calculated by author using the *2011 International Energy Outlook*'s reference case. EIA's projections assume a lower growth rate for China's emissions through 2030 than that of the past decade. Its projections for growth in U.S. emissions have declined substantially in recent years, likely because of the slowing of emissions growth attributed to the economic downturn.

24. Calculated by author using historical emissions data from the Carbon Dioxide Information Analysis Center (CDIAC), "Fossil-Fuel CO2 Emissions," Oak Ridge National Laboratory and the U.S. Department of Energy, 2011. http://cdiac.ornl.gov/trends/emis/meth_reg.html.

25. *World Energy Outlook 2007*.

26. Shares are for 2009. National Bureau of Statistics, *Zhongguo nengyuan tongji nianjian* [China Energy Statistical Yearbook] (Beijing: China Statistics Press, 2010); REN21, *Renewables Global Status Report 2011* (Paris: REN21, 2011).

27. "Zhongguo chengshi renkou shouci chaoguo nongcun renhou" [China's Urban Population Exceeded Rural Population for the First Time], *BBC Chinese News*, January 17, 2012. http://www.bbc.co.uk/zhongwen/simp/chinese_news/2012/01/120117_china_urban.shtml.

28. In China the average efficiency of coal power plants is rapidly catching up to that of developed countries as new, larger units come online and smaller, less efficient units are shut down. It is estimated that the average efficiency of China's coal-fired fleet was 32 percent in 2005, but it is expected to approach 40 percent by 2030 as more large, supercritical units come online and older subcritical units are phased out. In comparison, the majority of existing coal plants in the United States were built before 1989 using subcritical pulverized coal technology. International Energy Agency, *Cleaner Coal in China* (Paris: OECD/IEA, 2009), 51; Erik Shuster, "Tracking New Coal Fired Power Plants," National Energy Technology Laboratory, June 30, 2008 (updated July 12, 2011), http://www.netl.doe. gov/coal/refshelf/ncp.pdf.

29. Industry was 72 percent of total energy consumption in 2009. National Bureau of Statistics, *2010 Zhongguo nengyuan tongji nianjian* [2010 China Energy Statistical Yearbook] (Beijing: China Statistics Press, 2011). See also World Steel Association, "Statistics," 2012, http://www.worldsteel.org; International Aluminum Institute, "Statistics," 2012, http://www.world-aluminum.org. .

30. Depending on the methodology used, estimates of CO2 emissions embodied in China's exports as a percent of total domestic CO2 emissions range from 15 to 60 percent, with most studies estimating around 30 percent. Ming Xu et al., "CO2 Emissions Embodied in China's Exports from 2002 to 2008: A Structural Decomposition Analysis," *Energy Policy* 39, no. 11 (2011): 7381–88

31. Guri Bang, Gorild Heggelund, and Jonas Vevatne, *Shifting Strategies in the Global Climate Negotiations* (CICERO, 2005), http://www.cicero.uio.no/media/3079.pdf.

32. Xinhua, "Chinese Foreign Ministry Sets Up Climate Change Int'l Working Group," *China View*, September 5, 2007, http://news.xinhuanet.com/english/2007–09/05/content_6667432.htm.

33. Xinhua, "CAS Outlines Strategic Plan for China's Energy Development over Next 40 Years," *Beijing Review*, September 25, 2007, http://www.bjreview.com.cn/science/txt/2007-09/25/content_77642.htm.

34. "National Assessment Report on Climate Change Released" (Ministry of Science and Technology press release, December 31, 2006), http://www.most.gov.cn/eng/pressroom/200612/ t20061231_39425.htm.

35. National Development and Reform Commission, "China's National Climate Change Programme," June 2007; Chris Buckley, "Exclusive: China Preparing National Plan for Climate Change," *Reuters*, February 6, 2007, http://www.planetark.com/dailynewsstory.cfm/newsid/40197/story.htm; Richard McGregor, "China Delays Climate Change Plan Indefinitely," *Financial Times*, April 23, 2007, http://www.ft.com/cms/s/be763e8c-f1d6-11db-b5b6-000b5df10621.html.

36. Sebastian Oberthur and Herman E. Ott, *The Kyoto Protocol: International Climate Policy for the 21st Century* (Berlin: Springer, 1999), 24.

37. "China to Watch Others on Climate Change Action," *Reuters*, June 15, 2005, http:// www.enn.com/today.html?Id=7959.

38. P. Parameswaran, "Rich Nations Must Honor Climate Change Pledge: Developing Countries," *Agence France-Presse*, September 25, 2007.

39. People's Republic of China, "Guowuyuan changwuhui yanjiu jueding woguo kongzhi wenshiqiti paifang mubiao" [Standing Committee of China State Council to Study the Decision to Control Greenhouse Gas Emissions Targets], 2009, http://www.gov.cn/ldhd/2009-11/26/content_1474016.htm.

40. White House Press Office, "President to Attend Copenhagen Climate Talks: Administration Announces U.S. Emission Target for Copenhagen," November 25, 2009.

41. Paul Eckert and Claudia Parsons, "China's Hu Vows to Cut Carbon Output per GDP by 2020," *Reuters UK*, September 22, 2009, http://uk.reuters.com/article/2009/09/22/us-climate-china-idUKRE58L4XE20090922; Emma Graham-Harrison, "Hu's Carbon Commitment Marks New Era for China," *Reuters UK*, September 24, 2009, http://www.reuters.com/article/2009/09/24/us-china-climate-analysis-idUSRE58N15W20090924.

42. Coalition of Rainforest Nations, "Reducing Emissions from Deforestation in Developing Countries: Approaches to Stimulate Action" (Submission of Views of Seventeen Parties to the Eleventh Conference of the Parties to the United Nations Framework Convention on Climate Change, January 30, 2007), http://unfccc.int/files/methods_and_science/lulucf/application/ pdf/bolivia.pdf.

43. The language agreed to in December 2010 in Cancun states that "developing country Parties will take nationally appropriate mitigation actions in the context of sustainable development, supported and enabled by technology, financing and capacity-building, aimed at achieving a deviation in emissions relative to business as usual emissions in 2020." UNFCCC, *Outcome of the Work of the Ad Hoc Working Group on Long-term Cooperative Action Under the Convention, Draft Decision -/CP.16*," advance unedited version (December 2010).

44. United States General Accounting Office, *Selected Nations' Reports on Greenhouse Gas Emissions Varied in Their Adherence to Standards* (General Accounting Office, December 2003), http://www.gao.gov/new.items/d0498.pdf; David Streets et al., "Recent Reductions in China's Greenhouse Gas Emissions," *Science* 294, no. 5548 (November 30, 2001): 1835–37.

45. UNFCCC, *Outcome of the Work of the Ad Hoc Working Group*.

46. UNFCCC, *The United Nations Framework Convention on Climate Change*, 1992, article 3, principle 1, http://unfccc.int/essential_background/convention/background/items/1349.php.

47. Chinese Office of the National Coordination Committee on Climate Change, "Zhongguo CDM xiangmu yunxing guanli banfa" [Measures for Operation and Management of Clean Development Mechanism Projects in China] (October 12, 2005).

48. Bang, Heggelund, and Vevatne, *Shifting Strategies*.

49. Tauna Szymanski, "China's Take on Climate Change," *Sustainable Development, Ecosystems and Climate Change Committee Newsletter of the American Bar Association*, May 2006.

50. UNFCCC, *CDM: Registration*, January 20, 2012, http://cdm.unfccc.int/Statistics/Registration/AmountOfReductRegisteredProjPieChart.html.

51. Joanna I. Lewis, "The Evolving Role of Carbon Finance in Promoting Renewable Energy Development in China," *Energy Policy* 38, no. 6 (June 2010): 2875–86.

52. "UN Stops Approving China Wind Projects, Official Says," *Bloomberg*, December 2, 2009, http://www.bloomberg.com/apps/news?pid=newsarchive&sid=aU4wzufvXMpY.

53. UNFCCC, *Establishment of an Ad Hoc Working Group on the Durban Platform for Enhanced Action Draft Decision* (CP.17), December 2011.

54. Xinhua, "China to Cap Energy Use at 4 Bln Tonnes of Coal Equivalent by 2015," *China Radio International (CRI)*, March 4, 2011, http://english.cri.n/6909/2011/03/04/1461s624079.htm.

55. Lynn Price and Xuejun Wang, *Constraining Energy Consumption of China's Largest Industrial Enterprises Through Top-1000 Energy-Consuming Enterprise Program* (Berkeley: Lawrence Berkeley National Laboratory, 2007).

56. Jiabao Wen, "Report on the Work of the Government. Delivered at the Fourth Session of the Eleventh China National People's Congress," March 5, 2011, http://blogs.wsj.com/chinarealtime/2011/03/05/china-npc-2011-reports-full-text/.

57. "China's Energy Consumption per Unit of GDP Down 3.46 Percent in First 3 Quarters," *Xinhua News*, December 13, 2008, http://news.xinhuanet.com/english/2008-12/13/content_10497268.htm. Note that monetary values mentioned in this book are described in the same currency in which originally reported to avoid confusion with fluctuating exchange rates, except in some charts where monetary values have been converted to real rather than nominal values to allow for time-series comparisons.

58. "Key Targets of China's 12th Five-Year Plan," *Xinhua*, March 5, 2011, http://www.chinadaily.com.cn/xinhua/2011-03-05/content_1938144.html; PRC, "Zhonghua

renmin gongheguo guomin jingji he shehui fazhan di shier ge wunian guihua gangyao" [Twelfth Five-Year Plan for Economic and Social Development], March 16, 2011, http://news.xinhuanet.com/politics/2011-03/16/c_121193916.htm.

59. Jiang Lin et al., *Taking Out One Billion Tons of CO2: The Magic of China's 11th Five Year Plan?* (Berkeley: Lawrence Berkeley National Laboratory, 2007).

60. PRC, "Guowuyuan changwuhui yanjiu jueding woguo kongzhi wenshiqiti paifang mubiao" [Standing Committee of China State Council to Study the Decision to Control Greenhouse Gas Emissions Targets], 2009, http://www.gov.cn/ldhd/2009-11/26/content_1474016.htm.

61. David Cohen-Tanugi, "Putting It into Perspective: China's Carbon Intensity Target" (NRDC White Paper, October 2010).

62. National Development and Reform Commission (NDRC), "Guojia fazhan gaige wei bang gong ting guanyu kaizhan tan paifangquan jiaoyi shi dian gongzuo de tongzhi" [NDRC Notice on Pilot Trading Programs for the Development of Carbon Emissions Rights], Notice 2601, October 2011, http://www.ndrc.gov.cn/zcfb/zcfbtz/2011tz/t20120113_456506.htm.

63. The Tianjin Climate Exchange (TCX) is a joint venture of China National Petroleum Corporation Assets Management Co. Ltd. (CNPCAM), the Chicago Climate Exchange (CCX), and the city of Tianjin.

64. Implementation of China's national carbon-intensity target across all sectors and regions of the country is being supported by government-funded analysis led by the country's leading academic and research institutions.

65. Wen, "Report on the Work of the Government."

66. *International Energy Outlook.*

67. PRC, National People's Congress, "Zhonghua renmin gongheguo kezaisheng nengyuan fa" [Renewable Energy Law of the People's Republic of China], February 28, 2005.

68. The Renewable Energy Law and its related mechanisms are discussed further in chapter 3.

69. Sara Schuman, *Improving China's Existing Renewable Energy Legal Framework: Lessons from the International and Domestic Experience*, White Paper (NRDC, October 2010), 12–13; National People' Congress Standing Committee, "China Renewable Energy Law Decision," 2009, http://www.npc.gov.cn/huiyi/cwh/1112/2009-12/26/content_1533217.htm.

70. *Renewable Energy World*, "2010 Clean Energy Investment Hits a New Record," *Renewable Energy World*, January 11, 2011. http://www.renewableenergyworld.com/rea/news/article/2011/01/2010-clean-energy-investment-hits-a-new-record.

71. Ernst & Young, "Renewable Energy Country Attractiveness Indices: Country Focus—China," November 2011, http://www.ey.com/GL/en/Industries/Power–Utilities/RECAI–China.

72. "Zhang Guobao: Shi er wu mo lizheng feihuashi nengyuan zhan yici nengyuan bizhong 11.4%" [Zhang Guobao: "Twelfth Five" Push to Nonfossil Energy to Account

for 11.4 Percent Share of Primary Energy], *people.com.cn*, January 6, 2011, http://energy.people.com.cn/GB/13670716.html; PRC, "China Announces 16 Pct Cut in Energy Consumption per Unit of GDP by 2015," www.gov.cn, March 5, 2011, http://www.gov.cn/english/2011–03/05/content_1816947.htm.

73. PRC, "Guowuyuan tongguo jiakuai peiyu he fazhan zhanluexing xinxing chanye de jueding" [Decision on Speeding Up the Cultivation and Development of Emerging Strategic Industries], September 8, 2010, http://www.gov.cn/ldhd/2010-09/08/content_1698604.htm.

74. HSBC, "China's Next 5-Year Plan: What It Means for Equity Markets," October 2010.

75. Ibid.

76. Shisen Xu, "Green Coal-Based Power Generation for Tomorrow's Power" (paper presented at the APEC Energy Working Group: Expert Group on Clean Fossil Energy, Thermal Power Research Institute, Lampang, Thailand, February 24, 2006).

77. European Commission, *EU-China Summit: Joint Statement*, September 5, 2005, http://ec.europa.eu/comm/external_relations/china/summit_0905/index.htm; "UK Department of Environmental, Food and Rural Affairs," 2005, http://www.defra.gov.uk/environment/climatechange/internat/devcountry/china.htm.

78. "Carbon Capture Milestone in China," *Science Daily*, August 4, 2008, http://www.sciencedaily.com/releases/2008/07/080731135924.htm.

79. Yu, "Special Representative for Climate Change Negotiations of the Ministry of Foreign Affairs Yu Qingtai Receives Interview of the Media." He went on to say that "if the developing countries cannot maintain economic and social progress, eliminate poverty or raise people's living standards, the material foundation for coping with climate change will not exist, not mentioning the capacity to fight against climate change."

3. China in the Global Wind Power Innovation System

1. U.S. Department of Energy, "History of Wind Energy," September 12, 2005, http://www1.eere.energy.gov/wind/wind_history.html; Vaclav Smil, *Energy Transitions: History, Requirements, Prospects* (Santa Barbara: Praeger, 2010), 51–52; Daniel Yergin, *The Quest: Energy, Security and the Remaking of the Modern World* (New York: Penguin, 2011), 590–92.

2. Yergin, *The Quest*, 593–94.

3. International Energy Agency (IEA), *Renewables for Power Generation: Status & Prospects* (Paris: OECD/IEA, 2003).

4. A much more detailed history of wind energy is presented in Paul Gipe, *Wind Energy Comes of Age* (New York: Wiley, 1995).

5. IEA, *Renewables for Power Generation*.

6. IEA, *Renewables for Power Generation*.

7. IEA, "IEA Energy Statistics OECD R&D Database" (Paris: OECD/IEA, 2010), http://www.iea.org/stats/rd.asp.

8. Calculated using ibid. and OECD GDP Statistics, OECD, "OECD Statistics," *OECD.StatExtracts*, 2010, http://stats.oecd.org/Index.aspx.

9. IEA, "IEA Energy Statistics OECD R&D Database."

10. IEA, "IEA Energy Statistics OECD R&D Database."

11. Joanna I. Lewis and Ryan H. Wiser, "Fostering a Renewable Energy Technology Industry: An International Comparison of Wind Industry Policy Support Mechanisms," *Energy Policy* 35, no. 3 (March 2007): 1844–57; Joanna I. Lewis, "Technology Acquisition and Innovation in the Developing World: Wind Turbine Development in China and India," *Studies in Comparative International Development* 42, no. 3–4 (October 2007): 208–32.

12. See chapter 4 for further details about the origins of Vestas.

13. Rinie van Est, *Winds of Change: A Comparative Study of the Politics of Wind Energy Innovation in California and Denmark* (Utrecht: International Books, 1999).

14. Ryan Wiser and Mark Bolinger. *2010 Wind Technologies Market Report* (Berkeley: LBNL, June 2011), fig. 3.

15. The differences in the outcomes of the wind industries of Denmark and the Netherlands are explored at length in Linda Manon Kamp, "Learning in Wind Turbine Development: A Comparison Between the Netherlands and Denmark" (Ph.D. dissertation, University of Utrecht, 2002).

16 Van Est, *Winds of Change*.

17. Chapter 4 details the story of GE and the U.S. wind industry.

18. Wei Xie and Guisheng Wu, "Differences Between Learning Processes in Small Tigers and Large Dragons. Learning Processes of Two Color TV (CTV) Firms Within China," *Research Policy* 32 (2003): 1463–79.

19. Bengt-Åke Lundvall, *National Systems of Innovation: Toward a Theory of Innovation and Interactive learning* (London: Anthem Press, 2010).

20. Xielin Liu and Steven White, "Comparing Innovation Systems: A Framework and Application to China's Transitional Context," *Research Policy* 30 (2001): 1091–1114.

21. Chris Freeman, "The 'National System of Innovation' in Historical Perspective," *Cambridge Journal of Economics* 19, no. 1 (February 1995): 5–24; Sanjaya Lall, "Technological Capabilities in Emerging Asia," *Oxford Development Studies* 26, no. 2 (1998): 213–43; Alice H. Amsden, *The Rise of the Rest: Challenges to the West from Late-Industrializing Economies* (New York: Oxford University Press, 2001); Dan Breznitz, *Innovation and the State: Political Choice and Strategies for Growth in Israel, Taiwan, and Ireland* (New Haven: Yale University Press, 2007).

22. Liu and White, "Comparing Innovation Systems."

23. Pari Patel, "Localized Production of Technology for Global Markets," *Cambridge Journal of Economics* 19, no. 1 (1995): 141–53.

24. Daniele Archibugi and Jonathan Michie, "The Globalization of Technology: A New Taxonomy," *Cambridge Journal of Economics* 19, no. 1 (February 1995): 121–40; Daniele Archibugi and Carlo Pietrobelli, "The Globalization of Technology and Its

Implications for Developing Countries: Windows of Opportunity or Further Burden?" *Technological Forecasting & Social Change* 70 (2003): 861–83.

25. Rainier Walz, "The Role of Regulation for Sustainable Infrastructure Innovations: The Case of Wind Energy," *International Journal of Public Policy* 2, nos. 1–2 (2007): 57–88; Anna Bergek et al., "Analysing the Functional Dynamics of Technological Innovation Systems: A Scheme of Analysis," *Research Policy* 37, no. 3 (April 2008): 407–29; Simona O. Negro, Roald A. A. Suurs, and Marko P. Hekkert, "The Bumpy Road of Biomass Gasification in the Netherlands: Explaining the Rise and Fall of an Emerging Innovation System," *Technological Forecasting & Social Change* 75, no. 1 (January 2008): 57–77; Franco Malerba and Sunil Mani, eds., *Sectoral Systems of Innovation and Production in Developing Countries: Actors, Structure and Evolution* (London: Edward Elgar, 2009); Bo Carlsson and Rikard Stankiewicz, "On the Nature, Function and Composition of Technological Systems," *Journal of Evolutionary Economics* 1, no. 2 (1991): 93–118; Anna Bergek, Marko Hekkert, and Staffan Jacobsson, "Functions in Innovation Systems: A Framework for Analysing Energy System Dynamics and Identifying System Building Activities by Entrepreneurs and Policy Makers," in *Innovation for a Low Carbon Economy: Economic, Institutional and Management Approaches*, ed. Timothy J. Foxon, Jonathan Kohler, and Christine Oughton, 79–111 (London: Edward Elgar, 2008).

26. Mark Dodgson, "Learning, Trust and Inter-Firm Technological Linkages: Some Theoretical Associations," in *Technological Collaboration: The Dynamics of Cooperation and Industrial Innovation*, ed. Rod Coombs et al., 375–77 (Cheltenham: Edward Elgar, 1996).

27. Ikujiro Nonaka and Hirotaka Takeuchi, *The Knowledge Creating Company: How Japanese Companies Create the Dynamics of Innovation* (New York: Oxford University Press, 1995).

28. Kamp, "Learning in Wind Turbine Development."

29. Kamp, "Learning in Wind Turbine Development"; Raghu Garud, "On the Distinction Between Know-How, Know-Why, and Know-What," *Advances in Strategic Management* 14 (1997): 81–101.

30. AnnaLee Saxenian, *Regional Advantage: Culture and Competition in Silicon Valley and Route 128* (Cambridge: Harvard University Press, 1996); Yu Zhou, *The Inside Story of China's High-Tech Industry: Making Silicon Valley in Beijing* (Lanham: Rowman & Littlefield, 2008); Philip Cook and Olga Memedovic, *Strategies for Regional Innovation Systems: Learning Transfer and Applications*, Policy Papers (Vienna: United Nations Industrial Development Organization, 2003); Leonard Dudley, "Learning and the Interregional Transfer of Technology," *Southern Economic Journal* 40, no. 4 (April 1, 1974): 56370; Dieter Ernst, "A New Geography of Knowledge in the Electronics Industry? Asia's Role in Global Innovation Networks," *Policy Studies* 54 (Honolulu: East West Center, 2009).

31. Van Est, *Winds of Change*; Kamp, Smits, and Andriesse, "Notions on Learning Applied to Wind Turbine Development in the Netherlands and Denmark"; Peter

Karnoe, "Technological Innovation and Industrial Organization in the Danish Wind Industry," *Entrepreneurship and Regional Development* 2, no. 2 (April 1990): 105–24.

32. The global learning networks of the emerging wind turbine manufacturers of China, India, and South Korea are discussed further in chapter 6.

33. Amsden, *The Rise of the Rest*; Keun Lee and Chaisung Kim, "Technological Regimes, Catching Up, and Leapfrogging: Findings from the Korean Industries," *Research Policy* 30, no. 3 (March 2001): 459–83; Keun Lee, "Making a Technological Catch Up: Barriers and Opportunities," *Asian Journal of Technology Innovation* 13, no. 2 (2005): 97–131; Kelly Sims Gallagher, "Limits to Leapfrogging in Energy Technologies: Evidence from the Chinese Automobile Industry," *Energy Policy* 34, no. 4 (March 2006): 383–94; Richard R. Nelson, "The Changing Institutional Requirements for Technological and Institutional Catch Up," *International Journal of Technological Learning, Innovation and Development* 1, no. 1 (2007): 4–12.

34. There is a substantial literature describing the challenges associated with technology transfer in general—for example, Albert O. Hirschman and Kermit Gordon, *Development Projects Observed* (Washington, D.C.: Brookings Institution Press, 1967); Nathan Rosenberg and Claudio Frischtak, eds., *International Technology Transfer: Concepts, Measures, and Comparisons* (New York: Praeger, 1985); M. Kranzberg, "The Technical Elements in International Technology Transfer: Historical Perspectives," in *The Political Economy of International Technology Transfer*, ed. John R. McIntyre and Daniel S. Papp, 31–46 (New York: Quorum Books, n.d.); Denis Goulet, *The Uncertain Promise: Value Conflicts in Technology Transfer* (New York: New Horizons, 1989). There are also many works detailing technology transfer challenges that are specific to China—for example, U.S. Congress, Office of Technology Assessment (U.S. OTA), *Technology Transfer to China* (Washington, D.C.: U.S. OTA, July 1987); E. Mansfield, "Intellectual Property Protection, Foreign Direct Investment, and Technology Transfer," Discussion Paper (International Finance Corporation [IFC], 1994); Turlough F. Guerin, "Transferring Environmental Technologies to China: Recent Developments and Constraints," *Technological Forecasting & Social Change* 67, no. 1 (May 2001): 55–75; Paul Miesing, Mark P. Kriger, and Neil Slough, "Towards a Model of Effective Knowledge Transfer Within Transnationals: The Case of Chinese Foreign Invested Enterprises," *Journal of Technology Transfer* 32 (2007): 109–22.

35. Ogunlade Davidson, Bert Metz, and Sascha van Rooijen, *Methodological and Technological Issues in Technology Transfer, International Panel on Climate Change (IPCC)* (New York: Cambridge University Press, 2000).

36. Gamesa Eólica was a joint venture formed between the Spanish turbine manufacturer Gamesa and the Danish manufacturer Vestas. In this arrangement Gamesa paid licensing feeds to Vestas that allowed Gamesa to manufacture turbines made with Vestas technology solely within the Spanish market. This arrangement (discussed further in chapter 4) was terminated when the companies split in December 2001. See Rolf Wustenhagen, "Sustainability and Competitiveness in the Renewable Energy Sector: The Case of Vestas Wind Systems," *Greener Management Internation-*

al, no. 44, Special Issue on Sustainability Performance and Business Competitiveness (December 2003), http://www.iwoe.unisg.ch/org/iwo/web.nsf/SysWebRessources/Wue_vestaspaper_03–07–09/$FILE/Vestas_Paper_final.pdf.

37. This technology transfer model is discussed further in chapter 6.

38. Lewis and Wiser, "Fostering a Renewable Energy Technology Industry."

39. Canadian Wind Energy Association, "Manufacturing Commercial Scale Wind Turbines in Canada" (Ottawa: Canadian Wind Energy Association, April 24, 2003).

40. Van Est, *Winds of Change*; Anna Bergek and Staffan Jacobsson, "The Emergence of a Growth Industry: A Comparative Analysis of the German, Dutch and Swedish Wind Turbine Industries," in *Change, Transformation and Development*, ed. J. Stan Metcalfe and Uwe Canter (Berlin: PhysicaVerlag, 2003).

41. Peter M. Connor, "National Innovation, Industrial Policy and Renewable Energy Technology," Proceedings of the British Institute of Energy Economics Academic Conference, Oxford, September 25–26, 2003.

42. Ibid.

43. Several studies have attempted to estimate the total jobs created by the wind industry; see, for example, European Wind Energy Association (EWEA), *Industry and Employment*, vol. 3 in *Wind Energy: The Facts* (Brussels: EWEA, 2004), http://www.ewea.org/fileadmin/ewea_documents/documents/publications/WETF/Facts_Volume_4.pdf; National Wind Coordinating Committee, *The Effect of Wind Energy Development on State and Local Economies*, NWCC Wind Energy Series (Washington, D.C.: National Wind Coordinating Committee, January 1997). http://old.nationalwind.org/publications/wes/wes05.htm.

44. Allen Consulting Group, "Sustainable Energy Jobs Report Wind Manufacturing Case Study" (paper prepared for the Sustainable Energy Development Authority, Government of Australia, 2003).

45. The installation of fewer machines to reach a certain installed capacity can result in cost savings. Even though larger machines are more expensive than smaller machines on a per-unit basis, a significant portion of the cost is saved through reduced tower, foundation, and construction costs. Maintenance costs may also be saved because there are fewer turbines to be maintained per unit area. Cost savings also result from increases in turbine productivity over time. For example, in Denmark the annual energy yield per square meter of rotor area has grown by 5 percent per year since 1980. Soren Krohn, *Creating a Local Wind Industry: Experience from Four European Countries* (Montreal: Helios Center for Sustainable Energy Strategies, May 4, 1998).

46. Ibid.

47. Canadian Wind Energy Association, "Manufacturing Commercial Scale Wind Turbines in Canada."

48. Shipping turbines from Denmark to the Asia/Pacific region takes approximately eight weeks. See Allen Consulting Group, "Sustainable Energy Jobs Report Wind Manufacturing Case Study."

204   3. China in the Global Wind Power Innovation System

49. Linda M. Kamp, Ruud E.H.M Smits, and Cornelis D. Andriesse, "Notions on Learning Applied to Wind Turbine Development in the Netherlands and Denmark," *Energy Policy* 32, no. 14 (September 2004): 1625–37.

50. Torgny Moller and Gail Rajgor, "Denmark Picked for Global Headquarters," *Windpower Monthly*, October 1, 2004.

51. One source estimates that permanent magnets will be used in 15 to 25 percent of wind turbines by 2013. Ros Davidson, "U.S. Considers Official Probe of Illegal Chinese Wind Subsidy," *Windpower Monthly*, December 1, 2010.

52. World Trade Organization, "Understanding the WTO: Standards and Safety," n.d., http://www.wto.org/english/thewto_e/whatis_e/tif_e/agrm4_e.htm#TRS. As discussed in subsequent chapters, policies shown to be protectionist in the wind power industry have been challenged under other WTO provisions, including the SCM Agreement.

53. See, for example, United Steelworkers, "United Steelworkers' Section 301 Petition Demonstrates China's Green Technology Practices Violate WTO Rules," September 9, 2010, http://assets.usw.org/releases/misc/section-301.pdf; "Japan Starts WTO Dispute with Canada on Clean Power," *Reuters*, September 13, 2010, http://www.reuters.com/article/idUSTRE68C2RN20100913; Kari Williamson, "U.S. Wind Turbine Tower Manufacturers Join U.S.-China Trade Dispute," *Renewable Energy Focus*, January 4, 2012, http://www.renewableenergyfocus.com/view/22952/us-wind-turbine-tower-manufacturers-join-uschina-trade-dispute/.

54. The Chinese word usually used for local manufacturing is *guochanhua*, which, literally translated, means domestically produced, or made in China, but has frequently been (inaccurately) translated as "reverse engineering."

55. In recent years some companies have attempted to develop locally manufactured wind turbines that are also localized for Chinese conditions; for example, high-altitude wind turbines for the plateau regions of Southwest China, or turbines that function well in cold weather to withstand the cold temperatures of Inner Mongolia. Goldwind recently provided the wind turbines that were "tailor made" for a high-altitude wind farm in Yunnan province, "featuring larger blades and greater capacities for insulation, heat dissipation and lightening protection, in order to cope with harsher geographical and climate conditions in plateau areas." "High-Altitude Wind Farm Constructed in SW China," *China Daily*, December 30, 2011, http://www.chinadaily.com.cn/bizchina/2011-12/30/content_14358803.htm.

56. Historians have long speculated as to why China did not undergo a scientific and technological revolution at the same time as Europe (beginning in the 1600s) even though it is believed to have been the source of many inventions that proved important in Europe's scientific development. Many point to a difference in the cultural constructs of scientific development between the West and the East, although this has been a topic of much debate. This subject was addressed most comprehensively in Joseph Needham et al., *Science and Civilisation in China* (Cambridge: Cambridge University Press, 1954).

57. Valerie J. Karplus, *Innovation in China's Energy Sector*, Working Paper (Program on Energy and Sustainable Development, Center for Environmental Science and Policy, Stanford University, March 2007).

58. Liu and White, "Comparing Innovation Systems;" Kenneth Lieberthal, *Governing China: From Revolution Through Reform* (New York: Norton, 1995).

59. OECD, "Science, Technology and Innovation for the 21st Century: Meeting of the OECD Committee for Scientific and Technological Policy at Ministerial Level (Final Communique)" (OECD, January 30, 2004), http://www.oecd.org/document/1/0,3746,en_2649_34269_25998799_1_1_1_1,00&&en-USS_01DBC.html.

60. SSTC and IDRC, *A Decade of Reform: Science and Technology Policy in China* (IDRC Canada, 1997), http://idl-bnc.idrc.ca/dspace/bitstream/10625/15019/1/106731.pdf.

61. Karplus, *Innovation in China's Energy Sector*.

62. SSTC and IDRC, *A Decade of Reform*.

63. SSTC and IDRC, *A Decade of Reform*.

64. CAS has reportedly established over five hundred commercial enterprises in the high-tech sector as part of a government program to develop "technical enterprises" as subsidiaries of existing research institutes. DFI International for the Bureau of Export Administration, "U.S. Commercial Technology Transfers to the People's Republic of China" (DFI International for the Bureau of Export Administration, 1998), http://www.globalsecurity.org/wmd/library/report/1999/techtransfer2prc.htm.

65. CAS Official, Beijing, interview by Joanna Lewis, July 22, 2011.

66. SSTC and IDRC, *A Decade of Reform*.

67. Karplus, *Innovation in China's Energy Sector*.

68. Karplus, *Innovation in China's Energy Sector*, 3; Richard P. Suttmeier and Cong Cao, "China Faces the New Industrial Revolution: Achievement and Uncertainty in the Search for Research and Innovation Strategies," *Asian Perspective* 23, no. 3 (1999): 153–200.

69. Xiaomei Tan and Zhao Gang, *An Emerging Revolution: Clean Technology Research, Development and Innovation in China*, Working Paper (World Resources Institute, December 2009), http://www.wri.org.

70. Karplus, *Innovation in China's Energy Sector*.

71. National Bureau of Statistics and Ministry of Science and Technology, *Zhongguo keji tongji nianjian* [China Statistical Yearbook on Science and Technology] (Beijing: China Statistics Press, 2010).

72. China's energy-efficiency programs are described in more detail in chapter 2.

73. The *China Statistical Yearbook on Science and Technology* provides a yearly update of R&D expenditures in Chinese R&D institutions, industries, and universities and for targeted national R&D programs. Energy R&D data are not disaggregated by energy sector with the exception of "fossil energy." Industry data are reported in two categories: state-owned and "other" enterprises. National Bureau of Statistics and Ministry of Science and Technology, *Zhongguo keji tongji nianjian*; Ruud Kempener,

Laura A. Anadon, and Jose Candor, *Governmental Energy Innovation Investments, Policies, and Institutions in the Major Emerging Economies: Brazil, Russia, India, Mexico, China, and South Africa*, Discussion Paper, Energy Technology Innovation Policy Discussion Paper Series (Cambridge: Belfer School for Science and International Affairs, Kennedy School, Harvard University, November 2010).

74. David C. Mowery and Joanne E. Oxley, "Inward Technology Transfer and Competitiveness: The Role of National Innovation Systems," *Cambridge Journal of Economics* 19 (1995): 67–93.

75. Kempener, Anadon, and Candor, *Governmental Energy Innovation Investments*; OECD, *OECD Reviews of Innovation Policy: China Synthesis Report* (Paris: OECD, 2007).

76. The formation of the NEC was approved by the National People's Congress in January 2008, but it was not formally established until January 2010.

77. Before the NDRC was established in 2003, renewable energy policy was controlled by the State Development Planning Commission (SDPC) and the State Economic and Trade Commission (SETC). The NEC replaced the former National Energy Leading Group, and the NEA replaced the former Energy Bureau. Kempener, Anadon, and Candor, *Governmental Energy Innovation Investments*; Erica Downs, "China's 'New' Energy Administration," *China Business Review*, December 2008.

78. Downs, "China's 'New' Energy Administration."

79. A Ministry of Energy was established in 1988 but was disbanded five years later, reportedly because its administrative functions overlapped with other departments. Zhihong Wan, "Wen Heads 'Super Ministry' for Energy," *China Daily*, January 28, 2010, http://www.chinadaily.com.cn/china/2010-01/28/content_9388039.htm. According to Erica Downs, "China's fragmented energy bureaucracy has impeded energy governance because there is no single institution, such as a ministry of energy, with the authority to coordinate the interests of the various stakeholders. Turf battles among various energy institutions have often resulted in energy laws that fail to specify agencies responsible for the content of those laws, delaying or preventing implementation." Downs, "China's 'New' Energy Administration."

80. "Ke zai sheng nengyuan zhong changqi fazhan hua" [Medium- and Long-Term Plan for Renewable Energy Development in China], NDRC (September 2007); "Ke zai sheng nengyuan fazhan shiyi wu jihua" [Eleventh Five-Year Renewable Energy Development Plan], NDRC (March 2008); "Ke zai sheng nengyuan fazhan shier wu jihua" [Twelfth Five-Year Renewable Energy Development Plan], NDRC (December 2011).

81. Wenqiang Liu, Lin Gan, and Xiliang Zhang, "Cost Competitive Incentives for Wind Energy Development in China: Institutional Dynamics and Policy Changes," *Energy Policy* 30 (2002): 753–65.

82. A 1999 Circular issued by MOST and SDPC further stipulated that grid administrators must allow interconnection and purchase all interconnected power if grid capacity allows, and that projects using domestically made equipment receive a low-interest loan and a prolonged period of loan repayment. MOST and SDPC, "Circular

on Further Supporting the Development of Renewable Energy," SDPC (January 12, 1999), no. 44.

83. MOST, SDPC, and SETC, "Evaluation of Policies Designed to Promote the Commercialization of Wind Power Technology in China" (Energy Foundation China Sustainable Energy Program, May 15, 2002).

84. Ibid.

85. Debra J. Lew, "Alternatives to Coal and Candles: Wind Power in China," *Energy Policy* 28, no. 4 (April 2000): 271–86.

86. Karl-Eugen Feifel, "Wind Power in China, a German Company's Experience," *Industry & Environment* (2001).

87. While not elaborated here, there were extensive reports of gaming in the concession system, including the state-owned power companies bidding unreasonably low prices in an attempt to win projects and then renegotiating the tariffs to a higher price after the fact. There were also widespread claims that foreign developers and turbine suppliers were not being fairly considered for the projects, despite the fact that many of them could meet local content requirements.

88. In 2006 the NDRC issued a notice that proposed that prices for grid-connected wind power projects were to be set by the price control department of the State Council and directly informed by the prices resulting from the concession bids. "Ke zai sheng nengyuan dian jia fujia shou ru diao pei zan xing banfa" [Interim Measures on Renewable Energy Electricity Prices and Cost Sharing Management], NDRC (January 2006), no. 7.

89. An early concession project included an all-time low bid of 0.38 yuan per kWh, though the developer eventually renegotiated the tariff price.

90. Chapter 4 explores this phenomenon in more detail.

91. "Fagaiwei guanyu fengdian jianshe guanli youguan yaoqiu de tongzhi" [Notice on the Relevant Requirements for the Administration of the Construction of Wind Farms], NDRC (July 2005), no. 1204.

92. Ibid.

93. Jurisdiction for approval of projects under 50 MW may lie with the Development and Reform Commission of a province, autonomous region, or city, depending on the project. Ibid.

94. "Zhonghua renmin gongheguo ke zai sheng nengyuan fa" [Renewable Energy Law of the People's Republic of China]. National People's Congress, February 28, 2005. The law went into effect January 1, 2006.

95. "Ke zai sheng nengyuan dian jia fujia shou ru diao pei zan xing banfa" [Interim Measures on Renewable Energy Electricity Prices and Cost Sharing Management], NDRC (January 2006), no. 7.

96. The Interim Measures directed the NDRC to set a nationwide renewable surcharge levied on electricity users at a uniform rate based on the users' consumption of electricity. In 2006 the surcharge was set at RMB 0.001 per kWh, but it has doubled every two years for industrial users, who, as of 2009, were paying RMB 0.004

per kWh (excluding Sichuan province because of earthquake recovery efforts). The amount of renewable surcharges used to fund renewable-power generation totaled more than $490 million in the first half of 2009, up from $272 million in the previous six-month period. Sara Schuman, *Improving China's Existing Renewable Energy Legal Framework: Lessons from the International and Domestic Experience*, White Paper, NRDC (October 2010).

97. "Ke zai sheng nengyuan dian jia fujia shouru diao pei zan xing banfa" [Interim Measures on Revenue Allocation from the Renewable Surcharge], NDRC (2007), no. 44.

98. Under the 2007 Interim Measures, the surcharges were to be levied directly from the end-users' electricity bill by the grid company. Ultimately each grid company should receive the amount to which it is entitled, on the basis of the renewable power generated in its coverage area. In the first half of 2009, RMB 1.836 billion was exchanged through this program. There have, however, been widespread reports of challenges in implementating this program as designed. Schuman, *Improving China's Existing Renewable Energy Legal Framework*.

99. NDRC. "Medium- and Long-Term Plan."

100. In 2007 a Chinese Academy of Engineering study estimated China's expected wind power development would reach 120 GW in 2020, 270 GW in 2030, and 500 GW in 2050. After this and numerous other studies showing even more aggressive wind power development was possible, the official targets were increased. Xiangwan Du, *Research on the Development Strategy of China Renewable Energy* (Beijing: Science in China Press, 2008); "China Raises Wind Power Target to 1,000 GW by 2050," *People's Daily*, January 10, 2012, http://www.peoplesdaily-online.com/business/economy/27846-china-raises-wind-power-target-to-1000gw-by-2050.

101. 1 GW out of the total of 30 GW targeted for 2020. NDRC, "Medium and Long-Term Plan."

102. NDRC, "Medium and Long-Term Plan."

103. Junfeng Li, Pengfei Shi, and Hu Gao, *2010 China Wind Power Outlook* (GWEC and Greenpeace, 2010), http://www.greenpeace.org/eastasia/press/reports/wind-power-report-english-2010.

104. "China Raises Wind Power Target to 1,000 GW by 2050," *People's Daily*, January 10, 2012, http://www.peoplesdaily-online.com/business/economy/27846-china-raises-wind-power-target-to-1000gw-by-2050.

105. "Ke zai sheng nengyuan fadian youguan guanli guiding" [Management Rules on the Administration of Power Generation from Renewable Energy], NDRC (2006), no. 13.

106. The increase in feed-in tariff supposedly reflected the increasing cost of equipment and raw materials, as well as interest rates. The Guangdong feed-in tariff did not include the cost of grid connection. If project developers bore the grid connection costs, they could add RMB 0.01 per kWh to the benchmark prices for connection up to 50 km; RMB 0.02 for connection between 50 and 100 km; and RMB 0.03 for connection beyond 100 km. Li, Shi, and Gao, *China Wind Power Outlook 2010*.

107. "Guanyu wanshan fengli fadian shangwang dian jia zhengce de tongshi" [Notice on Improving Grid-Connected Wind Power Tariff Policy], NDRC (2009), no. 1906.

108. China ratified the Kyoto Protocol in August 2002, and the agreement entered into force on February 16, 2005. NDRC issued the "Measures for Operation and Management of Clean Development Mechanism Projects in China" in October 2005, as discussed in chapter 2. "Zhongguo CDM xiangmu yunxing guanli banfa" [Measures for Operation and Management of Clean Development Mechanism Projects in China], NDRC (October 2005).

109. Li Junfeng, Shi Pengfei, and Gao Hu, *China Wind Power Outlook 2010* (Beijing: Chinese Renewable Energy Industries Association, Global Wind Energy Council, and Greenpeace, October 2010).

110. Even though parties to the Kyoto Protocol agreed to extend it beyond 2012, these uncertainties still persist. For a more complete discussion, see chapter 2, as well as Joanna I. Lewis, "The Evolving Role of Carbon Finance in Promoting Renewable Energy Development in China," *Energy Policy* 38, no. 6 (June 2010): 2875–86.

111. Schuman, *Improving China's Existing Renewable Energy Legal Framework*, 12–13; National People' Congress Standing Committee, "Zhonghua renmin gongheguo ke zai sheng nengyuan fa (xiu zheng an)" [Renewable Energy Law of the People's Republic of China (Amended)], December 2009.

112. MOST, SDPC, and SETC, *Evaluation of Policies*; Liu, Gan, and Zhang, "Cost Competitive Incentives for Wind Energy Development in China."

113. "Fengli fadian shebei chanye hua zhuan xiang zijin guanli zan xing banfa" [Interim Measures on the Management of Special Project Funds for the Industrialization of Wind Power Generation Equipment], MOF (2008), no. 476. Note that this regulation was the target of scrutiny by the Office of the U.S. Trade Representative (USTR) under a WTO investigation. Chinese subsidies from this program have reportedly run into the hundreds of millions of dollars since its inception in 2008, and USTR claims that this violates WTO policy because the grants awarded are contingent on Chinese wind power manufacturers using parts and equipment manufactured domestically. Office of the United States Trade Representative, "WTO Dispute Settlement Proceedings: Subsidies on Wind Power Equipment, China," *Federal Register* 75, no. 249 (December 29, 2010): 82130–32; Andrew Krulewitz, "WTO Wind Industry Throwdown: U.S. vs. China," *Greentech Media*, December 28, 2010, http://www.greentechmedia.com/articles/read/wto-wind-industry-throwdown-US-vs-china1/.

114. Li, Shi, and Gao, *China Wind Power Outlook 2010*.

115. "Caizhengbu guojia shui wu zongju guanyu bufen ziyuan zonghe liyong jiqi ta chanpin zengzhishui zhengce wenti de tongzhi" [Circular of the Ministry of Finance and the State Administration of Taxation on Issues Concerning Value-Added Tax Policies for Certain Products Produced by Comprehensive Utilization of Resources and Other Goods], MOF and State Administration of Taxation (2001), no. 198.

116. "Guojia ke zai sheng nengyuan chanye fazhan zhidao mulu" [National Guidance Catalogue for Renewable Energy Industry Development], NDRC (2005), no. 2517.

117. "Guanyu tiaozheng zhongda jishu zhuangbei jinkou shuishou zhengce zan hanggui ding youguan qingdan de tongzhi" [Circular on Adjusting Relevant Lists of Tentative Provisions on Tax Policies for Import of Major Technical Equipment], MOF and State Administration of Taxation (April 2010), no. 17.

118. "Guanyu quxiao fengdian gongcheng xiangmu caigou shebei guochanhua lu yaoqiu de tongzhi" [Notice on Abolishing the Localization Rate Requirement for Equipment Procurement in Wind Power Projects]. NDRC (November 2009), no. 2991.

119. In 2009 there were reportedly eighty-three wind producers with annual production capacity exceeding 50 GW (based only on company estimates, not on actual production). Annual demand that year was closer to 10 GW.

120. "Feng dian shebei zhizao hangye zhun ru biaozhun (zhengqiu yijian gao)" [Wind Power Equipment Manufacturing Industry Access Standards-Draft], MIIT (March 25, 2010).

121. The necessity and efficacy of such regulations were widely disputed in the industry, with some stakeholders claiming that the regulations were not really about improving the health of the sector but rather a case of the government showing preference for the top three Chinese wind turbine manufacturers in an attempt to thwart competition from second-tier manufacturers. All three of the leading Chinese wind turbine manufacturers at this time were at least partially state-owned. As of early 2012 the requirements had not yet been fully implemented.

122. "Haizhang feng dian kaifa jianshe guanli zan xing banfa" [Measures for the Administration of Offshore Wind Power Development], State Oceanic Administration and NDRC Energy Bureau (2010), no. 29.

123. While foreign-owned developers were not excluded from participating in the earlier onshore concession projects, no foreign developers (with the exception of Hong Kong developers) were awarded any projects.

124. This is discussed further in chapter 4.

125. Yuanyuan Liu, "China Releases Technical Standards for Wind Power Industry." *Renewable Energy World*, September 22, 2011. http://www.renewableenergyworld.com/rea/news/article/2011/09/china-releases-technical-standards-for-wind-power-industry.

126. "Guanyu jiaqiang fengdianchang bingwang yunxing guanli de tongzhi" [Notice on Strengthening the Management of Wind Power Plant Grid Integration], NEA (July 2011), no. 182; "Guanyu yin fa fengdianchang gongli yuce yubao guanli zan xing banfa de tongzhi" [Operation and Provisional Management Methods for Wind Power Forecasting], NEA (June 2011), no. 177.

127. There were several high profile grid failures in 2011, most notably at Gansu's Jinquan wind farm (one of the largest in China), that have been attributed to a lack of

low-voltage ride through capability of the wind turbines. LVRT is therefore one of the key elements of the new grid codes.

128. The estimates from the 1980s were based on wind resources at 10 meters while the 2009 estimates were based on wind resources at 50 meters, better reflecting the hub height of modern wind turbines. Rong Zhu, "Study on Wind Resource Potential for Large Scale Development of Wind Power in China" (National Climate Center, China Meterological Administration, April 14, 2010). A recent study led by Azure International estimates a total offshore electricity generation potential of 11,580 terawatt hours (TWh). Azure International, "China's Annual Off-shore Wind Potential Estimated at 11,580 TWh," *Azure International*, n.d., http://www.azure-international.com/en/public-outreach/news/177-growth-potential-for-offshore-wind-2010.

129. Shi Pengfei in Eric Prideaux and Wu Qi, "China Wind Power 2010 Conference Report: Chinese Wind Sector Takes on the World," *Windpower Monthly*, December 1, 2010.

130. One student thesis characterizes the Chinese wind power innovation model as "learning by using," evolving from 1996–2002 into the wider utilization of "research, development and demonstration" to push the technology forward (Su Peng, "Wind Power Manufacturing in China," Master's thesis, Tsinghua University, 2005). This can be contrasted to Denmark's "learning by interacting" wind innovation model, with a focus on "knowledge transfer between turbine producers, turbine owners and researchers" (Kamp, Smits, and Andriesse, "Notions on Learning"); or to the U.S. wind industry, marked by a lack of collaboration and information flow among firms, but where learning by using the turbines was also important, particularly in California in the 1980s (Gipe, *Wind Energy Comes of Age*).

131. While foreign wind turbine manufacturers comfortably dominated annual wind turbine sales as recently as 2004—holding a 75 percent market share while Chinese-owned companies held just 25 percent—by 2009 their share had fallen to 25 percent, and by 2010 to just 10 percent (see chapter 4, fig. 4.11).

132. Vestas, "Vestas Opens China Technology R&D Center to Spearhead Chinese Wind Power Innovation," local press release no. 21/2010, Vestas China and Vestas Technology R&D, October 12, 2010, http://www.vestas.com/en/media/news/news-display.aspx?action=3&NewsID=2412; Andrea Fenn, "Gusting with Wind Power," *China Daily*, August 26, 2011, http://www.chinadailyapac.com/article/gusting-wind-power.

133. "China Ming Yang Wind Power Group Limited Reports Third Quarter 2010 Results," *MarketWatch*, November 15, 2010, http://www.marketwatch.com/story/china-ming-yang-wind-power-group-limited-reports-third-quarter-2010-results-2010-11-15.

134. Estimated from author's own database of historical wind turbine installations in China.

135. R. Wiser et al., "Wind Energy," in *IPCC Special Report on Renewable Energy Sources and Climate Change Mitigation*, ed. O. Edenhofer et al. (Cambridge: Cambridge University Press, 2011).

136. Ryan Wiser and Mark Bolinger, "2009 Wind Technologies Market Report" (U.S. DOE, August 2010), http://eetd.lbl.gov/ea/emp/reports/lbnl-3716e.pdf.

137. Ibid. Wind turbine costs are determined by a variety of inputs, including raw materials, components, and labor. In markets like the United States a large share of wind turbine equipment is imported, foreign exchange rates can also influence relative technology costs. Experts believe that U.S. wind turbine costs increased from 2004 to 2009 because of equipment and materials shortages, rising materials costs, and a weak U.S. dollar.

138. Wiser et al., "Wind Energy."

139. Wiser et al., "Wind Energy."

4. The Role of Foreign Technology in China's Wind Power Industry Development

1. "Major Merger in the Works, Nordtank and Micon," *Windpower Monthly*, June 1, 1997.

2. "NEG Micon Takes over Wind World," *Windpower Monthly*, September 1, 1998.

3. "China Develops Inner Mongolia 21 MW Project," *Windpower Monthly*, December 1, 1998.

4. "A Man of Clear Signals," *Windpower Monthly*, November 1, 1999.

5. "Another Gear Box Failure Warning—Red Flag Up for All Nordtank 600 kW Turbines," *Windpower Monthly*, December 1, 2000.

6. Goutou employee, Beijing, interview by Joanna Lewis, January 2004.

7. Vestas employee, Beijing, interview by Joanna Lewis, September 16, 2004.

8. Former NEG Micon China employee, Beijing, interview by Joanna Lewis, January 2004.

9. Rolf Wustenhagen, "Sustainability and Competitiveness in the Renewable Energy Sector: The Case of Vestas Wind Systems," *Greener Management International*, no. 44, Special Issue on Sustainability Performance and Business Competitiveness (December 2003).

10. Ibid.

11. "Vestas Wind," n.d., http://www.vestas.com//; Wustenhagen, "Sustainability and Competitiveness."

12. The first turbines that Vestas developed for the Danish market in 1981 had a rotor diameter of 15 meters, a 22 meter tower, and a 55 kW rating. Today Vestas's largest turbine has a 90 meter rotor and a 3 MW rating, meaning that the rotor diameter is six times bigger but the power rating is more than fifty times higher.

13. Vestas's first experience with an offshore wind farm, which involved ten 500 kW turbines in the Baltic Sea, was in 1995.

14. Wustenhagen, "Sustainability and Competitiveness."

15. Wustenhagen, "Sustainability and Competitiveness."

16. Wustenhagen, "Sustainability and Competitiveness."

17. "China's State Grid Energy Research Institute and Vestas Concludes First Part of Joint Study," July 29, 2010, http://www.vestas.com/en/media/news/news-display.aspx?action=3&NewsID=2368.

18. Jens Olsen, "Vestas Asia/Pacific" (paper presented at the China Senior Wind Energy Strategy Forum, Beijing, September 20, 2004).

19. The order, received from Chinese company Jiangsu Unipower Wind Power Co., Ltd., was said to have a value to Vestas of approximately 61 million euros. Vestas, "Vestas Receives Large Order for China and Establishes Local Blade Factory," *Vestas Stock Exchange Announcement No. 50/2004*, December 31, 2004, http://www.vestas.com/uk/news/press/newsDetails_UK.asp?ID=115.

20. According to the chief executive officer of Vestas at that time, Svend Sigaard, the decision to establish a blade factory in China was part of the company's strategy to expand manufacturing capacity close to the new markets and also served to expand the Vestas manufacturing base related to the U.S. dollar, as the increasing instability between the dollar and the euro makes European wind technology seem expensive to countries using currencies tied to the dollar, such as China. "Vestas Wind."

21. Vestas, *Annual Report 2009*, n.d., http://www.vestas.com/Admin/Public/DWSDownload.aspx?File=%2FFiles%2FFiler%2FEN%2FInvestor%2FCompany_announcements%2F2010%2F100210-CA_UK_AR.pdf.

22. "HRH Prince Joachim to Open Vestas' Blade Factory in China" (Vestas press release, June 8, 2006), http://www.vestas.com/files//Filer/EN/Press_releases/WS/2006/060608PMUK03Tianjin.pdf.

23. "Customer Day at Vestas' Manufacturing Facilities in Tianjin, China" (Vestas Asia Pacific A/S press release no. 1/2007, September 20, 2007), http://www.vestas.com/files//Filer/EN/Press_releases/Local/2007/AP070920LPMUK01.pdf; "Vestas Tianjin", n.d., http://www.vestas.com/en/jobs/work-locations/china/tianjin.aspx.

24. "Windmill Giant Vestas Opens Factory in Hohhot" (Embassy of Denmark in China, n.d.), http://www.ambbeijing.um.dk/en/menu/TheEmbassy/News/WindmillGiantVestasOpensFactoryInHohhot.htm.

25. Vestas, "Made in China, Made for China: Vestas Introduces New V60–850 kW Wind Turbine" (Vestas China in Hohhot, Inner Mongolia, press release, April 16, 2009), http://www.vestas.com/files//Filer/EN/Press_releases/Local/2009/CH_090416_LPR_UK_01.pdf.

26. Ibid.

27. As discussed in chapter 3, in early 2010 the Chinese Ministry of Industry and Information Technology released the Draft Access Conditions for Wind Power Equipment Industry, which restricted the operation of wind turbine manufacturers that did not have the capability to produce a 2.5 MW or larger turbine, did not have at least five years of experience in a related industry, and did not meet various financial, R&D and quality-control requirements.

28. "China's State Grid Energy Research Institute and Vestas Concludes First Part of Joint Study," *AllBusiness.com*, July 30, 2010, http://www.allbusiness.com/energy-utilities/utilities-industry-electric-power-power/14871446–1.html.

29. "Vestas China President Meets with Chinese Premier Wen Jiabao," *Vestas Wind*, April 30, 2010, http://vestas-iwt.it/en/media/news/news-display.aspx?action=3&NewsID=2389.

30. Zhihong Wan, "Vestas Opens $50m Beijing R&D Center," *China Daily*, October 13, 2010, http://www.chinadaily.com.cn/business/2010-10/13/content_11404459.htm; "Vestas Invests RMB350 mln to Set Up R&D Center in China," *istockanalyst.com*, October 12, 2010, http://www.istockanalyst.com/article/viewiStockNews/articleid/4574778.

31. Vestas, *Annual Report 2009*.

32. "From Out of Nowhere to the Top of Spain," *Windpower Monthly*, December 1, 2000.

33. Ibid.

34. Lopez Gandasegui, the former chief executive officer of Gamesa, has stated, "Why reinvent the wheel? Vestas's world market leadership is the result of verified performance, and Gamesa's aim is to offer precisely that." Despite the desire to use Vestas's years of experience to back its product, Gandasegui also recognizes that calling Gamesa merely a Spanish manufacturer of Vestas technology "understates the real depth of Gamesa." "From Out of Nowhere to the Top of Spain."

35. "From Out of Nowhere to the Top of Spain."

36. "Two Wind Giants Go Head to Head—Vestas and Gamesa Split," *Windpower Monthly*, January 1, 2002.

37. One of China's first wind power technology joint ventures was between Spain-based Made Renewable Technology and Luoyang-based China YiTuo Group. By 2000 the company was producing 330 kW and 660 kW turbines, with the latter 60 percent locally sourced. Yituo-Made is now defunct. Made filed for bankruptcy in 2003 and was bought by Spanish turbine manufacturer Gamesa Eólica. Debra J. Lew, "Alternatives to Coal and Candles: Wind Power in China," *Energy Policy* 28, no. 4 (April 2000): 271–86; Ginger Gardiner, "High Wind in China," *Composites World*, July 1, 2007, http://www.compositesworld.com/articles/high-wind-in-china.

38. "Gamesa: Global Presence," *Gamesacorp.com*, 2010; "History," *Gamesacorp.com*, n.d., http://www.gamesacorp.com/en/gamesaen/history/nowadays.html.

39. "Gamesa Shows Off Its New G10X-4.5 MW Wind Turbine and Other Innovations," *Renewbl*, April 23, 2010, http://www.renewbl.com/2010/04/23/gamesa-shows-off-its-new-g10x-4-5-mw-wind-turbine-and-other-innovations.html; "History."

40. "New Contracts to Boost Chinese Wind Industry," *Renewable Energy Focus*, October 29, 2010, http://www.renewableenergyfocus.com/view/13552/new-contracts-to-boost-chinese-wind-industry/.

41. Ibid.; "History."

42. "Nordex Company Information," n.d., http://www.nordex-online.com/en.

43. Thomas Richterich, "Letter to the Shareholders of Nordex AG," January 10, 2005, http://www.nordex-online.com/_e/aktionaersbrief_en.pdf.

44. Ibid.

45. "Nordex Increases Profitability," *Windpower Monthly*, September 1, 2010, http://www.windpowermonthly.com/news/login/1024798/.

46. "Nordex Company Information."

47. Karl-Eugen Feifel, "Wind Power in China, a German Company's Experience," *Industry & Environment* (2001).

48. The Ride the Wind Program is discussed in more detail in chapter 3. There was one other joint venture established as part of the program—that between Spanish turbine manufacturer Made (currently owned by Gamesa) and China's Luoyang First Tractor Factory, a commercial wing of the Chinese Ministry of Machinery.

49. "China Selects Partner to Ride the Wind, Joint Venture for 200 MW," *Windpower Monthly*, July 1, 1997.

50. Ibid.

51. "Wind Parks to Be Integrated into the Electricity Supply Network in China," *Renewable Energy Made in Germany*, 2005, http://www.german-renewable-energy.com/www/main.php?tplid=39& PHPSESSID =3ce725f848a24f28b12f5f2f65eccc5b.

52. Feifel, "Wind Power in China."

53. Feifel, "Wind Power in China."

54. Feifel, "Wind Power in China."

55. Feifel, "Wind Power in China."; former Nordex employee, Beijing, interview by Joanna Lewis, December 14, 2003.

56. Gardiner, "High Wind in China."

57. Ningxia Electric Power Group and the Ningxia Tianjing Electric Energy Development Group own 40 percent and 10 percent of the venture, respectively, while Nordex assumes the remaining 50 percent. Representing a total investment of $12 million, the company produced its first 1.5 MW turbines in November 2006 for its first major order of more than 130 turbines, or 200 MW. Ibid.

58. Kathryn Kranhold, "China's Price for Market Entry: Give Us Your Technology, Too," *Wall Street Journal*, February 26, 2004.

59. Ibid.

60. This project has gained some international recognition as the first wind farm in the world to be registered as a Clean Development Mechanism project.

61. Gipe, *Wind Energy Comes of Age*.

62. "Face Change for American Industry—Enron Wind Moves On," *Windpower Monthly*, March 1, 2000.

63. "Kenetech Files for Bankruptcy Protection, Meantime Corporation Stands Firm and Hopes for Clemency," *Windpower Monthly*, June 1, 1996.

64. Ibid.

65. Zond's founder and many former executives have since started their own company, Clipper Windpower. Peter Asthmus, "Gone with the Wind: How California Is Losing Its Clean Power Edge," *ReFocus*, 2002.

66. "Not Just a Marriage of Marketing Convenience," *Windpower Monthly*, June 1, 1997.

216   4. The Role of Foreign Technology in China's Wind Power Industry Development

67. "American Giant Moves into European Market," *Windpower Monthly*, November 1, 1997.

68. Ibid.

69. "Patent History with Bizarre Interludes," *Windpower Monthly*, May 1, 2003.

70. Ibid.

71. Robert D. Richardson and William L. Erdman, "Variable Speed Wind Turbine," Patent no. 5083039, USPTO, filed February 1, 1991, issued January 21, 1992.

72. "Patent History with Bizarre Interludes."

73. "Europe Patent Office Rules on Prior Art—Variable Speed Patent Withdrawn but GE Considers Appeal," *Windpower Monthly*, July 1, 2003; "European Interim Ruling on Patent Validity—GE Wind Sues Enercon and Its Agents in Britain and Canada," *Windpower Monthly*, May 1, 2003; "Combatants Call Truce in Patent War—End of Trans-Atlantic Dispute to Allow Technology to Progress," *Windpower Monthly*, June 1, 2004.

74. "Patent History with Bizarre Interludes."

75. "Patent History with Bizarre Interludes."

76. "Combatants Call Truce in Patent War—End of Trans-Atlantic Dispute to Allow Technology to Progress"; "Europe Patent Office Rules on Prior Art—Variable Speed Patent Withdrawn but GE Considers Appeal."

77. Diane Bartz, "ITC Finds for Mitsubishi in GE Wind Turbine Dispute," *Reuters*, January 8, 2010, http://www.reuters.com/article/idUSTRE6075DV20100108.

78. Paul Glader, "GE Sues over Wind Turbine Patent," *Wall Street Journal*, February 11, 2010, http://online.wsj.com/article/SB10001424052748703382904575059714259721350.html.

79. "MHI Sues GE Energy Alleging Wind Turbine Patent Infringement," *POWER-GEN WorldWide*, May 20, 2010, http://www.powergenworldwide.com/index/display/articledisplay.articles.powergenworldwide.renewables.wind.2010.05.ge-mitsubishi.QP129867.dcmp=rss.page=1.html.

80. A turbine is any motor in which air, wind, or steam spins blades on a shaft to create energy or perform a task—from windmills that pump water to hydroelectric generators that use the force of water behind a dam to make electricity.

81. Kranhold, "China's Price for Market Entry."

82. Kranhold, "China's Price for Market Entry."

83. Kranhold, "China's Price for Market Entry."

84. Kranhold, "China's Price for Market Entry."

85. Kranhold, "China's Price for Market Entry."

86. Kvaerner Hangfa, an equity joint venture established in 1995, is owned 61 percent by Norwegian Kvaerner Energy a.s. and 39 percent by Chinese state-owned Hangzhou Electric Equipment Works (Hangfa).

87. "GE Acquires One of China's Leading Suppliers of Hydropower Generation Equipment: Largest Acquisition for GE Power Systems in China" (GE Power Systems press release, February 4, 2003), http://www.gepower.com/about/press/en/2003_press/020403b.htm.

88. GE employee, Beijing, interview by Joanna Lewis, April 8, 2004.
89. Ibid.
90. "GE in China" (paper by GE employee presented at the GE Wind Energy China SOA, coorganized with the China Wind Energy Association, GE China Technology Center, Shanghai, China, September 20, 2004); "General Electric: Our Company: What Is Six Sigma?", n.d., http://www.ge.com/en/company/companyinfo/quality/whatis.htm.
91. GE employee, Shanghai, interview by Joanna Lewis, September 2004.
92. GE is currently constructing its fifth GE Global Research center, in Rio de Janeiro, expected to be completed in 2012. The $100 million center will be located on the Ilha do Bom Jesus peninsula and, when fully operational, will employ two hundred researchers and engineers. Work at the center will focus on advanced technologies for the oil and gas, renewable energy, mining, rail, and aviation industries. "GE Global Research," n.d., http://ge.geglobalresearch.com/.
93. Diane Brady, "Reaping the Wind: GE's Energy Initiative Is a Case Study in Innovation Without Borders," *BusinessWeek*, October 11, 2004, http://www.businessweek.com/magazine/content/04_41/b3903465.htm.
94. *MAKE Consulting Report*, July 2010.
95. *MAKE Consulting Report*, July 2010.
96. Research group of Prof. Su Jun, Tsinghua University School of Public Policy and Management's Institute on Science, Technology, and Innovation Policy, Beijing, interview by Joanna Lewis, June 25, 2010.
97. GE, "GE Drivetrain Technologies Signs LOIs with A-Power to Supply 900 Wind Turbine Gearboxes and Establish Joint Venture to Build Wind Turbine Assembly Facility" (GE press release, January 12, 2009), http://www.gedrivetrain.com/assets/PDFs/GET-A-PowerUSJan11Release.pdf.
98. GE, "GE Drivetrain Technologies and Chongqing XinXing Gear Finalize Joint Venture, Launch Gear Business for Wind Turbine Industry" (GE press release, August 20, 2009), http://www.getransportation.com/resources/doc_download/173-ge-drivetrain-technologies-and-chonqing-xinxing-gear-finalize-joint-venture.html.
99. GE, "GE and Harbin Electric Form Joint Venture to Capture Growth Opportunities in the World's Largest Wind Turbine Sales Territory" (GE press release, September 28, 2010), http://site.ge-energy.com/about/press/en/2010_press/092710c.htm.
100. Ibid.
101. Jesse Broehl, "U.S. Ready to Fight Hard with China's Trade 'Cheats,'" *Windpower Monthly*, August 1, 2010, http://www.windpowermonthly.com/windalert/news/login/1019208/.
102. Sumit Bose, "GE Global Research Blog," *GE's Next Generation Offshore Wind Project with DOE*, May 19, 2006, http://ge.geglobalresearch.com/blog/ges-next-generation-offshore-wind-project-with-doe/.
103. The top Chinese-owned wind turbine manufacturers, including Goldwind, are using direct-drive technology, for reasons discussed in more detail in chapter 5.

104. The exception is perhaps Clipper Windpower, which is led by former executives from Zond and Enron Wind.

105. These strategies are discussed in more detail in chapter 3.

106. Suzlon is discussed in more detail in chapter 6.

107. The technology transfer models of Chinese firms in the wind industry are examined in chapter 5.

108. See, e.g. Kelly Sims Gallagher, *China Shifts Gears: Automakers, Oil, Pollution, and Development* (Cambridge: MIT Press, 2006), 24–28.

109. Although some WTO regulations are clearly being ignored, others—particularly those concerning technology with potential military applications—are more stringently enforced. Enforcement has been mainly limited to preventing certain sensitive technology from being transferred to China; for example, although China reportedly has pressed U.S. technology firm Intel to build a Chinese plant for its most sophisticated silicon wafer for years, the U.S. government has prevented it, likely fearing resulting military applications.

110. "U.S. Commercial Technology Transfers to the People's Republic of China" (DFI International for the Bureau of Export Administration, 1998), http://www.globalsecurity.org/wmd/library/report/1999/techtransfer2prc.htm.

111. Ibid.

112. "U.S. Commercial Technology Transfers to the People's Republic of China."

113. Kranhold, "China's Price for Market Entry."

114. Gallagher, *China Shifts Gears*.

115. United Steelworkers, "United Steelworkers' Section 301 Petition Demonstrates China's Green Technology Practices Violate WTO Rules," September 9, 2010, http://assets.usw.org/releases/misc/section-301.pdf.

116. Ibid.

117. Paul Glader, "GE Chief Slams U.S. on Energy," *Wall Street Journal*, September 24, 2010, http://online.wsj.com/article/SB10001424052748703384204575509760331620520.html?mod=WSJ_business_whatsNews.

118. AWEA, "Statement on China's Trade Practices" (AWEA press release, n.d.), http://www.awea.org/newsroom/releases/09_10_10_China_Trade_Practices.html.

119. Doug Palmer, "U.S. Challenges China Wind Power Aid at WTO," *Reuters*, December 22, 2010, http://www.reuters.com/article/idUSTRE6BL3EU20101222.

120. Office of the USTR, "China Ends Wind Power Equipment Subsidies Challenged by the United States in WTO Dispute," June 7, 2011, http://www.ustr.gov/about-us/press-office/press-releases/2011/june/china-ends-wind-power-equipment-subsidies-challenged.

121. WTO, *Agreement on Subsidies and Countervailing Measures*, April 1994, http://www.wto.org/english/docs_e/legal_e/24-scm_01_e.htm.

122. Office of the USTR, "China Ends Wind Power Equipment Subsidies."

123. United Steelworkers, "USW Applauds Obama Administration Success Ending China Subsidies in Wind Energy Sector Following Section 301 Petition," June 7, 2011, http://www.usw.org/media_center/releases_advisories?id=0391.

4. The Role of Foreign Technology in China's Wind Power Industry Development   219

124. Yiyu Liu, "China to Halt Wind Turbine Subsidies," *China Daily*, September 15, 2011, http://www.chinadaily.com.cn/cndy/2011–06/08/content_12654114.htm.

125. "EU Joins Japan in Attacking Ontario's Renewables Tariff," *Environmental Finance*, August 15, 2011, http://www.environmental-finance.com/news/view/1917.

126. The U.S.-China dynamic in bilateral and multilateral forums is discussed further in chapter 7.

127. Matthew L. Wald, "Solyndra, Solar Firm Aided by Federal Loans, Shuts Doors," *New York Times*, August 31, 2011, http://www.nytimes.com/2011/09/01/business/energy-environment/solyndra-solar-firm-aided-by-federal-loans-shuts-doors.html?pagewanted=all; Brad Plummer, "Five Myths About the Solyndra Collapse," *Washington Post*, September 14, 2011, http://www.washingtonpost.com/blogs/ezra-klein/post/five-myths-about-the-solyndra-collapse/2011/09/14/gIQAfkyvRK_blog.html.

128. See chapter 6 for a more extensive discussion of the joint development model for wind power technology development between foreign and local firms.

129. American Superconductor Corporation employee, interview by Joanna Lewis, October 5, 2009.

130. "AMSC Issues Update Regarding Its Anticipated Fourth Quarter and Fiscal Year 2010 Financial Results" (AMSC press release, April 5, 2011), http://www.amsc.com.

131. "Debate on Overcapacity Blows Up in China's Wind Power Sector," *China Daily*, November 24, 2009, http://www.chinadaily.com.cn/business/2009–11/24/content_9034757_3.htm.

132. "China Fine-Tunes Wind Turbine Industry with New Guidelines," *CRI English*, May 10, 2011, http://english.cri.cn/6826/2011/05/10/1461s636641.htm.

133. "Grid Issue Taking Wind Out of Energy Plan's Sails," *China Daily*, February 16, 2011, http://www.china.org.cn/2011–02/16/content_21933267.htm.

134. Demi Zhu, Bloomberg New Energy Finance, Beijing, interview by Joanna Lewis, July 19, 2011.

135. Zachary Tracer and Andrew Herndon, "American Superconductor Plunges, Ex-Employee Jailed for Code Theft—Bloomberg," *Bloomberg*, September 15, 2011, http://www.bloomberg.com/news/2011–09–15/amsc-to-seek-sinovel-payments-on-lost-turbine-parts-contract.html; Joel Kirkland, "Case of Clean-Tech Theft Simmers in Washington ahead of China Talks," *ClimateWire*, January 27, 2012, http://www.eenews.net/public/climatewire/2012/01/27/1.

136. Kirkland, "Case of Clean-Tech Theft Simmers in Washington."

137. "AMSC & Sinovel Expand Partnership," *Renewable Energy World*, May 26, 2010, http://www.renewableenergyworld.com/rea/news/article/2010/05/amsc-sinovel-expand-partnership?cmpid=rss.

138. "Sinovel Website Banner: 'China No. 1 Is Just Our Starting Point,'"n.d., http://www.sinovel.com/en/images/banner-en.swf.

139. Paul Miesing, Mark P. Kriger, and Neil Slough, "Towards a Model of Effective Knowledge Transfer Within Transnationals: The Case of Chinese Foreign Invested Enterprises," *Journal of Technology Transfer* 32 (2007): 109–22.

140. Ibid.

141. Organisation for Economic Co-Operation and Development (OECD), *OECD Reviews of Innovation Policy: China Synthesis Report* (Paris: OECD, 2007).

142. Ibid.

143. There is still widespread disagreement in the business literature about the costs and benefits of moving innovative activities abroad. See, e.g., Julian Birkinshaw and Neil Hood, "Unleash Innovation in Foreign Subsidiaries," *Harvard Business Review* 79, no. 3 (2001): 131–38; John Seely Brown and John Hagel, "Innovation Blowback: Disruptive Management Practices from Asia," *McKinsey Quarterly* (2005).

144. For example, several Danish wind turbine component manufacturers pooled their resources to enter the Chinese market in mid-2010, opening a factory outside Beijing. Toke Christensen, "Danish Manufacturers Pool Resources to Break into China," *Windpower Monthly*, August 1, 2010.

5. Goldwind and the Emergence of the Chinese Wind Industry

1. The company was renamed the Goldwind Science and Technology Company Limited in 2001; in Chinese it is called *Xinjiang Jinfeng Keji Gufen Youxian Gongsi* or "*Jinfeng*."

2. Wu Gang, Goldwind CEO, Los Angeles, interview by Joanna Lewis, October 8, 2009.

3. "China's Most Powerful People 2009: Wu Gang," *Bloomberg BusinessWeek*, November 2009, http://images.businessweek.com/ss/09/11/1113_business_stars_of_china/28.htm.

4. The list is a ranking of companies by total annual wind power capacity sales.

5. Tianrun Investment is Goldwind's own investment arm for wind farm development, Tianyuan Service provides O&M, and Tianyun logistics provides other related services.

6. Goldwind employee, Goldwind manufacturing facility, Beijing, interview by Joanna Lewis, June 28, 2010.

7. "Xinjiang Goldwind Wins RMB 4.77-Billion Wind Turbine Orders," *HKTDC*, December 7, 2010, http://www.hktdc.com/info/vp/a/wr/en/2/5/1/1X078GN1/Xinjiang-Goldwind-Wins-RMB-4-77-Billion-Wind-Turbine-Orders.htm.

8. Kennix Chim, "China's Goldwind Raises $917 Million in HK IPO," *Reuters*, October 4, 2010, http://uk.reuters.com/article/idUKTRE6931TP20101004. Based on early reports of 2011 performance, Goldwind appears to have experienced losses in the middle of that year. "China Wind Market Outlook," *Bloomberg New Energy Finance*, November 23, 2011.

9. W. M. Yu and G. Wu, "The Development of China's Wind Turbine Manufacturing Industry and the Strategy of Goldwind Co.," in *Proceedings of the World Wind Energy Congress* (Beijing, 2004).

10. Ibid.

11. "REpower Unternehmen: History/Milestones," n.d., http://www.repower.de/index.php?id=39&L=1&period=1994–2001.

12. REpower has since licensed its wind turbine technology to many other companies, including Chinese companies Windey and Dongfang. REpower was later purchased by India's Suzlon, as discussed in chapter 6.

13. The 62 refers to the rotor length in meters, while 1.2 MW is the rated power output.

14. "Acquisition of REpower by Suzlon Is Important Step in International Cooperation," *World Wind Energy Association*, June 5, 2007, http://www.wwindea.org/home/index.php?option=com_content&task=view&id=175&Itemid=40.

15. Component suppliers include Techwin Electric, Jinli Magnetics, LM Glasfiber, ABB, and SKF, among others.

16. Wu Gang interview, October 8, 2009.

17. As part of the same deal, Goldwind also reportedly plans to purchase a 25 percent stake in Golden Concord of Xilinhot from Golden Concord Wind Equipment Holdings. "Xinjiang Goldwind Science and Technology to Acquire Two Companies," *OffshoreWind.biz*, November 26, 2010, http://www.offshorewind.biz/2010/11/26/xinjiang-goldwind-science-and-technology-to-acquire-two-companies-china/.

18. According to industry reports, Goldwind also hopes to obtain advanced technology for the design and production of wind turbine blades and to lower its production costs from the acquisition. Ibid.

19. Evan Osnos, "China's 863 Program, a Crash Program for Clean Energy," *New Yorker*, December 21, 2009, http://www.newyorker.com/reporting/2009/12/21/091221fa_fact_osnos.

20. Wu was quoted as saying that early turbine design attempts were a "terrible failure," and that "whole blades dropped off." Ibid.

21. Goldwind, "Company Overview" (Beijing, April 2010). R&D estimates are based on calculations from information presented in Goldwind annual reports. *MAKE Consulting Report*, July 2010.

22. "China's Goldwind Plans to Start Production of 6MW Turbines—Wind—Renewable Energy News—Recharge—Wind, Solar, Biomass, Wave/Tidal/Hydro and Geothermal," *ReCharge*, August 17, 2010, http://www.rechargenews.com/energy/wind/article226322.ece.

23. "Goldwind: Summary of Proposed Investment," International Finance Corporation, October 14, 2010, http://www.ifc.org/ifcext/spiwebsite1.nsf/projects/BA6FA1C1B81191498525777A0056668C.

24. The set-up costs for the center were provided by both the central government and the Xinjiang government, contributing RMB 5 million and 3 million, respectively. Daphne Ngar-yin Mah and Peter Hills, "The State-Industry-University Collaboration for Wind Energy: Local Diversity, Achievements and Limits in China," in *Proceedings of China Windpower 2010* (paper presented at the China Windpower 2010 Conference and Exhibition, Beijing, 2010).

25. Ibid.

26. Goldwind, "Goldwind Launches Goldwind University," *Goldwind Global: Company News*, September 28, 2011, http://www.goldwindglobal.com/web/news.do.

27. "Xinjiang Goldwind Science and Technology to Acquire Two Companies."

28. "The Gearbox Challenge," *Windpower Monthly*, November 1, 2005.

29. Chapter 4 discusses the gearbox problems of several Danish wind turbine manufacturers.

30. Having lighter-weight turbines allows developers to use smaller cranes; a lack of access to large construction cranes can be a significant barrier to wind development in some developing countries.

31. The avoidance of mechanical energy losses associated with gearboxes and couplings, as well as the passive air cooling design, results in these turbines being 3 to 5 percent more efficient than conventional drive systems, resulting in more electricity generated. In addition, the technology reportedly enables better grid connectivity with superior low-voltage ride through capabilities and high levels of reactive power control. Goldwind, "Company Overview."

32. The significance of the first wind concession projects in China is discussed further in chapter 3.

33. The increasing dominance of the big-five power companies in wind power development in China is discussed further in chapter 3.

34. "20% of Power Supply for Beijing's Olympic Venues to Be Wind-Generated," *China Association for Science and Technology*, July 24, 2008, http://english.cast.org.cn/n1181872/n1182018/n1182078/11107599.html.

35. Zhihong Wan, "Nation Eyes Offshore Wind Power," *China Daily*, December 10, 2007, http://www2.chinadaily.com.cn/bizchina/2007-12/10/content_6309294.htm.

36. "Goldwind's First Contracts for 10 GW Development," *Windpower Monthly*, August 1, 2008.

37. "Xinjiang Goldwind Wins RMB 4.77-Billion Wind Turbine Orders."

38. Goldwind, "Company Overview."

39. Wu Gang interview, October 8, 2009.

40. While it was widely reported that Goldwind's Uilk wind farm in Minnesota represented the first Chinese wind turbines installed on U.S. soil, Tang Energy installed Chinese wind turbines in Texas a few years earlier in a much less high-profile project. "U.S.-China Energy Cooperation Program, Founding Member Company Directory: Tang Energy", n.d., http://www.uschinaecp.org/members/detail/32; Wu Qi, "Goldwind Installs and Connects First Chinese Turbines on U.S. Soil," *Windpower Monthly*, February 2, 2010.

41. Ehren Goossens and Baldave Singh, "Goldwind Wins Power Supply Deal for U.S. Wind Project," *Bloomberg Businessweek*, December 20, 2010, http://www.businessweek.com/news/2010-12-20/goldwind-wins-power-supply-deal-for-u-s-wind-project.html; Harry Tournemille, "Goldwind Acquires Montana's Musselshell Wind Project from Volkswind," *Energy Boom*, January 19, 2012, http://www.energyboom

.com/wind/goldwind-acquires-montanas-musselshell-wind-project-volkswind; Ros Davidson, "Goldwind Purchases 106.5MW Illinois Project," *Windpower Monthly*, December 21, 2010; Ehren Goossens, "China's Goldwind Expanding in U.S. as Rivals Cut Back," *Bloomberg*, January 20, 2012. http://www.bloomberg.com/news/2012-01-19/goldwind-buys-two-10-megawatt-montana-wind-farms-from-volkswind.html.

42. Goldwind, "Company Overview."

43. "Goldwind's Wind Turbines Exported to Cuba," *China Real News*, November 30, 2008, http://chinarealnews.typepad.com/chinarealnews/2008/11/goldwinds-wind-turbines-exported-to-cuba.html; "China's Goldwind Wins 76.5 MW Wind Farm Project in Adama, Ethiopia," January 6, 2011, http://nazret.com/blog/index.php/2011/01/06/china-s-goldwind-wins-76-5-mw-wind-farm-projet-in-adama-ethiopia; Goldwind, "Company Overview."

44. According to an article in *USA Today*, "Honda's Ohio-built Accord is 70 percent domestic parts. Toyota's Corolla is made in a California plant alongside General Motors models. Ford's hit Fusion sedan is made in Mexico; only half its parts are from the USA or Canada. GM pitches its small HHR sport utility and giant Suburban straight at the American market, but they, too, are built in Mexico. HHR has only 41 percent American and Canadian parts." Chris Woodyard, "How Do You Tell Which Car Is More American?" *USA Today*, March 23, 2007, http://www.usatoday.com/money/autos/2007-03-22-american-usat_N.htm.

45. See Nikki Gloudeman, "'Buy American' Debate Goes Green," *Change.org*, April 13, 2010, http://environment.change.org/blog/view/buy_american_debate_goes_green.

46. Davidson, "Goldwind Purchases 106.5MW Illinois Project."

47. "Tim Rosenzweig Joins Goldwind USA, Inc. as Chief Executive Officer" (Goldwind press release, May 18, 2010).

48. Nie Peng, "Goldwind Gets $6b Credit Line to Fuel Its Int'l Drive," *China Daily*, May 21, 2010, http://www.chinadaily.com.cn/business/2010-05/21/content_9878442.htm.

49. Wu Gang interview, October 8, 2009.

50. "Goldwind Adds South America to Growing Global Wind Ambitions." *ReCharge*, July 7, 2011, http://www.rechargenews.com/energy/wind/article265993.ece.

51. As reported in chapter 3, Chinese wind companies were selling their turbines for 34 percent less than foreign companies were selling their products in the Chinese market in 2011.

52. Wu Gang interview, October 8, 2009.

53. This of course assumes a rigorous enforcement of intellectual property law in China. There are reports of Chinese manufacturers making modest modifications to their turbines in an attempt to circumvent license restrictions.

54. Wu Gang interview, October 8, 2009.

55. Estimated from author's database of historical wind turbine installations in China.

56. Goldwind employee interview, June 28, 2010.

57. Hanne May and Nicole Weinhold, "New Turbines on Offer," *New Energy*, 2009, http://www.newenergy.info/index.php?id=1542. AMSC's tumultuous relationship with Sinovel is described in chapter 4.

58. "About DEC, *Dongfang Electric*, 2010, http://www.dongfang.com.cn/index.php/aboutdecs/.

59. A-Power conducts operations through two Chinese companies: Liaoning GaoKe Energy Group Company Limited (GaoKe Energy), a wholly owned subsidiary of Head Dragon Holdings, and Liaoning High-Tech Energy Saving and Thermo-electricity Design Research Institute (GaoKe Design), 51 percent of which is owned by GaoKe Energy. The two companies are collectively referred to as Gaoke, which A-Power gained control of when it acquired Head Dragon Holdings and made it its wholly owned subsidiary. A-Power entered into an agreement with Norwin A/S of Denmark (Norwin) that gives GaoKe the exclusive right to produce and sell Norwin's 750 kW and 225 kW wind turbines in China. As part of the agreement with Norwin, a joint-venture company was established in Shenyang, China, that is 80 percent owned by GaoKe. GaoKe and Norwin also established a joint R&D facility in Shenyang to develop new wind turbine technology for both the China and the international markets. To secure these rights, GaoKe has reportedly agreed to pay Norwin a licensing fee of $3.5 million. "A-Power Signs Second Wind Turbine Deal, Enters Exclusive Agreement with Norwin A/S to Manufacture and Sell 750 kW and 225 kW Wind Turbines in China," *Electric Energy*, January 20, 2008, http://www.electricenergyonline.com/?page=show_news&id=82409.

60. Osnos, "China's 863 Program."

61. Wu Gang interview, October 8, 2009.

62. MOF, "Regulation No. 476," 2008. Note that this regulation was the target of scrutiny by the office of the USTR under a WTO investigation. Chinese subsidies from this program have reportedly run into the hundreds of millions of dollars since its inception in 2008, and the USTR claimed that this violates WTO policy because the grants awarded are contingent on Chinese wind power manufacturers using parts and equipment manufactured domestically. Andrew Krulewitz, "WTO Wind Industry Throwdown: U.S. vs. China," *Greentech Media*, December 28, 2010, http://www.greentechmedia.com/articles/read/wto-wind-industry-throwdown-u.s.-vs-china1/; Office of the United States Trade Representative, "WTO Dispute Settlement Proceedings: Subsidies on Wind Power Equipment, China," *Federal Register* 75, no. 249 (December 29, 2010): 82130–32.

63. Wu Qi, "Sinovel Denies Quality Control Behind Turbine Collapse," *Windpower Monthly*, November 11, 2010, http://www.windpowermonthly.com/news/login/1040576/.

64. National Academy of Engineering et al., "The Power of Renewables: Opportunities and Challenges for China and the United States" (National Academies Press, 2010).

65. Li, Shi, and Gao, *2010 China Wind Power Outlook*.

66. This project is discussed further in chapter 7. See also Rebecca Smith, "Chinese-Made Turbines to Fill U.S. Wind Farm," October 30, 2009, http://online.wsj.com/article/SB125683832677216475.html; Alex Pasternack, "Chinese Wind Farm in Texas: Green Jobs Fail?"*TreeHugger.com*, November 2, 2009, http://www.treehugger.com/files/2009/11/chinese-wind-farm-texas-green-jobs-fail.php.

67. Keith Bradsher, "China Wins in Wind Power, by Its Own Rules," *New York Times*, December 14, 2010, http://www.nytimes.com/2010/12/15/business/global/15chinawind.html?ref=keithbradsher.

68. Local content requirements are just one form of protectionist policy used in the wind industry. Even IPR can be used for protectionism. For example, GE has been accused of using lawsuits over patent infringements to prevent competitors from entering the U.S. market, as discussed in chapter 4.

69. Richard Blackwell, "Japan Takes Issue with Ontario's Green Energy Plan," *Globe and Mail*, September 13, 2010, http://www.theglobeandmail.com/report-on-business/japan-takes-issue-with-ontarios-green-energy-plan/article1705239/.

70. "B.C. Too Late to Use Local Content Rules for Developing Wind Power," *Douglas Magazine*, December 16, 2010, http://www.douglasmagazine.com/component/k2/item/3539.html.

71. Tamar Wilner, "Asian Firms Vie for Slice of Shrinking U.S. Wind Energy Pie," *Windpower Monthly*, August 1, 2010.

72. "Q3 2011 China Wind Market Outlook." *Bloomberg New Energy Finance*, November 23, 2011.

73. As discussed in chapter 3, many wind turbine manufacturers never make it this far.

74. Eric Prideaux and Wu Qi, "China Wind Power 2010 Conference Report: Chinese Wind Sector Takes on the World," *Windpower Monthly*, December 1, 2010.

6. Wind Energy Leapfrogging in Emerging Economies

1. Portions of this chapter are adapted from Joanna I. Lewis, "Building a National Wind Turbine Industry: Experiences from China, India and South Korea," *International Journal of Technology and Globalisation* 5, no. 3/4 (2011): 281–305; and Lewis, "Technology Acquisition and Innovation in the Developing World: Wind Turbine Development in China and India," *Studies in Comparative International Development* 42, no. 3–4 (October 2007): 208–32.

2. The history of modern wind power technology and the role of different nations and national innovation systems is discussed in more detail in chapter 3.

3. See further discussion of the literature in chapter 3. Chris Freeman, "The 'National System of Innovation' in Historical Perspective," *Cambridge Journal of Economics* 19, no. 1 (February 1995): 5–24; Sanjaya Lall, "Technological Capabilities in Emerging Asia," *Oxford Development Studies* 26, no. 2 (1998): 213–43; Alice H. Amsden, *The*

Rise of the Rest: Challenges to the West from Late-Industrializing Economies (New York: Oxford University Press, 2001); Keun Lee and Chaisung Kim, "Technological Regimes, Catching Up, and Leapfrogging: Findings from the Korean Industries," *Research Policy* 30, no. 3 (March 2001): 459–83; Keun Lee, "Making a Technological Catch Up: Barriers and Opportunities," *Asian Journal of Technology Innovation* 13, no. 2 (2005): 97–131;

4. Rinie van Est, *Winds of Change: A Comparative Study of the Politics of Wind Energy Innovation in California and Denmark* (Utrecht: International Books, 1999); Linda M. Kamp, Ruud E.H.M. Smits, and Cornelis D. Andriesse, "Notions on Learning Applied to Wind Turbine Development in the Netherlands and Denmark," *Energy Policy* 32, no. 14 (2004): 1625–37; Peter Karnoe, "Technological Innovation and Industrial Organization in the Danish Wind Industry," *Entrepreneurship and Regional Development* 2, no. 2 (April 1990): 105–24.

5. Binu Parthan and Xavier Lemaire, "Wind Power in India, Behind the Success Story" (paper presented at CSD 15, United Nations, New York, May 8, 2007).

6. *Indian Wind Energy Outlook 2009* (GWEC), http://www.gwec.net/fileadmin/documents/test2/GWEO_A4_2008_India_LowRes.pdf.

7. Ibid.

8. B. Rajsekhar, F. Van Hulle, and J. C. Jansen, "Indian Wind Energy Programme: Performance and Future Directions," *Energy Policy* 27, no. 11 (October 29, 1999): 669–78.

9. Emi Mizuno, "Cross-Border Transfer of Climate Change Mitigation Technologies: The Case of Wind Energy from Denmark and Germany to India" (Ph.D. dissertation, Massachusetts Institute of Technology, 2007).

10. Hanne May and Nicole Weinhold, "Going Global," *New Energy*, 2009, http://www.newenergy.info/index.php?id=1430.

11. *Indian Wind Energy Outlook 2009*. According to this study, some of the foreign turbine manufacturers source more than 80 percent of the components for their turbines in India and export them around the world to the United States, Europe, Australia, China, and Brazil.

12. *Indian Wind Energy Outlook 2009*.

13. Placement of Suzlon's international headquarters in Denmark was particularly strategic in 2004 since many former workers for the leading Danish wind companies, Vestas and NEG Micon, had been recently laid off after the streamlining that took place in conjunction with the merger of the two companies. T. Moller and G. Rajgor, "Denmark Picked for Global Headquarters," *Windpower Monthly*, September 2004.

14. "About Suzlon," *Suzlon.com*, January 2012.

15. *Red Herring Prospectus* (Suzlon Energy Limited, September 12, 2005), http://www.sebi.gov.in/dp/suz.pdf.

16. In late 2009 Suzlon divested some of its interest in Hansen, reportedly to reduce debt it acquired in its takeover of REpower, but it still holds a 26 percent interest in the company.

17. This is reportedly due to German regulations that require Suzlon to buy out the remaining shareholders in order to exercise full control and integrate the two companies. See Paul Beckett, "Suzlon Aims for Full Control of Repower," *Wall Street Journal*, December 10, 2009, http://online.wsj.com/article/SB10001424052748703514 404574587753147027832.html.

18. *Red Herring Prospectus.*

19. Suzlon Energy Limited Tianjin produces rotor blades, nacelles, nacelle covers, control panels, hubs, generators, and nose cones for wind turbines. It has an annual production capacity of 600 to 800 MW. As of September 2010 the Tianjin factory employed about eight hundred people. Ginger Gardiner, "High Wind in China," *Composites World*, July 1, 2007, http://www.compositesworld.com/articles/high-wind-in-china.

20. "Suzlon Completes Blade Retrofit Program," *India PRwire*, October 13, 2009, http://www.indiaprwire.com/pressrelease/information-technology/2009101335732.htm.

21. "GE Energy to Re-enter India's Wind Energy Market," *Forbes India*, October 26, 2009, http://www.moneycontrol.com/news/business/wherewind-blows_420818.html.

22. Suzlon, *Suzlon Annual Report 2009–2010: Sustaining Development Across 5 Continents, 25 Countries*, 2010, 11.

23. According to Suzlon's annual reports, the company spent 29.17 Rs crore on R&D, certification and product development, and quality assurance in 2009, and 73.57 Rs crore on these same categories in 2010. It spent 63.11 Rs crore on design change and technological upgrade charges in 2009, and 81.07 Rs crore on this category in 2010. (1 Rs crore is 10 million rupees.) Ibid., 106.

24. "Windicator," *Windpower Monthly*, January 2012, 64.

25. Choong-Yul Son, "The Status and Prospects of Wind Energy in Korea" (paper presented at the Wind Power Asia 2010, Beijing, June 23, 2010).

26. International Energy Agency, *Korea Goes for "Green Growth"* (September 11, 2008), http://www.iea.org/papers/Roundtable_SLT/korea_oct08.pdf.

27. "South Korea Shares Plans for Massive 2,500 MW of Wind Power Generation," *My Wind Power System*, November 19, 2011, http://www.mywindpowersystem.com/2011/11/south-korean-shares-plans-for-massive-2500-mw-of-wind-power-generation/.

28. "National Energy Plan 2009" (Korean Energy Economics Institute), http://www.keei.re.kr/main.nsf/index_en.html?open&p=/web_keei/en_Issues01.nsf/vicw04/59641581090E81B349256E2900483FAF&s=%3FOpenDocument.

29. International Energy Agency, *IEA Wind Annual Report 2007* (Paris: IEA, 2007), http://www.ieawind.org/AnnualReports_PDF/2007/CountryChapters/Korea-South.pdf.

30. Ibid.

31. Jong-Heon Lee, "South Korea's Doosan Goes All Out for Clean Energy," *UPI Asia.com*, http://www.upiasia.com/Economics/2009/11/20/south_koreas_doosan_goes_all_out_for_clean_energy/7639/.

32. MAKE Consulting, "South Korean Players Open New Front in Global WTG Battle," Research Note (December 2009), http://www.make-consulting.com/fileadmin/pdf/2009/091218_SouthKorean_players.pdf.

33. Power Engineering, "Samsung, Ontario Sign $3 Billion Wind, Solar Deal," *Renewable Energy World*, August 12, 2011, http://www.renewableenergyworld.com/rea/news/article/2011/08/samsung-ontario-sign-3-billion-wind-solar-deal.

34. Windtech International, "AAER Sells Wind Turbine to Hyundai Heavy Industries," March 16, 2009, http://www.windtech-international.com/content/view/2263/2/.

35. "AMSC and Hyundai Heavy Industries Expand Wind Power Strategic Alliance" (AMSC-Windtec press release, June 9, 2010), http://www.amsc-windtec.com/pdf/HHI%205MW%20Alliance%20060610%20Final.pdf.

36. "China's Fuxin to Purchase South Korea's Wind Turbines," August 14, 2009, http://en.sxcoal.com/NewsDetail.aspx?cateID=176&id=24143.

37. Lewis, "Technology Acquisition and Innovation in the Developing World."

38. Windtec was founded in Austria in 1995 and acquired by Germany's Pfleiderer Group in 2001. It became an independently owned company in 2005 and then became a wholly-owned subsidiary of AMSC in 2007. "About AMSC Windtec, a Wholly-Owned Subsidiary of AMSC," (AMSC-Windtec press release, 2010), http://www.amsc-windtec.com/about_amsc_windtec.html.

39. While Japan has historically not promoted the use of wind energy, the Japanese government has been rethinking the country's energy strategy in the wake of the Fukushima nuclear disaster, and wind energy (particularly offshore wind energy) is part of the discussion. See Yoko Kubota, "Japan Plans Floating Wind Power for Fukushima Coast," *Reuters*, September 13, 2011, http://www.reuters.com/article/2011/09/13/us-japan-wind-idUSTRE78C41M20110913.

40. Goldwind, one of the few Chinese firms with several years of operating experience, experienced major failures in hundreds of wind turbines it had installed across China, which were later traced to a material defect, as discussed in chapter 5. While Goldwind was able to repair the turbines and recover from this setback, unexpected technical failures can be extremely costly and can threaten the financial stability of a company.

41. See Peter Ford, "The World's First Carbon Positive City Will Be . . . in China? The Mayor of Baoding Is on a Crusade to Make It a Hub of Renewable Energy," *Christian Science Monitor*, August 16, 2009, http://abcnews.go.com/International/JustOneThing/story?id=8327868&page=1; "China Green Energy Technology Fund Was Created in Tianjin, Raising RMB1 to 1.5 Billion," *Zero2ipo.com.cn*, February 5, 2009, http://www.zero2ipo.com.cn/en/n/2009-2-5/200925140015.shtml.

7. Engaging China on Clean Energy Cooperation

1. Portions of this chapter are adapted from earlier papers written by the author, including Lewis, "The State of U.S.-China Relations on Climate Change"; and Nation-

al Academy of Engineering et al., "The Power of Renewables: Opportunities and Challenges for China and the United States" (Washington, D.C.: National Academies Press, 2010), chap. 7.

2. Samuel S. Kim, "Thinking Globally in Post-Mao China," *Journal of Peace* 27, no. 2 (May 1990): 191–209; Swaran Singh, "China's Quest for Multilateralism: Perspectives from India," *Procedia— Social and Behavioral Sciences* 2, no. 5 (2010): 7290–98; Gilbert Rozman, "Chinese Strategic Thinking on Multilateral Regional Security in Northeast Asia," *Orbis* 55, no. 2 (2011): 298–313.

3. Mark Lynas, "How Do I Know China Wrecked the Copenhagen Deal? I Was in the Room," *Guardian*, December 22, 2009. http://www.guardian.co.uk/environment/2009/dec/22/copenhagen-climate-change-mark-lynas; Zhongxiang Zhang, "Is It Fair to Treat China as a Christmas Tree to Hang Everybody's Complaints?" *Energy Economics* 32 (Supplement 1) (September 2010): S47–S56.

4. Lewis, "The State of U.S.-China Relations on Climate Change."

5. Orville Schell, Banning Garrett, Joanna Lewis, Jonathan Adams, Eileen Claussen, and Elliot Diringer, *A Roadmap for U.S.-China Cooperation on Energy and Climate Change* (New York: Asia Society Center for U.S.-China Relations and Pew Center on Global Climate Change, January 2009); Kenneth G. Lieberthal and David B. Sandalow, *Overcoming Obstacles to U.S.-China Cooperation on Climate Change* (Washington, D.C.: Brookings Institution, 2009).

6. White House Press Office, "U.S.-China Forum on Environment and Development, Co-Chaired by Vice President Al Gore and Premier Zhu Rongji," April 8, 1999, http://clinton6.nara.gov/1999/04/1999-04-08-fact-sheet-on-vice-president-and-premeir-zhrongji-forum.html.

7. U.S. Department of Treasury, "Joint U.S.-China Fact Sheet: U.S.-China Ten Year Energy and Environment Cooperation Framework," press release, June 18, 2008, http://www.ustreas.gov/press/releases/reports/uschinased10yrfactsheet.pdf.

8. U.S. Department of State, "U.S.-China Action Plans (Clean Water Action Plan; Clean and Efficient Transportation Action Plan; Nature Reserves and Protected Areas; Energy Efficiency Action Plan; Clean, Efficient, and Secure Electricity Production and Transmission Action Plan; Clean Air Action Plan; Wetlands Cooperation)," 2008, http://www.state.gov/g/oes/env/tenyearframework/index.htm.

9. U.S. Department of State, "U.S.-China Memorandum of Understanding to Enhance Cooperation on Climate Change, Energy and the Environment," July 28, 2009, http://www.state.gov/r/pa/prs/ps/2009/July/126592.htm.

10. The strategic component of the S&ED was transferred to the State Department and includes discussions on energy and climate change cooperation between the two countries. During the first meeting in July 2009, Treasury Secretary Timothy F. Geithner and Secretary of State Hillary Rodham Clinton were joined for the dialogue by their respective Chinese cochairs, State Councilor Dai Bingguo (for the strategic track) and Vice Premier Wang Qishan (for the economic track). U.S. Department of Treasury, "U.S.-China Strategic and Economic Dialogue," July 2009, http://www.ustreas.gov/initiatives/us-china/.

11. U.S. Department of State, "U.S.-China Joint Statement on Energy Security Cooperation," May 25, 2010, http://www.state.gov/r/pa/prs/ps/2010/05/142179.htm.

12. U.S. Department of Energy (DOE), "U.S.-China Clean Energy Research Center Announced," July 15, 2009, http://www.energy.gov/news2009/7640.htm.

13. U.S. DOE, "U.S.-China Clean Energy Announcements," November 17, 2009, http://www.energy.gov/news2009/8292.htm.

14. U.S. DOE, "Fact Sheet: U.S.-China Clean Energy Research Center," November 17, 2009, http://www.energy.gov/news2009/documents2009/U.S.-China_Fact_Sheet_CERC.pdf.

15. The subsequent Technology Management Plans (TMPs) agreed to by the CERC participants in 2011 further stipulate the specifics of the IPR arrangements to result from the joint research activities.

16. U.S. DOE, "Protocol Between the Department of Energy of the United States of America and the Ministry of Science and Technology and the National Energy Administration of the People's Republic of China for Cooperation on a Clean Energy Research Center," November 17, 2009, http://www.state.gov/documents/organization/135105.pdf.

17. U.S. DOE, "Secretary Chu Announces $37.5 Million Available for Joint U.S.-Chinese Clean Energy Research," press release, March 29, 2010, http://energy.gov/news/8804.htm.

18. U.S. DOE, "Fact Sheet: U.S.-China Electric Vehicles Initiative," November 17, 2009, http://www.energy.gov/news2009/documents2009/US-China_Fact_Sheet_Electric_Vehicles.pdf.

19. U.S. DOE, "U.S.-China Electric Vehicles (EV) Forum, September 28–29, 2009, Beijing, China," 2009, http://www.pi.energy.gov/122.htm; U.S. DOE, "Fact Sheet: U.S.–China Electric Vehicles Initiative."

20. U.S. DOE, "Fact Sheet: U.S.-China Energy Efficiency Action Plan," November 17, 2009, http://www.energy.gov/news2009/documents2009/US–China_Fact_Sheet_Efficiency_Action_Plan.pdf.

21. U.S. DOE, "Fact Sheet: U.S.-China Renewable Energy Partnership," November 17, 2009, http://www.energy.gov/news2009/documents2009/US–China_Fact_Sheet_Renewable_Energy.pdf.

22. U.S. DOE, "Fact Sheet: U.S.-China Cooperation on 21st Century Coal," November 17, 2009, http://www.energy.gov/news2009/documents2009/US–China_Fact_Sheet_Coal.pdf.

23. U.S. DOE, "Fact Sheet: U.S.-China Shale Gas Resource Initiative," November 17, 2009, http://www.energy.gov/news2009/documents2009/US–China_Fact_Sheet_Shale_Gas.pdf.

24. You-ling Wang, "First United States Renewable Energy Industry Forum and the Sino-U.S. Forum of Advanced Biofuels," *Xinhua News*, May 26, 2010, http://finance.sina.com.cn/j/20100526/20368007637.shtml.

25. "Boeing and Chinese Energy Officials Announce Sustainable Biofuel Initiatives," *PRNewswire-First Call*, May 27, 2010, http://www.prnewswire.com/news-releases/

boeing-and-chinese-energy-officials-announce-sustainable-biofuel-inititives-95007509.html.

26. "Applied Materials and China New Energy Leader CECEP to Collaborate on Advancing Solar PV," *BusinessWire Online*, May 26, 2010, http://www.marketwatch.com/story/correcting-and-replacing-applied-materials-and-china-new-energy-leader-cecep-to-collaborate-on-advancing-solar-pv-2010–05–26.

27. "CSEP Renewable Energy Program," China Sustainable Energy Program, 2009, http://www.efchina.org/FProgram.do?act=list&type=Programs&subType=3.

28. Beijing Energy Efficiency Center, http://www.beconchina.org.

29. For more details on the restructuring of FutureGen, see U.S. DOE, "U.S. and China Announce Cooperation on FutureGen and Sign Energy Efficiency Protocol at U.S.-China Strategic Economic Dialogue," December 15, 2006, http://www.energy.gov/news/4535.htm; U.S. DOE, "DOE Announces Restructured FutureGen Approach to Demonstrate Carbon Capture and Storage Technology at Multiple Clean Coal Plants," January 30, 2008, http://www.fossil.energy.gov/news/techlines/2008/08003-DOE_Announces_Restructured_FutureG.html.

30. Jeffrey Mervis, "Spending Bill Prohibits U.S.-China Collaborations," *Science Insider*, April 21, 2011, http://news.sciencemag.org/scienceinsider/2011/04/spending-bill-prohibits-us-china.html?etoc&sms_ss=email&at_xt=4db5967618586c24%2Co.

31. "Trade Sanctions Emerge as Tool to Force China and India to Curb Emissions," *Greenwire/E&E Daily*, March 21, 2007.

32. The requirement for a purchase of international reserve allowances amounts to a carbon allotment associated with the amount of carbon embedded in the imported product on a per unit basis. These border adjustments specifically target GHG-intensive products including iron, steel, aluminum, cement, bulk glass, and paper.

33. Houser et al., *Leveling the Carbon Playing Field*.

34. Chapters 4 and 5 provide an extended discussion of protectionism in the wind industry and recent WTO disputes between China and the United States on clean energy technologies.

35. Rebecca Smith, "Chinese-Made Turbines to Fill U.S. Wind Farm," October 30, 2009, http://online.wsj.com/article/SB125683832677216475.html; Pasternack, *Chinese Wind Farm in Texas*.

36. Michael Burnham, "China's A-Power to Build U.S. Wind Turbine Factory," *New York Times*, November 17, 2009, http://www.nytimes.com/gwire/2009/11/17/17greenwire-chinas-a-power-to-build-us-wind-turbine-factor-22742.html.

37. Brad Plumer, "Five Myths about the Solyndra Collapse," *Washington Post*, September 14, 2011, http://www.washingtonpost.com/blogs/ezra-klein/post/five-myths-about-the-solyndra-collapse/2011/09/14/gIQAfkyvRK_blog.html.

38. Mark Drajem and William McQuillen, "Solar-Panel Imports from China Said to Face U.S. Trade Complaint," *Bloomberg*, September 28, 2011, http://www.bloomberg.com/news/2011–09–28/solar-panel-imports-from-china-said-to-face-u-s-industry-trade-complaint.html.

39. Greentech Media, "U.S. Solar Energy Trade Assessment 2011: Trade Flows and Domestic Content for Solar Energy-Related Goods and Services in the United States" (Solar Energy Industries Association and GTM Research, August 2011), http://www.seia.org/galleries/pdf/GTM-SEIA_U.S._Solar_Energy_Trade_Balance_2011.pdf.

40. Ibid.

41. Joanna Lewis, "Energy and Climate Goals in China's 12th Five-Year Plan" (Pew Center on Global Climate Change, March 2011), http://www.pewclimate.org/docUploads/energy-climate-goals-china-twelfth-five-year-plan.pdf.

42. "China Orders Safety Drive After Environment Protests," *AFP*, September 29, 2011, http://www.google.com/hostednews/afp/article/ALeqM5iq5RwVArs9ynQPmgUsYkuQcJfpwA?docId=CNG.457606e6eafc0843f310eeff8abd8c70.251; Reuters, "China villagers protest solar plant pollution—Xinhua," *Reuters*, September 18, 2011, http://af.reuters.com/article/commoditiesNews/idAFL3E7KI01B20110918.

43. "Guanyu wanshan taiyangneng guangfu fadian shangwang dian jia zhengce de tongzhi" [NDRC Notice on Improving Electricity Tariff Policy for Grid Connected Solar PV] (NDRC, 2011), no. 1594.

44. "China Doubles Solar Power Target to 10 GW by 2015," *Reuters*, May 6, 2011, http://af.reuters.com/article/energyOilNews/idAFL3E7G554620110506.

45. "Will China Rescue the Solar Industry. What About Its Secret Over Capacity Problem," *Techpulse360.com*, December 10, 2009, http://techpulse360.com/2009/12/10/will-china-rescue-the-solar-industry-what-about-its-secret-over-capacity-problem/.

46. Michael Kanellos, "Will Solyndra, or Part of It, Get Bought?" *Greentech Media*, August 31, 2011, http://www.greentechmedia.com/articles/read/will-solyndra-or-part-of-it-get-bought/.

47. U.S. DOE, "SunShot Initiative: About," October 6, 2011, http://www1.eere.energy.gov/solar/sunshot/about.html.

48. McKinsey estimates a total GHG abatement potential of 2.8 Gt by 2030 in China's power sector, plus an additional 1 Gt from end-use electricity consumption efficiency gains. McKinsey & Company, "China's Green Revolution: Prioritizing Technologies to Achieve Energy and Environmental Sustainability," February 2009, http://origin.mckinsey.com/clientservice/sustainability/pdf/china_green_revolution.pdf.

49. For example, all users of wind energy today have benefited from the initial investments in R&D made by the Danish government in the 1970s and 1980s, as discussed in chapter 3.

50. Institute of Electrical Engineering, Chinese Academy of Sciences, "The Research and Proposals on Incentive Policy and Measurements of Chinese PV Market Development and the Acceleration" (prepared for the China Sustainable Energy Program, 2009).

51. "Patented Method for Improved Solar Cell Efficiency Expected to Lower Solar Energy Costs: Amtech Comments Further on Significance of Research Agreement with Yingli and the ECN to Develop High Efficiency N-Type Solar Cells," *azonano.com*, June 1, 2009, http://www.azonano.com/news.aspx?newsID=11800;"Yingli,

Amtech Partner in Next-Gen Solar Efficiency Research," *cleantech.com*, June 1, 2009, http://cleantech.com/news/4528/yingli-amtech-systems-partner-next.

52. Cadmium telluride (CdTe), copper indium gallium selenide (CIGS), and amorphous silicon (A-Si) are the main types of thin-film technologies.

53. Concentrating solar power systems use mirrors or lenses to concentrate a large area of solar thermal energy (sunlight) onto a small area. Electrical power is produced when the concentrated heat drives a steam turbine connected to an electrical power generator.

54. McKinsey & Company and Institute of Electrical Engineering, CAS, "CSP in China," 2009.

55. Greenpeace International, SolarPACES, and ESTELA, "Concentrating Solar Power: Global Outlook 2009," May 25, 2009, http://www.greenpeace.org/raw/content/international/press/reports/concentrating-solar-power-2009.pdf.

56. U.S. DOE, "Concentrating Solar Power Fact Sheet," 2009, http://apps1.eere.energy.gov/solar/cfm/faqs/third_level.cfm?name=Concentrating%20Solar%20Power/cat=Benefits.

57. These capital cost and cost of energy estimates are based on U.S. market estimates and take into account U.S. tax credits and therefore are not necessarily applicable to other national markets.

58. McKinsey & Company and Institute of Electrical Engineering, CAS, "CSP in China."

59. McKinsey & Company and Institute of Electrical Engineering, CAS, "CSP in China."

60. Ucilia Wang, "The Rise of Concentrating Solar Thermal Power," *Renewable Energy World*, June 6, 2011, http://www.renewableenergyworld.com/rea/news/article/2011/06/the-rise-of-concentrating-solar-thermal-power.

61. Ibid.

62. U.S. DOE, "DOE to Invest More Than $5 Million for Concentrating Solar Power," November 29, 2007, http://www.energy.gov/print/5752.htm.

63. The capacity factor (CF) defines the fraction of the rated power potential of a turbine that is actually realized over the course of a year given expected variations in wind speed. Studies have reported CFs of 23 percent compared with 34 percent in the United States (and some U.S. wind farms have a CF as high as 48 percent). Ecofys Beijing and Azure International, "The Value of Carbon in China" (WWF Hong Kong, July 2008), http://www.azure-international.com/images/stories/azure/Value%20of%20Carbon%20in%20China.pdf.

64. For example, if the average price of wind power in China is RMB 0.55 per KWh and the CF is increased to 34 percent, the price could be decreased to RMB 0.38 per kWh. Ecofys Beijing and Azure International, "The Value of Carbon in China."

65. NEA official, Beijing, interview by Joanna Lewis, July 2011.

Bibliography

AFP. "China Orders Safety Drive After Environment Protests." *AFP,* September 29, 2011. http://www.google.com/hostednews/afp/article/ALeqM5iq5RwVArs9ynQPmgUsYkuQcJfpwA?docId=CNG.457606e6eafc0843f310eeff8abd8c70.251.
AllBusiness.com. "China's State Grid Energy Research Institute and Vestas Concludes First Part of Joint Study." *AllBusiness.com,* July 30, 2010. http://www.allbusiness.com/energy-utilities/utilities-industry-electric-power-power/14871446-1.html.
Allen Consulting Group. "Sustainable Energy Jobs Report Wind Manufacturing Case Study." Paper prepared for the Sustainable Energy Development Authority, Government of Australia, 2003.
American Superconductor. "AMSC Issues Update Regarding Its Anticipated Fourth Quarter and Fiscal Year 2010 Financial Results." Press release. April 5, 2011. http://www.amsc.com.
AMSC-Windtec. "About AMSC Windtec, a Wholly-Owned Subsidiary of AMSC." Press release. 2010. http://www.amsc-windtec.com/about_amsc_windtec.html.
———. "AMSC and Hyundai Heavy Industries Expand Wind Power Strategic Alliance." Press release. June 9, 2010. http://www.amsc-windtec.com/pdf/HHI%205MW%20Alliance%200610%20Final.pdf.
Amsden, Alice H. *The Rise of the Rest: Challenges to the West from Late-Industrializing Economies.* New York: Oxford University Press, 2001.
Archibugi, Daniele, and Jonathan Michie. "The Globalization of Technology: A New Taxonomy." *Cambridge Journal of Economics* 19, no. 1 (1995): 121–40.

Archibugi, Daniele, and Carlo Pietrobelli. "The Globalization of Technology and Its Implications for Developing Countries: Windows of Opportunity or Further Burden?" *Technological Forecasting & Social Change* 70 (2003): 861–83.

Asthmus, Peter. "Gone with the Wind: How California Is Losing Its Clean Power Edge." *ReFocus* 3, no. 5 (2002): 38–40.

AWEA. "Statement on China's Trade Practices." Press release. September 10, 2011. http://www.awea.org/newsroom/releases/09_10_10_China_Trade_Practices .html.

Azure International. "China's Annual Off-shore Wind Potential Estimated at 11,580 TWh." *Azure International Report, Beijing*. 2010. http://www.azure-international .com/en/public-outreach/news/177-growth-potential-for-offshore-wind- 2010.

Bang, Guri, Gorild Heggelund, and Jonas Vevatne. *Shifting Strategies in the Global Climate Negotiations*. CICERO. 2005. http://www.cicero.uio.no/media/3079.pdf.

"Baogao chen qing cang gaoyuan bingchuan meinian tuisuo mi tu" [Tibetan Plateau Glaciers Retreat 7.8 m Each Year]. *Chinanews.com*, November 17, 2011.

Barton, John. *New Trends in Technology Transfer. Implications for National and International Policy*. Geneva: International Center for Trade and Sustainable Development (ICTSD), 2007.

Bartz, Diane. "ITC Finds for Mitsubishi in GE Wind Turbine Dispute." *Reuters*, January 8, 2010. http://www.reuters.com/article/idUSTRE6075DV20100108.

Beckett, Paul. "Suzlon Aims for Full Control of REpower." *Wall Street Journal Online*, December 10, 2009. http://online.wsj.com/article/SB10001424052748703514404574587753147027832.html.

Behr, Peter. "Tilting with Wind Turbines: A Legal War Slows Industry Growth." *ClimateWire*, August 4, 2010. http://www.eenews.net/public/climatewire/2010/ 08/04/1.

Bergek, Anna, Marko Hekkert, and Staffan Jacobsson. "Functions in Innovation Systems: A Framework for Analysing Energy System Dynamics and Identifying System Building Activities by Entrepreneurs and Policy Makers." In *Innovation for a Low Carbon Economy: Economic, Institutional and Management Approaches*, ed. Timothy J. Foxon, Jonathan Kohler, and Christine Oughton, 79–111. Cheltenham, UK: Edward Elgar, 2008.

Bergek, Anna, and Staffan Jacobsson. "The Emergence of a Growth Industry: A Comparative Analysis of the German, Dutch and Swedish Wind Turbine Industries." In *Change, Transformation and Development*, ed. J. Stan Metcalfe and Uwe Canter, 197–228. Heidelberg: Physica/Springer, 2003.

Bergek, Anna, Staffan Jacobsson, Bo Carlsson, Sven Lindmark, and Annika Rickne. "Analysing the Functional Dynamics of Technological Innovation Systems: A Scheme of Analysis." *Research Policy* 37, no. 3 (2008): 407–29.

Birkinshaw, Julian, and Neil Hood. "Unleash Innovation in Foreign Subsidiaries." *Harvard Business Review* 79, no. 3 (2001): 131–38.

Blackwell, Richard. "Japan Takes Issue with Ontario's Green Energy Plan." *Globe and Mail* (Toronto), September 13, 2010. http://www.theglobeandmail.com/report-on-business/japan-takes-issue-with-ontarios-green-energy-plan/article1705239/.
Bloem, Hans, et al. *Renewable Energy Snapshots 2010*. Luxembourg: European Commission Joint Research Centre, 2010.
Bloomberg. "UN Stops Approving China Wind Projects, Official Says." *Bloomberg*, December 2, 2009. http://www.bloomberg.com/apps/news?pid=newsarchive&sid=aU4wzufvXMpY.
Bloomberg BusinessWeek. "China's Most Powerful People 2009: Wu Gang." *Bloomberg BusinessWeek*, November 2009. http://images.businessweek.com/ss/09/11/1113_business_stars_of_china/28.htm.
Bloomberg New Energy Finance. "China Wind Market Outlook." *Bloomberg New Energy Finance*, November 23, 2011.
———. "Q3 2011 China Wind Market Outlook." *Bloomberg New Energy Finance*, November 23, 2011.
Bodansky, Daniel. "Preliminary Thoughts on the Copenhagen Accord." *Ethiopian Review*, December 21, 2009. http://www.ethiopianreview.com/index/17713.
Bose, Sumit. GE Global Research Blog. *GE's Next Generation Offshore Wind Project with DOE*. May 19, 2006. http://ge.geglobalresearch.com/blog/ges-next-generation-offshore-wind-project-with-doe/.
BP. *BP Statistical Review of World Energy*. London: BP, 2007.
Bradsher, Keith. "China Wins in Wind Power, by Its Own Rules." *New York Times*, December 14, 2010. http://www.nytimes.com/2010/12/15/business/global/15chinawind.html?ref=keithbradsher.
Brady, Diane. "Reaping the Wind: GE's Energy Initiative Is a Case Study in Innovation Without Borders." *BusinessWeek*, October 11, 2004. http://www.businessweek.com/magazine/content/04_41/b3903465.htm.
Breznitz, Dan. *Innovation and the State: Political Choice and Strategies for Growth in Israel, Taiwan, and Ireland*. New Haven: Yale University Press, 2007.
Breznitz, Dan, and Michael Murphree. *Run of the Red Queen: Government, Innovation, Globalization, and Economic Growth in China*. New Haven: Yale University Press, 2011.
Broehl, Jesse. "U.S. Ready to Fight Hard with China's Trade 'Cheats.'" *Windpower Monthly*, August 1, 2010. http://www.windpowermonthly.com/windalert/news/login/1019208/.
Brown, John Seely, and John Hagel. "Innovation Blowback: Disruptive Management Practices from Asia." *McKinsey Quarterly*. 2005.
BTM Consult. *International Wind Energy Development: World Market Update 2010*. Denmark: BTM/Navigant Consulting, 2011.
———. *International Wind Energy Development: World Market Update 2009*. Denmark: BTM/Navigant Consulting, 2010.

Buckley, Chris. "Exclusive: China Preparing National Plan for Climate Change." *Reuters*, February 6, 2007. http://www.planetark.com/dailynewsstory.cfm/newsid/40197/story.htm.

Burnham, Michael. "China's A-Power to Build U.S. Wind Turbine Factory." *New York Times*, November 17, 2009. http://www.nytimes.com/gwire/2009/11/17/17greenwire-chinas-a-power-to-build-us-wind-turbine-factor-22742.html.

BusinessWire Online. "Applied Materials and China New Energy Leader CECEP to Collaborate on Advancing Solar PV." *BusinessWire Online*, May 26, 2010. http://www.marketwatch.com/story/correcting-and-replacing-applied-materials-and-china-new-energy-leader-cecep-to-collaborate-on-advancing-solar-pv-2010-05-26.

Byrd, Robert, and Chuck Hagel. *S. Res. 98: Expressing the Sense of the Senate Regarding the Conditions for the United States Becoming a Signatory to Any International Agreement on Greenhouse Gas Emissions Under the United Nations Framework Convention on Climate Change*. Washington, D.C.: U.S. Senate, 1007.

Canadian Wind Energy Association. "Manufacturing Commercial Scale Wind Turbines in Canada." Ottawa: Canadian Wind Energy Association, April 24, 2003.

Carbon Dioxide Information Analysis Center. "Fossil Fuel CO2 Emissions." 2009. http://cdiac.ornl.gov/trends/emis/meth_reg.html.

———. "Fossil-Fuel CO2 Emissions." Oak Ridge National Laboratory and the U.S. Department of Energy, 2011. http://cdiac.ornl.gov/trends/emis/meth_reg.html.

Carlsson, Bo, and Rikard Stankiewicz. "On the Nature, Function and Composition of Technological Systems." *Journal of Evolutionary Economics* 1, no. 2 (1991): 93–118.

Chim, Kennix. "China's Goldwind Raises $917 Million in HK IPO." *Reuters*, October 4, 2010. http://uk.reuters.com/article/idUKTRE6931TP20101004.

China Association for Science and Technology. "20% of Power supply for Beijing's Olympic Venues to be Wind-Generated." July 24, 2008. http://english.cast.org.cn/n1181872/n1182018/n1182078/11107599.html.

China Coal Resource. "China's Fuxin to Purchase South Korea's Wind Turbines." *China Coal Resource*, August 14, 2009. http://en.sxcoal.com/NewsDetail.aspx?cateID=176&id=24143.

China Daily. "Debate on Overcapacity Blows Up in China's Wind Power Sector." *China Daily*, November 24, 2009. http://www.chinadaily.com.cn/business/2009-11/24/content_9034757_3.htm.

———. "Grid Issue Taking Wind Out of Energy Plan's Sails. *China Daily*, February 16, 2011. http://www.china.org.cn/2011-02/16/content_21933267.htm.

——— "High-Altitude Wind Farm Constructed in SW China." *China Daily*, December 30, 2011. http://www.chinadaily.com.cn/bizchina/2011-12/30/content_14358803.htm.

China Energy Group. *China Energy Databook Version 7.0 (CD-ROM)*. Berkeley: Lawrence Berkeley National Laboratory, 2008.

China Real News. "Goldwind's Wind Turbines Exported to Cuba." *China Real News*, November 30, 2008. http://chinarealnews.typepad.com/chinarealnews/2008/11/goldwinds-wind-turbines-exported-to-cuba.html.

China Sustainable Energy Program. "CSEP Renewable Energy Program." China Sustainable Energy Program, 2009. http://www.efchina.org/FProgram.do?act=list&type=Programs&subType=3.

Chinese Office of the National Coordination Committee on Climate Change. "Zhongguo CDM xiangmu yunxing guanli banfa" [Measures for Operation and Management of Clean Development Mechanism Projects in China]. October 12, 2005.

Christensen, Toke. "Danish Manufacturers Pool Resources to Break into China." *Windpower Monthly*, August 1, 2010.

Cleantech.com. "Yingli, Amtech Partner in Next-Gen Solar Efficiency Research." *cleantech.com*, June 1, 2009. http://cleantech.com/news/4528/yingli-amtech-systems-partner-next.

CNA. *National Security and the Threat of Climate Change*. Alexandria: CNA Corporation, 2007.

Coalition of Rainforest Nations. "Reducing Emissions from Deforestation in Developing Countries: Approaches to Stimulate Action." Submission of Views of Seventeen Parties to the Eleventh Conference of the Parties to the United Nations Framework Convention on Climate Change, January 30, 2007. http://unfccc.int/files/methods_and_science/lulucf/application/ pdf/bolivia.pdf.

Cohen-Tanugi, David. "Putting It into Perspective: China's Carbon Intensity Target." NRDC White Paper, October 2010.

Connor, Peter M. "National Innovation, Industrial Policy and Renewable Energy Technology." *Proceedings of the British Institute of Energy Economics Academic Conference*, Oxford, September 25–26, 2003.

Conway, Thomas, and Pieter Tans. "Trends in Atmospheric Carbon Dioxide." NOAA/ESRL, November 2011. http://www.esrl.noaa.gov/gmd/ccgg/trends/global.html.

Cook, Philip, and Olga Memedovic. *Strategies for Regional Innovation Systems: Learning Transfer and Applications*. Policy Papers. Vienna: United Nations Industrial Development Organization, 2003.

CRI English. "China Fine-Tunes Wind Turbine Industry with New Guidelines." *CRI English*, May 10, 2011. http://english.cri.cn/6826/2011/05/10/1461s636641.htm.

Davidson, Ogunlade, Bert Metz, and Sascha van Rooijen. *Methodological and Technological Issues in Technology Transfer: Special Report of the Intergovernmental Panel on Climate Change*. New York: Cambridge University Press, 2000.

Davidson, Ros. "Goldwind Purchases 106.5MW Illinois Project." *Windpower Monthly*, December 21, 2010.

———. "U.S. Considers Official Probe of Illegal Chinese Wind Subsidy." *Windpower Monthly*, December 1, 2010.

DFI International for the Bureau of Export Administration. "U.S. Commercial Technology Transfers to the People's Republic of China." 1998. http://www.globalsecurity.org/wmd/library/report/1999/techtransfer2prc.htm.

Diamant, Rachel, Helen Davison, and Meir P. Pugatch. *Promoting Technology Transfer in Developing Countries: Lessons from Public-Private Partnerships in the Field of Pharmaceuticals*. Stockholm: Stockholm Network, 2007.

Dodgson, Mark. "Learning, Trust and Inter-Firm Technological Linkages: Some Theoretical Associations." In *Technological Collaboration: The Dynamics of Cooperation and Industrial Innovation*, ed. Rod Coombs, Albert Richards, Pier Paolo Saviotti, and Vivien Walsh, 54–75. Cheltenham: Edward Elgar, 1996.

Douglas Magazine. "B.C. Too Late to Use Local Content Rules for Developing Wind Power." *Douglas Magazine*, December 16, 2010. http://www.douglasmagazine.com/component/k2/item/3539.html.

Downs, Erica. "China's 'New' Energy Administration." *China Business Review* (December 2008).

Drajem, Mark, and William McQuillen."Solar-Panel Imports from China Said to Face U.S. Trade Complaint." *Bloomberg*, September 28, 1011. http://www.bloomberg.com/news/2011-09-28/solar-panel-imports-from-china-said-to-face-u-s-industry-trade-complaint.html.

Du, Xiangwan. *Research on the Development Strategy of China Renewable Energy*. Beijing: Science in China Press, October 2008.

Dudley, Leonard. "Learning and the Interregional Transfer of Technology." *Southern Economic Journal* 40, no. 4 (April 1974): 563–70.

Eckert, Paul, and Claudia Parsons. "China's Hu Vows to Cut Carbon Output per GDP by 2020. *Reuters UK*, September 22, 2009. http://uk.reuters.com/article/2009/09/22/us-climate-china-idUKTRE58L4XE20090922.

Ecofys Beijing and Azure International. "The Value of Carbon in China." *WWF Hong Kong* (July 2008). http://www.azure-international.com/images/stories/azure/Value%20of%20Carbon%20in%20China.pdf.

Economy, Elizabeth. "China vs. Earth." *Nation*, May 7, 2007. http://www.thenation.com/doc/20070507/economy.

Electric Energy. "A-Power Signs Second Wind Turbine Deal, Enters Exclusive Agreement with Norwin A/S to Manufacture and Sell 750 kW and 225 kW Wind Turbines in China." *Electric Energy*, January 20, 2008. http://www.electricenergyonline.com/?page=show_news&id=82409.

Embassy of Denmark in China. "Windmill Giant Vestas Opens Factory in Hohhot." http://www.ambbeijing.um.dk/en/menu/TheEmbassy/News/WindmillGiantVestasOpensFactoryInHohhot.htm.

Energy Information Administration, U.S. Department of Energy. *International Energy Outlook 2011*. Washington, D.C.: Energy Information Administration, 2011.

———. *International Energy Outlook 2009*. Washington, D.C.: Energy Information Administration, 2009.

———. *International Energy Outlook 2006*. Washington, D.C.: Energy Information Administration, 2006.

———. *International Energy Outlook 2004*. Washington, D.C.: Energy Information Administration, 2004.

Energy Research Institute. "China Wind Power Development Towards 2030—Feasibility Study on Wind Power Contribution to 10% of Power Demand in China." Energy Foundation China Sustainable Energy Program, May 18, 2010.

Environmental Finance. "EU Joins Japan in Attacking Ontario's Renewables Tariff." *Environmental Finance,* August 15, 2011. http://www.environmental-finance.com/news/view/1917.

Ernst, Dieter. "China's Innovation Policy is a Wake-up Call for America." *AsiaPacific Issues* 100 (May 2011): 12.

———. "A New Geography of Knowledge in the Electronics Industry? Asia's Role in Global Innovation Networks." *Policy Studies* 54 (2009).

Ernst & Young. "Renewable Energy Country Attractiveness Indices: Country Focus—China." November 2011. http://www.ey.com/GL/en/Industries/Power–Utilities/RECAI–China.

van Est, Rinie. *Winds of Change: A Comparative Study of the Politics of Wind Energy Innovation in California and Denmark.* Utrecht: International Books, 1999.

European Commission. "EU-China Summit: Joint Statement." European Commission, September 5, 2006. http://ec.europa.eu/comm/external_relations/china/summit_0905/index.htm.

European Wind Energy Association. *Industry and Employment.* Volume 3 of *Wind Energy: The Facts.* Brussels: European Wind Energy Association, 2003. http://www.ewea.org/fileadmin/ewea_documents/documents/publications/WETF/Facts_Volume_4.pdf.

Fan, Daidu, and Congxian Li. "Complexities of China's Coast in Response to Climate Change." In *Advances in Climate Change Research* 2, Supplement 1 (2006): 54–58.

Feifel, Karl-Eugen. "Wind Power in China, a German Company's Experience." *Industry & Environment,* 2001.

Fellman, Joshua. "China to Hold Primary Energy Use to 4.2 Billion Tons in 2015, Xinhua Says." *Bloomberg,* October 20, 2010. http://www.bloomberg.com/news/2010-10-30/china-to-hold-primary-energy-use-to-4-2-billion-tons-in-2015-xinhua-says.html.

Fenn, Andrea. "Gusting with Wind Power." *China Daily,* August 26, 2011. http://www.chinadailyapac.com/article/gusting-wind-power.

Forbes India. "GE Energy to Re-enter India's Wind Energy Market." *Forbes India.* October 26, 2009. http://www.moneycontrol.com/news/business/wherewindblows_420818.html.

Ford, Peter. 2009. "The World's First Carbon Positive City Will Be . . . in China? The Mayor of Baoding Is on a Crusade to Make It a Hub of Renewable Energy." *Christian Science Monitor,* August 16, 2009. http://abcnews.go.com/International/JustOneThing/story?id=8327868&page=1.

Freeman, Chris. "The 'National System of Innovation' in Historical Perspective." *Cambridge Journal of Economics* 19. no. 1 (February 1995): 5–24.

Gallagher, Kelly Sims. *China Shifts Gears: Automakers, Oil, Pollution, and Development.* Cambridge: MIT Press, 2006.

———. "Limits to Leapfrogging in Energy Technologies: Evidence from the Chinese Automobile Industry." *Energy Policy* 34, no. 4 (2006): 383–94.

Gamesa. "Gamesa: Global Presence." *Gamesacorp.com.* 2010.

———. "History." *Gamesacorp.com*. n.d. http://www.gamesacorp.com/en/gamesaen/history/nowadays.html.
Gardiner, Ginger. "High Wind in China." *Composites World*. July 1, 2007. http://www.compositesworld.com/articles/high-wind-in-china.
Garud, Raghu. "On the Distinction Between Know-How, Know-Why, and Know-What." *Advances in Strategic Management* 14 (1997): 81–101.
"GE in China." Paper presented at the GE Wind Energy China SOA, co-organized with China Wind Energy Association, GE China Technology Center, Shanghai, September 20, 2004.
GE Global Research. http://ge.geglobalresearch.com/.
GE Power Systems."GE Acquires One of China's Leading Suppliers of Hydropower Generation Equipment: Largest Acquisition for GE Power Systems in China." Press release. February 4, 2003. http://www.gepower.com/about/press/en/2003_press/020403b.htm.
General Electric. "GE and Harbin Electric Form Joint Venture to Capture Growth Opportunities in the World's Largest Wind Turbine Sales Territory." Press release. September 28, 2010. http://site.ge-energy.com/about/press/en/2010_press/092710c.htm.
———. "GE Drivetrain Technologies and Chongqing XinXing Gear Finalize Joint Venture, Launch Gear Business for Wind Turbine Industry." Press release. August 20, 2009. http://www.getransportation.com/resources/doc_download/173-ge-drivetrain-technologies-and-chonqing-xinxing-gear-finalize-joint-venture.html.
———. "GE Drivetrain Technologies Signs LOIs with A-Power to Supply 900 Wind Turbine Gearboxes and Establish Joint Venture to Build Wind Turbine Assembly Facility." Press release. January 12, 2009. http://www.gedrivetrain.com/assets/PDFs/GET-A-PowerUSJan11Release.pdf.
———. "General Electric: Our Company: What Is Six Sigma?" 2012. http://www.ge.com/en/company/companyinfo/quality/whatis.htm
Gilchrist, Valerie J., and Robert L. Williams."Key Informant Interviews." In *Doing Qualitative Research*, ed. Benjamin F. Crabtree and William L. Miller, 71–88. 2nd ed. Thousand Oaks, Calif.: Sage, 1999.
Gill, Gerard J. *O.K., the Data's Lousy, but It's All We've Got (Being a Critique of Conventional Methods)*. Gatekeeper Series. London: Sustainable Agriculture Programme of the International Institute for Environment and Development, 1993.
Gipe, Paul. *Wind Energy Comes of Age*. New York: Wiley, 1995.
Glader, Paul. "GE Chief Slams U.S. on Energy." *Wall Street Journal*, September 24, 2010. http://online.wsj.com/article/SB10001424052748703384204575509760331620520.html?mod=WSJ_business_whatsNews.
———. "GE Sues over Wind Turbine Patent." *Wall Street Journal*, February 11, 2010. http://online.wsj.com/article/SB10001424052748703382904575059714259721350.html.
Global Wind Energy Council. *Global Wind Energy Outlook 2008*. Brussels: GWEC, 2008.

———. *Global Wind Report, Annual Market Update 2010*. Brussels: GWEC, 2011.
———. *Global Wind 2009 Report*. Brussels: GWEC, 2009.
———. *Indian Wind Energy Outlook 2009*. Brussels: GWEC, 2009.
Gloudeman, Nikki. "'Buy American' Debate Goes Green." *Change.org*, April 13, 2010. http://environment.change.org/blog/view/buy_american_debate_goes_green.
Goldemberg, Jose."Leapfrog Energy Technologies." *Energy Policy* 26, no. 10 (1998): 729–41.
Goldwind. "Company Overview." Beijing, April 2010.
———. "Goldwind Launches Goldwind University." *Goldwind Global: Company News*, September 28, 2011. http://www.goldwindglobal.com/web/news.do.
———. "Tim Rosenzweig Joins Goldwind USA, Inc. as Chief Executive Officer." Press release. May 18, 2010.
Goossens, Ehren. "China's Goldwind Expanding in U.S. as Rivals Cut Back." *Bloomberg*, January 20, 2012. http://www.bloomberg.com/news/2012-01-19/goldwind-buys-two-10-megawatt-montana-wind-farms-from-volkswind.html.
Goossens, Ehren, and Baldave Singh. "Goldwind Wins Power Supply Deal for U.S. Wind Project." *Bloomberg Businessweek*, December 20, 2010. http://www.businessweek.com/news/2010-12-20/goldwind-wins-power-supply-deal-for-u-s-wind-project.html.
Goulet, Denis. *The Uncertain Promise: Value Conflicts in Technology Transfer*. New York: New Horizons, 1989.
Graham-Harrison, Emma. "Hu's Carbon Commitment Marks New Era for China." *Reuters*, September 24, 2009. http://www.reuters.com/article/2009/09/24/us-china-climate-analysis-idUSTRE58N15W20090924.
Greenpeace International, SolarPACES, and ESTELA. "Concentrating Solar Power: Global Outlook 2009." May 25, 2009. http://www.greenpeace.org/raw/content/international/press/reports/concentrating-solar-power-2009.pdf.
Greentech Media. "U.S. Solar Energy Trade Assessment 2011: Trade Flows and Domestic Content for Solar Energy-Related Goods and Services in the United States." *Solar Energy Industries Association and GTM Research*, August 2011. http://www.seia.org/galleries/pdf/GTM-SEIA_U.S._Solar_Energy_Trade_Balance_2011.pdf.
GTZ. *BMU CDM-JI Initiative Country Study China of the CDM Service Unit China*. December 2008. http://gemmeronline.de/resources/CDM-CountryStudyChina.pdf.
Guerin, Turlough F. "Transferring Environmental Technologies to China: Recent Developments and Constraints." *Technological Forecasting & Social Change* 67, no. 1 (May 2001): 55–75.
Hirschman, Albert O., and Kermit Gordon. *Development Projects Observed*. Washington, D.C.: Brookings Institution Press, 1967.
HKTDC. "Xinjiang Goldwind Wins RMB 4.77-Billion Wind Turbine Orders." *HKTDC*, December 7, 2010. http://www.hktdc.com/info/vp/a/wr/en/2/5/1/1X078GN1/Xinjiang-Goldwind-Wins-RMB-4-77-Billion-Wind-Turbine-Orders.htm.

Houser, Trevor. "China's Energy Consumption and Opportunities for U.S.-China Cooperation to Address the Effects of China's Energy Use." Washington, D.C., June 14, 2007. www.uscc.gov/hearings/2007hearings/written_testimonies/07_06_14_1 5wrts/07_06_14_houser_statement.php.

Houser, Trevor, Rob Bradley, Britt Childs, Jacob Werksman, and Robert Heilmayr. *Leveling the Carbon Playing Field: International Competition and U.S. Climate Policy Design*. Washington, D.C.: Peterson Institute for International Economics and World Resources Institute, 2008.

HSBC. "China's Next 5-Year Plan: What It Means for Equity Markets." HSBC, October 2010.

India PRwire. "Suzlon Completes Blade Retrofit Program." *India PRwire*, October 13, 2009. http://www.indiaprwire.com/pressrelease/information-technology/2009101335732.htm.

Information Office of the State Council. *Zhongguo yingdui qihou bianhua de zhengce yu xingdong 2011 bai pi shu* [China's Policies and Actions for Addressing Climate Change, 2011 White Paper]. Beijing: State Council, 2011.

Institute of Electrical Engineering, Chinese Academy of Sciences. "The Research and Proposals on Incentive Policy and Measurements of Chinese PV Market Development and the Acceleration." Prepared for the China Sustainable Energy Program, 2009.

International Aluminum Institute. "Statistics." 2012. http://www.world-aluminum.org.

International Energy Agency. *Cleaner Coal in China*. Paris: OECD/IEA, 2009.

——. *IEA Energy Statistics OECD R&D Database*. Paris: OECD/IEA, 2010. http://www.iea.org/stats/rd.asp.

——. *IEA Wind Annual Report 2007*. Paris: IEA, 2007. http://www.ieawind.org/AnnualReports_PDF/2007/CountryChapters/KoreaSouth.pdf.

——. *Korea Goes for "Green Growth."* Paris: IEA, September 11, 2008. http://www.iea.org/papers/Roundtable_SLT/korea_octo8.pdf.

——. *Renewables for Power Generation: Status & Prospects*. Paris: OECD/IEA, 2003.

——. *World Energy Outlook 2007*. Paris: International Energy Agency/OECD, 2007.

International Finance Corporation. "Goldwind: Summary of Proposed Investment." October 14, 2010. http://www.ifc.org/ifcext/spiwebsite1.nsf/projects/BA6FA1C-1B81191498525777A0056668C.

Intergovernmental Panel on Climate Change (IPCC). "Summary for Policymakers." In *Managing the Risks of Extreme Events and Disasters to Advance Climate Change Adaptation*, ed. C. B. Field et al., 1–19. Special Report of Working Groups I and II. Cambridge: Cambridge University Press, 2012.

Kamp, Linda Manon. "Learning in Wind Turbine Development: A Comparison Between the Netherlands and Denmark." Ph.D. dissertation, University of Utrecht, 2002.

Kamp, Linda M., Ruud E.H.M Smits, and Cornelis D. Andriesse. "Notions on Learning Applied to Wind Turbine Development in the Netherlands and Denmark." *Energy Policy* 32, no. 14 (2004): 1625–37.

Kanellos, Michael. "Will Solyndra, or Part of It, Get Bought?" *Greentech Media*, August 31, 2011. http://www.greentechmedia.com/articles/read/will-solyndra-or-part-of-it-get-bought/.

Karnoe, Peter. "Technological Innovation and Industrial Organization in the Danish Wind Industry." *Entrepreneurship and Regional Development* 2, no. 2 (April 1990): 105–24.

Karplus, Valerie J. "Innovation in China's Energy Sector." Working Paper. Program on Energy and Sustainable Development, Center for Environmental Science and Policy, Stanford University, March 2007.

Kempener, Ruud, Laura A. Anadon, and Jose Candor. "Governmental Energy Innovation Investments, Policies, and Institutions in the Major Emerging Economies: Brazil, Russia, India, Mexico, China, and South Africa." Discussion Paper. *Energy Technology Innovation Policy Discussion Paper Series*. Belfer School for Science and International Affairs, Kennedy School, Harvard University, November 2010.

Kim, Eun-shik, Dong Kyun Park, Xueyong Zhao, Sun Kee Hong, Kang Suk Koh, Min Hwan Shu, and Young Sam Kim. "Sustainable Management of Grassland Ecosystems for Controlling Asian Dusts and Desertification in Asian Continent and a Suggestion of Eco-Village Study in China." *Ecological Research* 21, no. 6 (November 2006): 907–11.

Kim, Samuel S. "Thinking Globally in Post-Mao China." *Journal of Peace* 27, no. 2 (May 1990): 191–209.

Kirkland, Joel. "Case of Clean-Tech Theft Simmers in Washington ahead of China Talks." *ClimateWire*, January 27, 2012. http://www.eenews.net/public/climatewire/2012/01/27/1.

Kong, Xinxin. *Corporate R&D in China: The Role of Research Institutes*. Graduate School of the Chinese Academy of Social Sciences Institute of Industrial Economics, 2003.

Korean Energy Economics Institute. "National Energy Plan 2009." Korean Energy Economics Institute, 2009. http://www.keei.re.kr/main.nsf/index_en.html?open&p=/web_keei/en_Issues01.nsf/view04/59641581090E81B349256E2900483FAF&s=%3F OpenDocument.

Kranhold, Kathryn. "China's Price for Market Entry: Give Us Your Technology, Too." *Wall Street Journal*, February 26, 2004.

Kranzberg, M. "The Technical Elements in International Technology Transfer: Historical Perspectives." In *The Political Economy of International Technology Transfer*, ed. John R. McIntyre, Daniel S. Papp, 31–46. New York: Quorum Books, 1986.

Krohn, Soren. *Creating a Local Wind Industry: Experience from Four European Countries*. Montreal: Helios Center for Sustainable Energy Strategies, May 4, 1998.

Krulewitz, Andrew. "WTO Wind Industry Throwdown: U.S. vs. China." *Greentech Media*, December 28, 2010. http://www.greentechmedia.com/articles/read/wto-wind-industry-throwdown-u.s.-vs-china1/.

Kubota, Yoko. "Japan Plans Floating Wind Power for Fukushima Coast." *Reuters*, September 13, 2011. http://www.reuters.com/article/2011/09/13/us-japan-wind-idUSTRE78C41M20110913.

Lall, Sanjaya. "Technological Capabilities in Emerging Asia." *Oxford Development Studies* 26, no. 2 (1998): 213–43.

Lee, Jong-Heon. "South Korea's Doosan Goes All Out for Clean Energy." *UPI Asia.com*, November 20, 2009. http://www.upiasia.com/Economics/2009/11/20/south_koreas_doosan_goes_all_out_for_clean_energy/7639/.

Lee, Keun. "Making a Technological Catch Up: Barriers and Opportunities." *Asian Journal of Technology Innovation* 13, no. 2 (2005): 97–131.

Lee, Keun, and Chaisung Kim. "Technological Regimes, Catching Up, and Leapfrogging: Findings from the Korean Industries." *Research Policy* 30, no. 3 (March 2001): 459–83.

Levine, Mark D., and Lynn Price. "Assessment of China's Energy-Saving and Emission-Reduction Accomplishments and Opportunities During the 11th Five Year Plan: Findings and Recommendations." Paper presented at the ChinaFAQs Network for Climate and Energy Information Meeting, Washington, D.C., December 2009.

Lew, Debra J. "Alternatives to Coal and Candles: Wind Power in China." *Energy Policy* 28, no. 4 (April 2000): 271–86. http://dx.doi.org/10.1016/S0301–4215(99)00077–4.

Lewis, Joanna I. "China." In *Climate Change and National Security: A Country-Level Analysis*. Washington, D.C.: Georgetown University Press, 2011.

———. "China's Strategic Priorities in International Climate Negotiations." *Washington Quarterly* 31, no. 1 (2007): 155–74.

———. *Decoding China's Climate and Energy Policy Post-Copenhagen*. Policy brief. German Marshall Fund of the United States, June 2010. http://www.gmfus.org/publications/publication_view?publication.id=677.

———. "Energy and Climate Goals in China's 12th Five-Year Plan." Pew Center on Global Climate Change, March 2011. http://www.pewclimate.org/docUploads/energy-climate-goals-china-twelfth-five-year-plan.pdf.

———. Environmental Challenges: From the Local to the Global. In *China Today, China Tomorrow*, ed. Joseph Fewsmith. Lanham, Md.: Roman and Littlefield, 2010.

———. "The Evolving Role of Carbon Finance in Promoting Renewable Energy Development in China." *Energy Policy* 38, no. 6 (June 2010): 2875–86.

———. "The State of U.S.-China Relations on Climate Change: Examining the Bilateral and Multilateral Relationship." *China Environment Series*, no. 11 (December 2010): 7–39.

———. "Technology Acquisition and Innovation in the Developing World: Wind Turbine Development in China and India." *Studies in Comparative International Development* 42, nos. 3–4 (October 2007): 208–32.

Lewis, Joanna I., Jeffrey Logan, and Michael B. Cummings. "Understanding the Climate Challenge in China." In *Climate Change Science in Policy*, ed. Stephen H. Schneider, Armin Rosencranz, Michael Mastrandrea, and Kristin Kuntz-Duriseti. Washington, D.C.: Island Press, 2009.

Lewis, Joanna I., and Ryan H. Wiser. "Fostering a Renewable Energy Technology Industry: An International Comparison of Wind Industry Policy Support Mechanisms." *Energy Policy* 35, no. 3 (March 2007): 1844–57.

Li, Congxian, Daidu Fan, B. Deng, and D. J. Wang. "Some Problems of Vulnerability Assessment in the Coastal Zone of China." In *Global Change and Asian Pacific Coasts: Proceedings of the APN/SURVAS/LOICZ Joint Conference on Coastal Impacts of Climate Change and Adaptation in the Asia-Pacific Region*, Kobe, Japan, November 14–16, 2000. Asia Pacific Network for Global Change Research, 49–56.

Li Junfeng, Shi Pengfei, and Gao Hu. *China Wind Power Outlook 2010*. Beijing: Chinese Renewable Energy Industries Association, Global Wind Energy Council, and Greenpeace, October 2010.

Lieberthal, Kenneth. *Governing China: From Revolution Through Reform*. New York: Norton, 1995.

Lin, Erda, Wei Xiong, Hui Ju, Yue Li, Liping Bai, and Liyong Xie. "Climate Change Impacts on Crop Yield and Quality with CO_2 Fertilization in China." *Philosophical Transactions of Biological Sciences* 360, no. 1463 (November 29, 2005): 2149–54.

Lin, Erda, Yinlong Xu, Shaohong Wu, Hui Hu, and Shiming Ma. "Synopsis of China National Climate Change Assessment Report (II): Climate Change Impacts and Adaptation." *Advances in Climate Change Research* 3 (2007): 6–11.

Lin, Jiang, Nan Zhou, Mark Levine, and David Fridley. *Taking Out One Billion Tons of CO_2: The Magic of China's 11th Five Year Plan?* Berkeley: Lawrence Berkeley National Laboratory, 2007.

Liu, Wenqiang, Lin Gan, and Xiliang Zhang. "Cost Competitive Incentives for Wind Energy Development in China: Institutional Dynamics and Policy Changes." *Energy Policy* 30 (2002): 753–65.

Liu, Xielin, and Steven White. "Comparing Innovation Systems: A Framework and Application to China's Transitional Context." *Research Policy* 30 (2001): 1091–1114.

Liu, Yiyu. "China to Halt Wind Turbine Subsidies." *China Daily*, September 15, 2011. http://www.chinadaily.com.cn/cndy/2011-06/08/content_12654114.htm.

Liu, Yuanyuan. "China Releases Technical Standards for Wind Power Industry." *Renewable Energy World*, September 22, 2011. http://www.renewableenergyworld.com/rea/news/article/2011/09/china-releases-technical-standards-for-wind-power-industry.

———. "Prospects for China Wind Turbine Manufacturers Remain Gloomy." *Renewable Energy World*, October 25, 2011. http://www.renewableenergyworld.com/rea/news/article/2011/10/prospects-for-china-wind-turbine manufacturers-remain-gloomy.

Lundvall, Bengt-Åke. *National Systems of Innovation: Toward a Theory of Innovation and Interactive learning*. London: Anthem Press, 2010.

Lundvall, Bengt-Åke, et al. "National Systems of Production, Innovation and Competence Building." *Research Policy* 31, no. 2 (February 2002): 213–31.

Lynas, Mark. "How Do I Know China Wrecked the Copenhagen Deal? I Was in the Room." *Guardian*, December 22, 2009. http://www.guardian.co.uk/environment/2009/dec/22/copenhagen-climate-change-mark-lynas.

Mah, Daphne Ngar-yin, and Peter Hills. "The State-Industry-University Collaboration for Wind Energy: Local Diversity, Achievements and Limits in China." In *The Proceedings of China Windpower 2010*. Beijing, October 13, 2010.

MAKE Consulting. *MAKE Consulting Report*. July 2010.

———. "South Korean Players Open New Front in Global WTG Battle." Research Note. MAKE Consulting, December 2009. http://www.make-consulting.com/fileadmin/pdf/2009/091218_SouthKorean_players.pdf.

Malerba, Franco, and Sunil Mani, eds. *Sectoral Systems of Innovation and Production in Developing Countries: Actors, Structure and Evolution*. London: Edward Elgar, 2009.

Mansfield, Edwin. "Intellectual Property Protection, Foreign Direct Investment, and Technology Transfer." Discussion Paper. International Finance Corporation, 1994.

MarketWatch. "China Ming Yang Wind Power Group Limited Reports Third Quarter 2010 Results." *MarketWatch*, November 15, 2010. http://www.marketwatch.com/story/china-ming-yang-wind-power-group-limited-reports-third-quarter-2010-results-2010-11-15.

Mason, Jeff. "China to Watch Others on Climate Change Action." *Reuters*, June 15, 2005. http://www.enn.com/today.html?Id=7959.

May, Hanne, and Nicole Weinhold. "Going Global." *New Energy*, 2009. http://www.newenergy.info/index.php?id=1430.

———. "New Turbines on Offer." *New Energy*, 2009. http://www.newenergy.info/index.php?id=1542.

McElroy, Michael B., Xi Lu, Chris P. Nielsen, and Yuxuan Wang. "Potential for Wind-Generated Electricity in China." *Science* 325, no. 5946 (September 2009): 1378–80.

McGregor, Richard. "China Delays Climate Change Plan Indefinitely." *Financial Times*, April 23, 2007. http://www.ft.com/cms/s/be763e8c-f1d6-11db-b5b6-000b5df10621.html.

McKinsey & Company. *China's Green Revolution: Prioritizing Technologies to Achieve Energy and Environmental Sustainability*. February 2009. http://origin.mckinsey.com/clientservice/sustainability/pdf/china_green_revolution.pdf.

McKinsey & Company and Institute of Electrical Engineering, Chinese Academy of Sciences. *CSP in China*. Beijing, 2009.

Mervis, Jeffrey. "Spending Bill Prohibits U.S.-China Collaborations." *Science Insider*, April 21, 2011. http://news.sciencemag.org/scienceinsider/2011/04/spending-bill-prohibits-us-china.html?etoc&sms_ss=email&at_xt=4db5967618586c24%2Co.

Miesing, Paul, Mark P. Kriger, and Neil Slough. "Towards a Model of Effective Knowledge Transfer within Transnationals: The Case of Chinese Foreign Invested Enterprises." *Journal of Technology Transfer* 32 (2007): 109–22.

Ministry of Finance. "Fengli fadian shebei chanye hua zhuan xiang zijin guanli zan xing banfa" [Interim Measures on the Management of Special Project Funds for the Industrialization of Wind Power Generation Equipment]. 2008.

Ministry of Finance and State Administration of Taxation. "Caizhengbu guojia shui wu zongju guanyu bufen ziyuan zonghe liyong jiqi ta chanpin zengzhishui zhengce wenti de tongzhi" [Circular of the Ministry of Finance and the State Administration of Taxation on Issues Concerning Value-Added Tax Policies for Certain Products Produced by Comprehensive Utilization of Resources and Other Goods]. 2001.

Ministry of Industry and Information Technology. "Feng dian shebei zhizao hangye zhun ru biaozhun (zhengqiu yijian gao)" [Wind Power Equipment Manufacturing Industry Access Standards-Draft]. March 25, 2010.

Ministry of Science and Technology. "National Assessment Report on Climate Change Released." Press release. December 31, 2006. http://www.most.gov.cn/eng/pressroom/200612/ t20061231_39425.htm.

Ministry of Science and Technology, State Development Planning Commission, and State Economic and Trade Commission. *Evaluation of Policies Designed to Promote the Commercialization of Wind Power Technology in China*. Beijing: Energy Foundation China Sustainable Energy Program, May 15, 2002.

Mizuno, Emi. "Cross-border Transfer of Climate Change Mitigation Technologies: The Case of Wind Energy from Denmark and Germany to India." Ph.D. dissertation, Massachusetts Institute of Technology, 2007.

Moller, Torgny, and Gail Rajgor. "Denmark Picked for Global Headquarters." *Windpower Monthly*, October 1, 2004.

Moran, Daniel, ed., *Climate Change and National Security: A Country-Level Analysis*. Washington, D.C.: Georgetown University Press, 2011.

Mowery, David C., and Joanne E. Oxley. "Inward Technology Transfer and Competitiveness: The Role of National Innovation Systems." *Cambridge Journal of Economics* 19 (1995): 67–93.

National Academy of Engineering, National Research Council, Chinese Academy of Sciences, and Chinese Academy of Engineering. *The Power of Renewables: Opportunities and Challenges for China and the United States*. Washington, D.C.: National Academies Press, 2010.

National Bureau of Statistics. *China Energy Statistical Yearbook 2006*. Beijing: China Statistics Press, 2007.

——. *2010 Zhongguo nengyuan tongji nianjian* [2010 China Energy Statistical Yearbook]. Beijing: China Statistics Press, 2011.

——. *2010 Zhongguo tongji nianjian* [2010 China Statistical Yearbook]. Beijing: China Statistics Press, 2010.

National Bureau of Statistics and Ministry of Science and Technology. *Zhongguo keji tongji nianjian* [China Statistical Yearbook on Science and Technology]. Beijing: China Statistics Press, 2010.

National Development and Reform Commission. *China's National Climate Change Programme*, June 2007.

——. *Medium- and Long-Term Development Plan for Renewable Energy in China*, September 2007.

———. "Fagaiwei guanyu fengdian jianshe guanli youguan yaoqiu de tongzhi" [Notice on the Relevant Requirements for the Administration of the Construction of Wind Farms], July 2005.

———. "From the Chinese Government, a Requirement on Wind Farms," 2005. http://documents.nytimes.com/chinas-requirements-for-wind-farms?ref=global.

———. "Guanyu quxiao fengdian gongcheng xiangmu caigou shebei guochanhua lu yaoqiu de tongzhi" [Notice on Abolishing the Localization Rate Requirement for Equipment Procurement in Wind Power Projects]. November 2009.

———. "Guanyu wanshan fengli fadian shangwang dian jia zhengce de tongshi" [Notice on Improving Grid-Connected Wind Power Tariff Policy]. 2009.

———. "Guojia fazhan gaige wei bang gong ting guanyu kaizhan tan paifangquan jiaoyi shi dian gongzuo de tongzhi" [NDRC Notice on Pilot Trading Programs for the Development of Carbon Emissions Rights]. October 2011.

———. *Guojia ke zai sheng nengyuan chanye fazhan zhidao mulu* [National Guidance Catalogue for Renewable Energy Industry Development]. 2005.

———. "Ke zai sheng nengyuan dian jia fujia shou ru diao pei zan xing banfa" [Interim Measures on Renewable Energy Electricity Prices and Cost Sharing Management]. January 2006.

———. "Ke zai sheng nengyuan fadian youguan guanli guiding" [Management Rules on the Administration of Power Generation from Renewable Energy]. 2006.

———. "Ke zai sheng nengyuan fazhan shiyi wu ji hua" [Eleventh Five-Year Renewable Energy Development Plan]. March 2008.

———. "Management Rules Related to Renewable Energy Power Generation." January 5, 2006.

———. "Outline Measures for the Administration of Offshore Wind Power Development." 2010, no. 29.

———. Pricing Department. "NDRC Notice on Improving Solar PV Electricity Pricing Policy (No. 1594)." 2011. http://www.ndrc.gov.cn/zcfb/zcfbtz/2011tz/t20110801_426501.htm.

———. "Regulation No. 7." January 4, 2006.

———. "Regulation No. 44." January 11, 2007.

———. "Regulation No. 1906." July 2009.

———. "Zhongguo CDM xiangmu yunxing guanli banfa" [Measures for Operation and Management of Clean Development Mechanism Projects in China]. October 2005.

———. "Regulation No. 2517." November 29, 2005.

National People's Congress. "The Renewable Energy Law of the People's Republic of China." February 28, 2005.

National People's Congress Standing Committee. "Zhonghua renmin gongheguo ke zai sheng nengyuan fa (xiu zheng an)" [Renewable Energy Law of the People's Republic of China (Amended)]. December 2009.

National Wind Coordinating Committee. *The Effect of Wind Energy Development on State and Local Economies*. NWCC Wind Energy Series. Washington, D.C.:

National Wind Coordinating Committee, January 1997. http://old.nationalwind.org/publications/wes/wes05.htm.

Nazret.com. "China's Goldwind Wins 76.5 MW Wind Farm Project in Adama, Ethiopia." *Nazret.com.* January 6, 2011. http://nazret.com/blog/index.php/2011/01/06/china-s-goldwind-wins-76-5-mw-wind-farm-projet-in-adama-ethiopia.

Needham, Joseph, et al. *Science and Civilisation in China.* Cambridge: Cambridge University Press, 1954.

Negro, Simona O., Roald A. A. Suurs, and Marko P. Hekkert. "The Bumpy Road of Biomass Gasification in the Netherlands: Explaining the Rise and Fall of an Emerging Innovation System." *Technological Forecasting & Social Change* 75, no. 1 (January 2008): 57–77.

Nelson, Richard R. "The Changing Institutional Requirements for Technological and Institutional Catch Up." *International Journal of Technological Learning, Innovation and Development* 1, no. 1 (2007): 4–12.

Nemets, Alexandr, and Thomas Torda. "China's Guochanhua (Reverse Engineering)." June 13, 2002. http://archive.newsmax.com/archives/articles/2002/6/13/24549.shtml.

Netherlands Environmental Assessment Agency. "China Now No. 1 in CO2 Emissions; USA in Second Position." 2007. http://www.pbl.nl/en/dossiers/Climatechange/moreinfo/Chinanowno1inCO2emissionsUSAinsecondposition.html.

Nonaka, Ikujiro, and Hirotaka Takeuchi. *The Knowledge Creating Company: How Japanese Companies Create the Dynamics of Innovation.* New York: Oxford University Press, 1995.

Nordex. "Nordex Company Information." http://www.nordex-online.com/en.

Oberthur, Sebastian, and Herman E. Ott. *The Kyoto Protocol: International Climate Policy for the 21st Century.* Berlin: Springer, 1999.

O'Connell, Ric. "CSP in the United States." Paper presented at the Chinese Academy of Sciences, Beijing, October 23, 2008.

Office of the United States Trade Representative. "China Ends Wind Power Equipment Subsidies Challenged by the United States in WTO Dispute." June 7, 2011.

——. "WTO Dispute Settlement Proceedings: Subsidies on Wind Power Equipment, China." *Federal Register* 75, no. 249 (December 29, 2010): 82130–32. http://www.ustr.gov/about-us/press-office/press-releases/2011/june/china-ends-wind-power-equipment-subsidies-challenged.

OffshoreWind.biz. "Xinjiang Goldwind Science and Technology to Acquire Two Companies." *OffshoreWind.biz*, November 26, 2010. http://www.offshorewind.biz/2010/11/26/xinjiang-goldwind-science-and-technology-to-acquire-two-companies-china/.

Olsen, Jens. "Vestas Asia/Pacific." Paper presented at the China Senior Wind Energy Strategy Forum, September 20, 2004, Beijing.

Organization for Economic Co-operation and Development. *OECD Reviews of Innovation Policy: China Synthesis Report.* Paris: OECD, 2007.

———. *OECD Science, Technology and Industry Scoreboard 2009*. Paris: OECD, 2009.

———. "OECD Statistics." *OECD.StatExtracts*. http://stats.oecd.org/Index.aspx.

———. *Patent Database*. Paris: OECD, 2011.

———. *Science, Technology and Innovation for the 21st Century: Meeting of the OECD Committee for Scientific and Technological Policy at Ministerial Level (Final Communique)*. January 30, 2004. http://www.oecd.org/document/1/0,3746,en_2649_34269_25998799_1_1_1_1,00&&en-USS_01DBC.html.

Osnos, Evan. "China's 863 Program, a Crash Program for Clean Energy." *New Yorker*, December 21, 2009. http://www.newyorker.com/reporting/2009/12/21/091221fa_fact_osnos.

Palmer, Doug. "U.S. Challenges China Wind Power Aid at WTO." *Reuters*, December 22, 2010. http://www.reuters.com/article/idUSTRE6BL3EU20101222.

Parameswaran, P. "Rich Nations Must Honor Climate Change Pledge: Developing Countries." *Agence France-Presse*, September 25, 2007.

Parthan, Binu, and Xavier Lemaire. "Wind Power in India, Behind the Success Story." Paper presented at CSD 15, United Nations, New York, May 8, 2007.

Pasternack, Alex. "Chinese Wind Warm in Texas: Green Jobs Fail?" *TreeHugger.com*, November 2, 2009. http://www.treehugger.com/files/2009/11/chinese-wind-farm-texas-green-jobs-fail.php.

Patel, Pari. "Localized Production of Technology for Global Markets." *Cambridge Journal of Economics* 19, no. 1 (1995): 141–53.

"Patented Method for Improved Solar Cell Efficiency Expected to Lower Solar Energy Costs: Amtech Comments Further on Significance of Research Agreement with Yingli and the ECN to Develop High Efficiency N-Type Solar Cells." *Azonano.com*, June 1, 2009. http://www.azonano.com/news.aspx?newsID=11800.

Peng, Nie. "Goldwind Gets $6b Credit Line to Fuel Its Int'l Drive." *China Daily*, May 21, 2010. http://www.chinadaily.com.cn/business/2010-05/21/content_9878442.htm.

Peng, Su. "Wind Power Manufacturing in China." Master's thesis, Tsinghua University, June 2005.

People's Daily. "China Raises Wind Power Target to 1,000 GW by 2050." January 10, 2012. http://www.peoplesdaily-online.com/business/economy/27846-china-raises-wind-power-target-to-1000gw-by-2050.

People.com.cn. "Zhang Guobao: Shi er wu mo lizheng feihuashi nengyuan zhan yici nengyuan bizhong 11.4%" [Zhang Guobao: "Twelfth Five" Push to Nonfossil Energy to Account for 11.4 Percent Share of Primary Energy]. *People.com.cn*, January 6, 2011. http://energy.people.com.cn/GB/13670716.html.

People's Republic of China. "China Announces 16 Pct Cut in Energy Consumption per Unit of GDP by 2015." *www.gov.cn*, March 5, 2011. http://www.gov.cn/english/2011-03/05/content_1816947.htm.

———. "Guanyu tiaozheng zhongda jishu zhuangbei jinkou shuishou zhengce zan hanggui ding youguan qingdan de tongzhi" [Circular on Adjusting Relevant Lists of Tentative Provisions on Tax Policies for Import of Major Technical Equipment]. April 2010.

———. "Guowuyuan changwuhui yanjiu jueding woguo kongzhi wenshiqiti paifang mubiao" [Standing Committee of China State Council to Study the Decision to Control Greenhouse Gas Emissions Targets]. 2009. http://www.gov.cn/ldhd/2009-11/26/content_1474016.htm.

———. "Guowuyuan tongguo jiakuai peiyu he fazhan zhanluexing xinxing chanye de jueding" [Decision on Speeding Up the Cultivation and Development of Emerging Strategic Industries]. September 8, 2010. http://www.gov.cn/ldhd/2010-09/08/content_1698604.htm.

———. "Notice Regarding Policy Issues for Comprehensive Utilization of Some Natural Resources and Other Goods." December 1, 2001, no. 198.

———. "Regulation No. 476." August 11, 2008.

———. "Zhonghua renmin gongheguo guomin jingji he shehui fazhan di shier ge wunian guihua gangyao" [Twelfth Five-Year Plan for Economic and Social Development]. March 16, 2011. http://news.xinhuanet.com/politics/2011-03/16/c_121193916.htm.

———. "Zhonghua renmin gongheguo ke zai sheng nengyuan fa" [Renewable Energy Law of the People's Republic of China]. February 2005.

Plumer, Brad. "Five Myths About the Solyndra Collapse." *Washington Post*, September 14, 2011. http://www.washingtonpost.com/blogs/ezra-klein/post/five-myths-about-the-solyndra-collapse/2011/09/14/gIQAfkyvRK_blog.html.

Power Engineering. "Samsung, Ontario Sign $3 Billion Wind, Solar Deal." *Renewable Energy World*, August 12, 2011. http://www.renewableenergyworld.com/rea/news/article/2011/08/samsung-ontario-sign-3-billion-wind-solar-deal.

POWER-GEN WorldWide. "MHI Sues GE Energy Alleging Wind Turbine Patent Infringement." *POWER-GEN WorldWide*, May 20, 2010. http://www.powergenworldwide.com/index/display/articledisplay.articles.powergenworldwide.renewables.wind.2010.05.ge-mitsubishi.QP129867.dcmp=rss.page=1.html.

Price, Lynn, and Xuejun Wang. *Constraining Energy Consumption of China's Largest Industrial Enterprises Through Top-1000 Energy-Consuming Enterprise Program.* Berkeley: Lawrence Berkeley National Laboratory, 2007.

Prideaux, Eric, and Wu Qi. "China Wind Power 2010 Conference Report: Chinese Wind Sector Takes on the World." *Windpower Monthly*, December 1, 2010.

PRNewswire. "Boeing and Chinese Energy Officials Announce Sustainable Biofuel Initiatives." *PRNewswire-First Call*, May 27, 2010. http://www.prnewswire.com/news-releases/boeing-and-chinese-energy-officials-announce-sustainable-biofuel-initiatives-95007509.html.

Qi, Wu. "Goldwind Installs and Connects First Chinese Turbines on U.S. Soil." *Windpower Monthly*, February 2, 2010. http://windpowermonthly.com/news/981149/Goldwind-installs-connects-first-Chinese-turbines-US-soil/?DCMP=ILC-SEARCH.

———. "Sinovel Denies Quality Control Behind Turbine Collapse." *Windpower Monthly*, November 11, 2010. http://www.windpowermonthly.com/news/login/1040576/.

"Qihoubianhua guojia pinggu baogao weilai woguo hai pingmian jiang jixu shangsheng" [National Assessment Report on Climate Change: China's Sea Level Will Continue to Rise in Future]. *Xinhuanet.com*, November 16, 2011. http://news.xinhuanet.com/tech/2011-11/16/c_122286961.htm

Rajsekhar, B., F. Van Hulle, and J. C. Jansen. "Indian Wind Energy Programme: Performance and Future Directions." *Energy Policy* 27, no. 11 (October 29, 1999): 669–78.

ReCharge. "China's Goldwind Plans to Start Production of 6MW Turbines—Wind—Renewable Energy News—Recharge—Wind, Solar, Biomass, Wave/Tidal/Hydro and Geothermal." *ReCharge*, August 17, 2010. http://www.rechargenews.com/energy/wind/article226322.ece.

———. "Goldwind Adds South America to Growing Global Wind Ambitions." *ReCharge*, July 7, 2011. http://www.rechargenews.com/energy/wind/article265993.ece.

Recknagel, Paul. "Mapping WTG Manufacturers in China." GTZ Renewable Energy Program, 2010.

Red Herring. *Red Herring Prospectus: Suzlon Energy Limited*. Prospectus. Suzlon Energy Limited, September 12, 2005. http://www.sebi.gov.in/dp/suz.pdf.

Renewbl. "Gamesa Shows Off Its New G10X-4.5 MW Wind Turbine and Other Innovations." *Renewbl*, April 23, 2010. http://www.renewbl.com/2010/04/23/gamesa-shows-off-its-new-g10x-4-5-mw-wind-turbine-and-other-innovations.html.

Renewable Energy Focus. "New Contracts to Boost Chinese Wind Industry." *Renewable Energy Focus*, October 29, 2010. http://www.renewableenergyfocus.com/view/13552/new-contracts-to-boost-chinese-wind-industry/.

Renewable Energy Made in Germany. "Wind Parks to Be Integrated into the Electricity Supply Network in China." *Renewable Energy Made in Germany*. 2005. http://www.german-renewable-energy.com/www/main.php?tplid=39& PHPSESSID=3ce725f848a24f28b12f5f2f65eccc5b.

Renewable Energy World. "AMSC & Sinovel Expand Partnership." *Renewable Energy World*, May 26, 2010. http://www.renewableenergyworld.com/rea/news/article/2010/05/amsc-sinovel-expand-partnership?cmpid=rss.

———. "2010 Clean Energy Investment Hits a New Record," *Renewable Energy World*, January 11, 2011. http://www.renewableenergyworld.com/rea/news/article/2011/01/2010-clean-energy-investment-hits-a-new-record.

REN21. *Renewables Global Status Report 2011*. Paris: REN21 Secretariat.

REPower. *REpower Unternehmen: History/Milestones*. http://www.repower.de/index.php?id=39&L=1&period=1994–2001.

Reuters. "China Doubles Solar Power Target to 10 GW by 2015." *Reuters*, May 6, 2011. http://af.reuters.com/article/energyOilNews/idAFL3E7G554620110506.

———. "China Villagers Protest Solar Plant Pollution—Xinhua." *Reuters*, September 18, 2011. http://af.reuters.com/article/commoditiesNews/idAFL3E7KI01B20110918.

———. "Japan Starts WTO Dispute with Canada on Clean Power." *Reuters*, September 13, 2010. http://www.reuters.com/article/idUSTRE68C2RN20100913.

Richardson, Robert D., and William L. Erdman. "Variable Speed Wind Turbine." Patent no. 5083039, USPTO, filed February 1, 1991, issued January 21, 1992.

Richterich, Thomas. "Letter to the Shareholders of Nordex AG." January 10, 2005. http://www.nordex-online.com/_e/aktionaersbrief_en.pdf.

Rosenberg, Nathan, and Claudio Frischtak, eds. *International Technology Transfer: Concepts, Measures, and Comparisons*. New York: Praeger, 1985.

Rozman, Gilbert. "Chinese Strategic Thinking on Multilateral Regional Security in Northeast Asia." *Orbis* 55, no. 2 (2011): 298–313.

Saxenian, AnnaLee. *Regional Advantage: Culture and Competition in Silicon Valley and Route 128*. Cambridge: Harvard University Press, 1996.

Schell, Orville, Banning Garrett, Joanna Lewis, Jonathan Adams, Eileen Claussen, and Elliot Diringer. *A Roadmap for U.S.-China Cooperation on Energy and Climate Change*. New York: Asia Society Center for U.S.-China Relations and Pew Center on Global Climate Change, January 2009.

Schuman, Sara. *Improving China's Existing Renewable Energy Legal Framework: Lessons from the International and Domestic Experience*. White Paper. Natural Resources Defense Council, October 2010.

Science Daily. "Carbon Capture Milestone in China." *Science Daily*, August 4, 2008. http://www.sciencedaily.com/releases/2008/07/080731135924.htm.

Segal, Adam. *Advantage: How American Innovation Can Overcome the Asian Challenge*. New York: Norton, 2011.

Shi, Pengfei. "2008 Wind Power Market Shares in China." 2009.

Shirk, Susan L. *The Political Logic of Economic Reform in China*. Berkeley: University of California Press, 1993.

Shuster, Erik. "Tracking New Coal Fired Power Plants." National Energy Technology Laboratory, June 30, 2008 (updated July 12, 2011). http://www.netl.doe.gov/coal/refshelf/ncp.pdf.

Simon, Denis Fred, and Cong Cao. *China's Emerging Technological Edge*. Cambridge: Cambridge University Press, 2009.

Singh, Swaran. "China's Quest for Multilateralism: Perspectives from India." *Procedia— Social and Behavioral Sciences* 2, no. 5 (2010): 7290–98.

Sinton, Jonathan E. "Accuracy and Reliability of China's Energy Statistics." *China Economic Review* 12 (2001): 373–83.

Smil, Vaclav. *Energy Transitions: History, Requirements, Prospects*. Santa Barbara: Praeger, 2010.

Smith, Rebecca. "Chinese-Made Turbines to Fill U.S. Wind Farm." *Wall Street Journal*, October 30, 2009. http://online.wsj.com/article/SB125683832677216475.html.

Solomon, S., D. Qin, M. Manning, Z. Chen, M. Marquis, K. B. Averyt, M. Tignor, and H. Miller, eds. "Summary for Policymakers." In *Climate Change 2007: The Physical Science Basis. Contribution of Working Group I to the Fourth Assessment Report* of the Intergovernmental Panel on Climate Change, 1–17. Cambridge: Cambridge University Press, 2007.

Son, Choong-Yul. "The Status and Prospects of Wind Energy in Korea." Paper presented at Wind Power Asia 2010, Beijing, June 23, 2010.

South Centre and Center for International Environmental Law. The Technology Transfer Debate in the UNFCCC: Politics, Patents and Confusion. IP Quarterly Update (2008).

"South Korea Shares Plans for Massive 2,500 MW of Wind Power Generation." My Wind Power System, November 19, 2011. http://www.mywindpowersystem.com/2011/11/south-korean-shares-plans-for-massive-2500-mw-of-wind-power-generation/.

State Oceanic Administration and NDRC Energy Bureau. "Haizhang feng dian kaifa jianshe guanli zan xing banfa" [Measures for the Administration of Offshore Wind Power Development]. 2010.

State Science and Technology Commission (SSTC) and International Development Research Center (IDRC). A Decade of Reform: Science and Technology Policy in China. IDRC Canada, 1997. http://idl-bnc.idrc.ca/dspace/bitstream/10625/15019/1/106731.pdf.

Steinfeld, Edward S. Playing Our Game: Why China's Economic Rise Doesn't Threaten the West. Oxford: Oxford University Press, 2010.

Streets, David, Kejun Jiang, Xiulian Hu, and Jonathan E. Sinton. "Recent Reductions in China's Greenhouse Gas Emissions." Science 294, no. 5548 (November 30): 1835–37.

Suttmeier, Richard P., and Cong Cao. "China Faces the New Industrial Revolution: Achievement and Uncertainty in the Search for Research and Innovation Strategies." Asian Perspectives 23, no. 3 (1999). 153–200.

Suzlon. "About Suzlon." Suzlon.com, January 2012.

———. Suzlon Annual Report 2009–2010: Sustaining Development Across 5 Continents, 25 Countries. Pune, India, 2010.

Szymanski, Tauna. "China's Take on Climate Change." Sustainable Development, Ecosystems and Climate Change Committee Newsletter of the American Bar Association, May 2006.

Tan, Xiaomei, and Zhao Gang. An Emerging Revolution: Clean Technology Research, Development and Innovation in China. Working paper. Washington, D.C.: World Resources Institute, December 2009. http://www.wri.org.

Titus, J. G. "Greenhouse Effect, Sea Level Rise, and Land Use." Land Use Policy 7, no. 2 (April 1990): 138–53.

Tournemille, Harry. "Goldwind Acquires Montana's Musselshell Wind Project from Volkswind." Energy Boom, January 19, 2012. http://www.energyboom.com/wind/goldwind-acquires-montanas-musselshell-wind-project-volkswind.

Tracer, Zachary, and Andrew Herndon. "American Superconductor Plunges, Ex-Employee Jailed for Code Theft—Bloomberg." Bloomberg, September 15, 2011. http://www.bloomberg.com/news/2011-09-15/amsc-to-seek-sinovel-payments-on-lost-turbine-parts-contract.html.

"Trade Sanctions Emerge as Tool to Force China and India to Curb Emissions." *Greenwire/E&E Daily*, March 21, 2007.

United Nations Framework Convention on Climate Change (UNFCCC). *CDM: Registration.* January 20, 2012. http://cdm.unfccc.int/Statistics/Registration/AmountOfReductRegisteredProjPieChart.html.

———. *Decision 1/CP.1 The Berlin Mandate: Review of the Adequacy of Article 4, Paragraph 2(a) and (b), of the Convention, Including Proposals Related to a Protocol and Decisions on Follow-up.* 1995.

———. *Establishment of an Ad Hoc Working Group on the Durban Platform for Enhanced Action Draft Decision* (CP.17). December 2011.

———. *Outcome of the Work of the Ad Hoc Working Group on Long-term Cooperative Action under the Convention, draft decision -/CP.16* (advance unedited version). December 2010.

———. *Parties & Observers.* 2009. http://unfccc.int/parties_and_observers/items/2704.php.

———. *The United Nations Framework Convention on Climate Change.* 1992. http://unfccc.int/essential_background/convention/ background/items/1349.php.

United States Congress. Office of Technology Assessment. *Technology Transfer to China.* Washington, D.C.: U.S. Congress, Office of Technology Assessment, July 1987.

United States Department of Energy. "Concentrating Solar Power Fact Sheet." 2009. http://apps1.eere.energy.gov/solar/cfm/faqs/third_level.cfm/name=Concentrating%20Solar%20Power/cat= Benefits.

———. "DOE Announces Restructured FutureGen Approach to Demonstrate Carbon Capture and Storage Technology at Multiple Clean Coal Plants." January 30, 2008. http://www.fossil.energy.gov/news/techlines/2008/08003-DOE_Announces_Restructured_FutureG.html.

———. "DOE to Invest More than $5 Million for Concentrating Solar Power." November 29, 2007. http://www.energy.gov/print/5752.htm.

———. "Fact Sheet: U.S.-China Clean Energy Research Center." November 17, 2009. http://www.energy.gov/news2009/documents2009/U.S.-China_Fact_Sheet_CERC.pdf.

———. "Fact Sheet: U.S.-China Cooperation on 21st Century Coal." November 17, 2009. http://www.energy.gov/news2009/documents2009/US-China_Fact_Sheet_Coal.pdf.

———. "Fact Sheet: U.S.-China Electric Vehicles Initiative." November 17, 2009. http://www.energy.gov/news2009/documents2009/US-China_Fact_Sheet_Electric_Vehicles.pdf.

———. "Fact Sheet: U.S.-China Energy Efficiency Action Plan." November 17, 2009. http://www.energy.gov/news2009/documents2009/US-China_Fact_Sheet_Efficiency_Action_Plan.pdf.

———. "Fact Sheet: U.S.-China Renewable Energy Partnership." November 17, 2009. http://www.energy.gov/news2009/documents2009/US-China_Fact_Sheet_Renewable_Energy.pdf.

———. "Fact Sheet: U.S.-China Shale Gas Resource Initiative." November 17, 2009. http://www.energy.gov/news2009/documents2009/US-China_Fact_Sheet_Shale_Gas.pdf.

———. "History of Wind Energy." September 12, 2005. http://www1.eere.energy.gov/wind/wind_history.html.

———. "Protocol Between the Department of Energy of the Untied States of America and the Ministry of Science and Technology and the National Energy Administration of the People's Republic of China for Cooperation on a Clean Energy Research Center." November 17, 2009. http://www.state.gov/documents/organization/135105.pdf.

———. "Secretary Chu Announces $37.5 Million Available for Joint U.S.-Chinese Clean Energy Research." Press release. March 29, 2010. http://energy.gov/news/8804.htm.

———. "SunShot Initiative: About." October 6, 2011. http://www1.eere.energy.gov/solar/sunshot/about.html.

———. "20 Percent Wind Energy by 2030." July 2008. http://www.20percentwind.org.

———. "U.S. and China Announce Cooperation on FutureGen and Sign Energy Efficiency Protocol at U.S.-China Strategic Economic Dialogue." December 15, 2006. http://www.energy.gov/news/4535.htm.

———. U.S.-China Electric Vehicles (EV) Forum, September 28–29, 2009, Beijing, China." 2009. http://www.pi.energy.gov/122.htm.

———. "Wind and Water Power Program: History of Wind Energy." September 12, 2005. http://www1.eere.energy.gov/windandhydro/wind_history.html.

United States Department of State. "U.S.-China Action Plans (Clean Water Action Plan; Clean and Efficient Transportation Action Plan; Nature Reserves and Protected Areas; Energy Efficiency Action Plan; Clean, Efficient, and Secure Electricity Production and Transmission Action Plan; Clean Air Action Plan; Wetlands Cooperation)." http://www.state.gov/g/oes/env/tenyearframework/index.htm.

———. "U.S.-China Joint Statement on Energy Security Cooperation." May 25, 2010. http://www.state.gov/r/pa/prs/ps/2010/05/142179.htm.

———. "U.S.-China Memorandum of Understanding to Enhance Cooperation on Climate Change, Energy and the Environment." July 28, 2009. http://www.state.gov/r/pa/prs/ps/2009/july/126592.htm.

United States Department of Treasury. "Joint U.S.-China Fact Sheet: U.S.-China Ten Year Energy and Environment Cooperation Framework." Press release. June 18, 2008. http://www.ustreas.gov/press/releases/reports/uschinased10yrfactsheet.pdf.

———. "US-China Strategic and Economic Dialogue," July 2009. http://www.ustreas.gov/initiatives/us-china/.

United States General Accounting Office. *Selected Nations' Reports on Greenhouse Gas Emissions Varied in Their Adherence to Standards*. Washington, D.C.: General Accounting Office, December 2003. http://www.gao.gov/new.items/d0498.pdf.

United Steelworkers. "United Steelworkers' Section 301 Petition Demonstrates China's Green Technology Practices Violate WTO Rules." September 9, 2010. http://assets.usw.org/releases/misc/section-301.pdf.

———. "USW Applauds Obama Administration Success Ending China Subsidies In Wind Energy Sector Following Section 301 Petition." June 7, 2011. http://www.usw.org/media_center/releases_advisories?id=0391.

US-China Energy Cooperation Program. *Founding Member Company Directory: Tang Energy.* http://www.uschinaecp.org/members/detail/32.

Vestas. *Annual Report 2009.* http://www.vestas.com/Admin/Public/DWSDownload.aspx?File=%2FFiles%2FFiler%2FEN%2FInvestor%2FCompany_announcements%2F2010%2F100210-CA_UK_AR.pdf.

———. "China's State Grid Energy Research Institute and Vestas Concludes First Part of Joint Study." July 29, 2010. http://www.vestas.com/en/media/news/news-display.aspx?action=3&NewsID=2368.

———. "HRH Prince Joachim to Open Vestas' Blade Factory in China." Press release. June 8, 2006. http://www.vestas.com/files//Filer/EN/Press_releases/VWS/2006/060608PMUK03Tianjin.pdf.

———. "Made in China, Made for China: Vestas Introduces New V60–850 kW Wind Turbine." Press release, Vestas China in Hohhot, Inner Mongolia. April 16, 2009. http://www.vestas.com/files//Filer/EN/Press_releases/Local/2009/CH_090416_LPR_UK_01.pdf.

———. "Vestas China President Meets with Chinese Premier Wen Jiabao." *Vestas Wind*, April 30, 2010. http://vestas-iwt.it/en/media/news/news-display.aspx?action=3&NewsID=2389.

———. "Vestas Receives Large Order for China and Establishes Local Blade Factory." *Vestas Stock Exchange Announcement No. 50/2004*, December 31, 2004. http://www.vestas.com/uk/news/press/newsDetails_UK.asp?ID=115.

———. *Vestas Tianjin.* http://www.vestas.com/en/jobs/work-locations/china/tianjin.aspx.

———. *Vestas Wind.* http://www.vestas.com.

Vestas Asia Pacific. "Customer Day at Vestas' Manufacturing Facilities in Tianjin, China." Press release no. 1/2007. September 20, 2007. http://www.vestas.com/files//Filer/EN/Press_releases/Local/2007/AP070920LPMUK01.pdf.

"Vestas Invests RMB350 Mln to Set Up R&D Center in China." *istockanalyst.com*. October 12, 2010. http://www.istockanalyst.com/article/viewiStockNews/articleid/4574778.

Wald, Matthew L. "Solyndra, Solar Firm Aided by Federal Loans, Shuts Doors." *New York Times*, August 31, 2011. http://www.nytimes.com/2011/09/01/business/energy-environment/solyndra-solar-firm-aided-by-federal-loans-shuts-doors.html?pagewanted=all.

Walz, Rainier. "The Role of Regulation for Sustainable Infrastructure Innovations: The Case of Wind Energy." *International Journal of Public Policy* 2, nos. 1–2 (2007): 57–88.

Wan, Zhihong. "Nation Eyes Offshore Wind Power." *China Daily*, December 10, 2007. http://www2.chinadaily.com.cn/bizchina/2007-12/10/content_6309294.htm.

———. "Vestas Opens $50m Beijing R&D Xenter." *China Daily*, October 13, 2010. http://www.chinadaily.com.cn/business/2010-10/13/content_11404459.htm.

———. "Wen Heads 'Super Ministry' for Energy." *China Daily*, January 28, 2010. http://www.chinadaily.com.cn/china/2010-01/28/content_9388039.htm.

Wang, Ucilia. "The Rise of Concentrating Solar Thermal Power." *Renewable Energy World*, June 6, 2011. http://www.renewableenergyworld.com/rea/news/article/2011/06/the-rise-of-concentrating-solar-thermal-power.

Wang, You-ling. "First United States Renewable Energy Industry Forum and the Sino-U.S. Forum of Advanced Biofuels." *Xinhua News*, May 26, 2010. http://finance.sina.com.cn/j/20100526/20368007637.shtml.

Wen, Jiabao. "Report on the work of the Government." Delivered at the Fourth Session of the Eleventh China National People's Congress, March 5, 2011. http://blogs.wsj.com/chinarealtime/2011/03/05/china-npc-2011-reports-full-text/.

White House Press Office. "President to Attend Copenhagen Climate Talks: Administration Announces U.S. Emission Target for Copenhagen." November 25, 2009.

———. "U.S.-China Forum on Environment and Development, Co-Chaired by Vice President Al Gore and Premier Zhu Rongji." April 8, 1999. http://clinton6.nara.gov/1999/04/1999-04-08-fact-sheet-on-vice-president-and-premeir-zhrongji-forum.html.

"Will China Rescue the Solar Industry? What About Its Secret Overcapacity Problem?" *Techpulse360.com*, December 10, 2009. http://techpulse360.com/2009/12/10/will-china-rescue-the-solar-industry-what-about-its-secret-over-capacity-problem/.

Williamson, Kari. "U.S. Wind Turbine Tower Manufacturers Join U.S.-China Trade Dispute," *Renewable Energy Focus*, January 4, 2012. http://www.renewableenergyfocus.com/view/22952/us-wind-turbine-tower-manufacturers-join-uschina-trade-dispute/.

Wilner, Tamar. "Asian Firms Vie for Slice of Shrinking U.S. Wind Energy Pie." *Windpower Monthly*, August 1, 2010. http://www.windpowermonthly.com/winduscan/news/login/1019245/.

Windpower Monthly. "American Giant Moves into European Market." *Windpower Monthly*, November 1, 1997.

———. "Another Gear Box Failure Warning—Red Flag Up for All Nordtank 600 kW Turbines." *Windpower Monthly*, December 1, 2000.

———. "China Develops Inner Mongolia 21 MW project." *Windpower Monthly*, December 1, 1998.

———. "China Selects Partner to Ride the Wind, Joint Venture for 200 MW." *Windpower Monthly*, July 1, 1997.

———. "Combatants Call Truce in Patent War—End of Trans-Atlantic Dispute to Allow Technology to Progress." *Windpower Monthly*, June 1, 2004.

———. "Europe Patent Office Rules on Prior Art—Variable Speed Patent Withdrawn but GE Considers Appeal." *Windpower Monthly*, July 1, 2003.

———. "European Interim Ruling on Patent Validity—GE Wind Sues Enercon and Its Agents in Britain and Canada." *Windpower Monthly*, May 1, 2003.
———. "Face Change for American Industry—Enron Wind Moves On." *Windpower Monthly*, March 1, 2000.
———. "From Out of Nowhere to the Top of Spain." *Windpower Monthly*, December 1, 2000.
———. "The Gearbox Challenge." *Windpower Monthly*, November 1, 2005.
———. "Goldwind Purchases 106.5MW Illinois Project." *Windpower Monthly*, December 21, 2010.
———. "Goldwind's First Contracts for 10 GW Development." *Windpower Monthly*, August 1, 2008.
———. "Kenetech Files for Bankruptcy Protection, Meantime Corporation Stands Firm and Hopes for Clemency." *Windpower Monthly*, June 1, 1996.
———. "Major Merger in the Works, Nordtank and Micon." *Windpower Monthly*, June 1, 1997.
———. "A Man of Clear Signals." *Windpower Monthly*, November 1, 1999.
———. "NEG Micon Takes over Wind World." *Windpower Monthly*, September 1, 1998.
———. "Nordex Increases Profitability." *Windpower Monthly*, September 1, 2010.
———. "Not Just a Marriage of Marketing Convenience." *Windpower Monthly*, June 1, 1997.
———. "Patent History with Bizarre Interludes." *Windpower Monthly*, May 1, 2003.
———. "Two Wind Giants Go Head to Head—Vestas and Gamesa Split." *Windpower Monthly*, January 1, 2002.
———. "Windicator." *Windpower Monthly*, January 2012.
Windtech International. "AAER Sells Wind Turbine to Hyundai Heavy Industries." March 16, 2009. http://www.windtech-international.com/project-and-contracts/aaer-sells-wind-turbine-to-hyundai-heavy-industries.
Wiser, Ryan, and Mark Bolinger. *2010 Wind Technologies Market Report*. Berkeley: LBNL, June 2011.
———. *Wind Technologies Market Report 2009*. U.S. Department of Energy, August 2010.
Wiser, Ryan, and Maureen Hand. "Wind Power: How Much, How Soon, and at What Cost?" In *Generating Electricity in a Carbon-Constrained World*, ed. Fereidoon P. Sioshansi, 241–70. Maryland Heights: Elsevier, 2009.
Wiser, R., Z. Yang, M. Hand, O. Hohmeyer, D. Infield, P.H. Jensen, V. Nikolaev, M. O'Malley, G. Sinden, and A. Zervos. "Wind Energy." In *IPCC Special Report on Renewable Energy Sources and Climate Change Mitigation*, ed. O. Edenhofer, R. Pichs, Y. Sokona, K. Seyboth, and P. Matschoss, 535–608. Cambridge: Cambridge University Press, 2011.
Woodyard, Chris. "How Do You Tell Which Car Is More American?" *USA Today*, March 23, 2007. http://www.usatoday.com/money/autos/2007-03-22-american-usat_N.htm.

World Bank. *World Development Indicators (Population)*. Washington, D.C.: World Bank, 2011.

World Health Organization. *Climate Change and Human Health: Risks and Responses*. Geneva: World Health Organization, 2003.

World Steel Association. "Statistics." 2012. http://www.worldsteel.org.

World Trade Organization. "Agreement on Subsidies and Countervailing Measures." April 1994. http://www.wto.org/english/docs_e/legal_e/24-scm_01_e.htm.

———. "Understanding the WTO: Standards and Safety." http://www.wto.org/english/thewto_e/whatis_e/tif_e/agrm4_e.htm#TRS.

World Wind Energy Association. "Acquisition of REpower by Suzlon Is Important Step in International Cooperation. World Wind Energy Association, June 5, 2007. http://www.wwindea.org/home/index.php?option=com_content&task=view&id=175&Itemid=40.

———. *World Wind Energy Report 2008*. World Wind Energy Association, 2008. http://www.wwindea.org/home/images/stories/worldwindenergyreport2008_s.pdf.

Wustenhagen, Rolf. "Sustainability and Competitiveness in the Renewable Energy Sector: The Case of Vestas Wind Systems." *Greener Management International*, no.44. Special Issue on Sustainability Performance and Business Competitiveness (December 2003). http://www.iwoe.unisg.ch/org/iwo/web.nsf/SysWebRessources/Wue_vestaspaper_03-07-09/$FILE/Vestas_Paper_final.pdf.

Xie, Wei, and Guisheng Wu. "Differences Between Learning Processes in Small Tigers and Large Dragons. Learning Processes of Two Color TV (CTV) Firms Within China." *Research Policy* 32 (2003): 1463–79.

Xinhua. "CAS Outlines Strategic Plan for China's Energy Development over Next 40 Years." *Beijing Review*, September 25, 2007. http://www.bjreview.com.cn/science/txt/2007-09/25/content_77642.htm.

———. "China to Cap Energy Use at 4 Bln Tonnes of Coal Equivalent by 2015." *China Radio International (CRI)*, March 4, 2011. http://english.cri.cn/6909/2011/03/04/1461s624079.htm.

———. "Chinese Foreign Ministry Sets Up Climate Change Int'l Working Group." *China View*, September 5, 2007. http://news.xinhuanet.com/english/2007-09/05/content_6667432.htm.

———. "Key Targets of China's 12th Five-Year Plan." *Xinhua*, March 5, 2011. http://www.chinadaily.com.cn/xinhua/2011-03-05/content_1938144.html.

Xinhua News. "China's Energy Consumption per Unit of GDP Down 3.46 Percent in First 3 Quarters." *Xinhua News*, December 13, 2008. http://news.xinhuanet.com/english/2008-12/13/content_10497268.htm.

Xu Ming, Ran Li, John C. Crittenden, and Yongsheng Chen. "CO2 Emissions Embodied in China's Exports from 2002 to 2008: A Structural Decomposition Analysis." *Energy Policy* 39, no. 11 (2011): 7381–88.

Xu, Shisen. "Green Coal-Based Power Generation for Tomorrow's Power." Paper presented at APEC Energy Working Group: Expert Group on Clean Fossil Energy, Thermal Power Research Institute, Lampang, Thailand, February 24, 2006.

Yergin, Daniel. *The Quest: Energy, Security and the Remaking of the Modern World*. New York: Penguin, 2011.

Yu, Qingtai. "Special Representative for Climate Change Negotiations of the Ministry of Foreign Affairs Yu Qingtai Receives Interview of the Media." September 22, 2007. http://www.chinaembassy.org.in/.

Yu, W. M., and G. Wu. "The Development of China's Wind Turbine Manufacturing Industry and the Strategy of Goldwind Co." In *Proceedings of the World Wind Energy Congress*. Beijing, November 31, 2004.

Zero2IPO. "China Green Energy Technology Fund Was Created in Tianjin, Raising RMB 1 to 1.5 Billion." *Zero2ipo.com.cn*, February 5, 2009. http://www.zero2ipo.com.cn/en/n/2009-2-5/200925140015.shtml.

Zhang, Zhongxiang. "Is It Fair to Treat China as a Christmas Tree to Hang Everybody's Complaints?" *Energy Economics* 32 (Supplement 1) (September 2010): S47–S56.

"Zhongguo chengshi renkou shouci chaoguo nongcun renhou" [China's Urban Population Exceeded Rural Population for the First Time]. *BBC Chinese News*, January 17, 2012. http://www.bbc.co.uk/zhongwen/simp/chinese_news/2012/01/120117_china_urban.shtml.

Zhou, Yu. *The Inside Story of China's High-Tech Industry: Making Silicon Valley in Beijing*. Lanham: Rowman and Littlefield, 2008.

Zhu, Rong. "Study on Wind Resources Potential for Large Scale Development of Wind Power." Presentation at Chinese Metrological Association, National Climate Center, April 13, 2010.

Zyuga, Pluvia, and Dominique Guellec. *Who Licenses Out Patents and Why? Lessons from a Business Survey*. Paris: OECD, 2009.

Index

Italic page numbers refer to figures and tables.

Aerodyn, 126, 157, 158, 161
Aerolaminates, 77
Aeronautica, 161
Aerpac B.V., 151
Agreement on High Energy Physics (1970), 171
Air China, 177
American Council on Renewable Energy, U.S.–China Program, 177
American Power Act, 180
American Superconductor Corporation (AMSC), 117–18, 135, 154, 157, 158, 161, 228*n*38
American Wind Energy Association (AWEA), 94, 116
Amtech, 184
Andersen, Lars, 83
A-Power, 104, 135, 158, 161, 165, 181, 224*n*59
Applied Materials, 177
Argentina, 77, 161
Asia Society, Initiative for U.S.–China Cooperation on Energy and Climate, 177
Australia: and Goldwind, 134, 137, 159; and Suzlon, 150; wind power industry in, 142
Australian Commonwealth Scientific and Industrial Research Organization, 24
automobile industry, 115, 131, 223*n*44
Avantis Energy, 154, 158, 161
avoided-deforestation targets, 15
Azure International, 211*n*128

Balke-Dürr Zhangjiakou Installation Company, 89
Beijing Arbitration Commission, 118
Beijing Beizhong, 157
Beijing Energy Efficiency Center (BECon), 178
Bingaman, Jeff, 180
biomass power technologies, 23, 24, 177
Boeing, 94, 176, 177
Boeing Research & Technology, 176
Bokuk Electric, 155
Bonus, 76, 123
Brazil: and clean energy collaboration, 168; and climate negotiations, 15; and common knowledge sources, 161; and Gamesa, 86; and General Electric, 217*n*92; and Goldwind, 131,

Brazil (*continued*)
135; and Suzlon, 150; wind power industry in, 142; and wind turbine technology, 38
British Aerospace, 94
Brookings Institution, 177
Bulgaria, 131

California: and learning networks, 34; wind power industry in, 80, 93; and wind turbine technology, 28, 32, 211*n*130
Canada: and Goldwind, 130, 131; and international trade tensions, 117; and patents, 98; and Samsung, 154, 164; wind power industry in, 142, 144; and wind turbine technology, 29, 38
Canadian Wind Energy Association, 41
Cancun Agreements, 20
capacity factor (CF), 186, 233*n*63, 233*n*64
cap-and-trade programs, China's pilot programs, 21
carbon capture and sequestration (CCS) technologies: China's development of, 24; and technology transfer, 166; and U.S.–China cooperation, 173, 179, 185
carbon dioxide concentrations, 6
carbon dioxide emissions: and China's coal-based energy sources, 11; and China's energy-efficiency programs, 19, 21; China's inventories of, 15–16, 20–21; and Chinese energy intensity, 9; Chinese rate of, 9–10, *10*, 194*n*19, 194*n*22, 195*n*30; and concentrating solar power, 184; of United States, 9, 10, 15. *See also* greenhouse gas (GHG) emissions
carbon intensity: and assessing compliance, 20; and China's reduction targets, 14–15, 17, 18, 20–22, 46, 181, 198*n*64; and provincial targets, 21
carbon trading: and carbon leakage, 180; and international allowances, 180. *See also* cap-and-trade programs
Carnegie Endowment for International Peace, 177
certified emissions reductions (CER): and China's Clean Development Mechanism involvement, 16, 17, 55, 70; and development countries, 16
Chicago Climate Exchange (CCX), 21, 198*n*63
China: as centrally planned economy, 33, 44, 45, 47, 66–67; climate politics in, 11–18; domestic energy strategies of, 1; economic growth rates of, 5–6; GDP of, 8, 10, 14–15, 20, 24, 46; and Group of 77, 13–16; mean surface temperatures in, 6, 193*n*5; and Near Zero Emission Coal partnership, 24; and negotiating position of, 17–18, 169; science and technology development in, 44–47, 204*n*56; solar resources of, 185. *See also* U.S.–China cooperation; wind power industry of China
China Classification Society Industrial Corporation, 132
China Development Bank, 131
China Energy Conservation and Environmental Protection Group, 177
China General Certification Center (CGC), 56, 132
China Hydropower Engineering Consulting Group Corporation, 140
China Meteorological Administration, 12, 59
China National Offshore Oil Corporation (CNOOC), 127, 129
China National Petroleum Corporation Assets Management Co. Ltd. (CNPCAM), 198*n*63
China National Standardization Commission, 58–59, 74
China PV Test Center, 184
China's Decision on Accelerating Scientific and Technological Progress (1995), 44–45
China's Decision on Reform of the Science and Technology Management System (1985), 44
China Shipbuilding Industry Corporation Haizhuang Windpower Equipment Co. Ltd. (CSIC Haizhuang), 157, 161
China Sustainable Energy Program (CSEP), 178
China Wind Energy Association, 117
Chinese Academy of Engineering, 208*n*100
Chinese Academy of Sciences (CAS): and climate change policy, 13; and concentrating solar power, 184, 185; and science and technology development, 45, 205*n*64; and U.S.–China cooperation, 176–77
Chinese Academy of Social Sciences, 13
Chinese Patent Office (State Intellectual Property Office), 119
Chinese wind turbine manufacturers: expansion of, 138–41, 164, 165; experience overseas, 64, 64, 67, 119–20, 130, 138, 140, 141–44, 193;

funding support for, 56, 69, 70, 72, 209*n*113; government support for, 56–57, 108, 111, 139, 143, 163, 165–66, 181, 210*n*121; and home markets, 36, 107–8, *107*, *108*, 139; and joint-venture partnerships, 51–52, 61, 68, 78–79, 88–92, 99–100, 104–5, 106, 107–11, *108*; and learning networks, 63–64; and localization models, 40–41, 43, 52, 60, 73, 139, 163, 181, 204*n*54, 204*n*55; map of manufacturing facilities, *63*; number and ownership of, 61, *62*, 67; price of turbines, 132, 223*n*51; R&D expenditures of, 62; reputation of, 139–40, 144; and technology transfers, 64, 75, 79, 109–10, 111, 157; and wind bases, 54, 64, 72, 139; and wind concession projects, 58, 65, 82; and wind turbine cost, 65–66, *66*, 138–39; and wind turbine size, 60–61, 67, 73. *See also specific companies*

Chongqing XinXing Fengneng Investment Company, 104

Chu, Steven, 173, 174

City Group, 150

CKD NOVÉ Energo, 124, 162

Clean Air Task Force, Asia Clean Energy project, 177

Clean Development Mechanism (CDM): and certified emissions reduction credits, 16, 17, 55, 70; and China's climate change policy, 16; and wind power industry of China, 55, 70, 115, 215*n*60

clean energy technologies: bilateral collaboration on, 168–69, 171–77; and U.S.–China cooperation, 4, 168, 169–70, 171, 173, 174, 176, 177, 178, 180–81, 187

climate change: and China's energy system, 1; China's mitigation strategies, 11–12, 13, 17, 18, 21–22, 24–25, 168, 183; China's policy on, 3, 12, 13, 14, 15, 16, 18, 168, 169; developing nations' capacity for fighting, 199*n*79; and environmental trends, 6; and greenhouse gas emissions, 1–2, 190–91*n*2; impacts facing China, 6–8; international cooperation on, 3, 12–16, 17, 18, 165, 168, 169, 171; multilateral cooperation on, 169, 171, 186–87; and U.S.–China cooperation, 4, 171–78, 186–87, 229*n*10; and U.S. Congress's legislative proposals, 180, 231*n*32

Clinton, Hillary Rodham, 173, 229*n*10

Clipper Windpower, 215*n*65, 218*n*104

coal: and China's carbon dioxide emissions, 11; China's development of zero-emission coal technologies, 24; and China's energy system, 5, 10, 169; and China's Twelfth Five-Year Plan, 23; concentrating solar power compared to, 185; efficiency of China's coal power plants, 11, 19, 195*n*28; and United States, 169; and U.S.–China cooperation, 172, 173, 174, 176, 179, 183. *See also* carbon capture and sequestration (CCS) technologies

combined heat and power (CHP) systems, 141

competition: and China's innovation system, 46; and China's Ride the Wind Program, 51; and Danish wind turbine manufacturers, 77; and domestic environments, 156; and foreign-owned wind turbine manufacturers in China, 113, 120; and General Electric, 98, 115; global competition in wind power industry, 36, 148; and joint development, 158; and technology transfer from second- or third-tier wind power companies, 36, 157; and U.S. and Chinese solar manufacturers, 117; and U.S.–China cooperation, 178, 179–82; in wind power industry of China, 138

Composite Technology Corporation, 154

Comprehensive R&D Plan on Green Technology of South Korea, 153

concentrating solar power (CSP), and U.S.–China cooperation, 183, 184–85

Conference of the Parties to the United Nations Framework Convention on Climate Change (UNFCCC): COP 15, Copenhagen, 15, 17–18, 20; COP 16, Cancun, 15, 196*n*43; COP 17, Durban, 17, 18. *See also* Kyoto Protocol; United Nations Framework Convention on Climate Change (UNFCCC)

Costa Rica, 15

Costas Computer Technology A/S, 81

"cost-plus-profit" formula, and wind energy policy in China, 51

Cryscapital, 150

CSR Zhushou, 157, 161

Cuba, and Goldwind, 130, *132*

Cultural Revolution, 44

Customs Administration, 52, 69

Dabancheng, 63, 123

Daewoo Shipbuilding and Marine, 153–54, 155, 158
Dai Bingguo, 173, 229n10
DanControl, 77
Danish Wind Technology (DWT), 81
Dansk Staalvindue Industri, 80
Darrieus Turbine design, 80
Datang Renewable Power Co. Ltd., 54, 86
DEG, 92
Delft University of Technology, Netherlands, 126
Deng Xiaoping, 44
Denmark: and benefits of local manufacturing, 203n45; and common knowledge sources, 161; and delivery lead times, 203n48; and Gamesa, 86; innovation model of, 60, 211n130; and learning networks, 34, 148; skilled labor force of, 42; and Suzlon, 150, 226n13; windmills for electricity generation in, 27; wind power industry in, 36, 41, 76, 77–86, 113, 232n49; and wind power industry of China, 22, 51, 61, 75–76, 77, 78–79, 82, 93, 105–6, 107, 109, 134, 220n144; and wind turbine technology development, 28, 29, 31–32, 34, 36, 60, 159
developed nations: and greenhouse gas emissions, 14; and national innovation systems, 33; and technology transfer, 36, 145–46, 157. See also specific nations
developing nations: and certified emissions reduction credits, 16; China as source of technology transfer, 3, 135; China in spokesperson role for, 169; China's status as, 10, 14, 15, 16, 24; and greenhouse gas emissions, 15, 16, 166, 196n43; and Group of 77, 14; industrial strategy in, 147–48; innovation capacity of, 24, 199n79; and international reserve allowances, 180; and national innovation systems, 33; and technology transfer, 36, 145–46, 157, 166–67; and wind turbine manufacturers, 36. See also specific nations
DeWind, 154, 157, 158
Dongfang Electric Corporation (DEC): and common knowledge sources, 161; and exports, 141; and General Electric, 104; and home-country markets, 36; and joint development, 158; and licensing agreements, 157; and REpower, 221n12; as wind turbine manufacturer, 129, 135, 138

Doosan Heavy Industries, 153, 154, 158, 161
Double Increase Program, 51, 68–69
Draft Access Conditions for Wind Power Equipment Industry of China, 213n27
Durban Platform, 17

economic development: and China as developing nation, 10; and China's centrally planned economy, 33, 47; and China's transition to market economy, 47, 67, 112; and effects of climate change in China, 8, 10, 12, 25, 169; and greenhouse gas emissions, 8, 13; structure of Chinese economy, 21; and wind power technology, 126
Ecuador, 132
863 Program (National High Tech R&D Program) of China, 44, 56, 68, 69, 71, 126, 185
electric grid: and China's renewable energy surcharges, 53, 56, 208n98; grid integration technology, 127, 135, 140–41, 163, 183, 186; and U.S.–China cooperation, 176; and variable-speed turbines, 97; and wind energy policy in China, 50–51, 52, 53, 55–56, 58–59, 68, 70, 71, 72, 118, 140–41, 163, 206n82, 207n88, 210–11n127; and wind power industry of India, 149; and wind power industry of South Korea, 153; and wind turbine technology, 28
electricity: and coal power plants, 10, 195n28; and concentrating solar power, 184–85, 233n53; and hydropower, 10, 22; and power sector reform, 28; and U.S.–China cooperation, 173; windmills generating, 26–27; and wind turbine technology, 29
Electricity Act of India (2003), 149
electricity sector of China: annual electricity production targets, 22; carbon footprint of, 2; and General Electric, 93; surcharge on electricity rates, 53, 71, 207–8n96, 208n98; and technology transfer, 115; value-added tax on wind electricity, 56–57, 69; and wind power electricity price, 69, 233n64
Eleventh Five-Year Plan for National Economic and Social Development of the People's Republic of China (Eleventh Five-Year Plan): and energy-efficiency programs, 18, 19, 20; and government stimulus funding, 24; and renewable energy, 50; Science

Index   269

and Technology Plan, 46, 56; and solar energy technology, 185; and wind turbine technology, 71
Eleventh Five-Year Plan of India, 148
Eleventh Five-Year Renewable Energy Development Plan of China, 54, 72
Elin EBG Motoren GmbH, 151
emissions. *See* carbon dioxide emissions; certified emissions reductions (CER); greenhouse gas (GHG) emissions
emissions trading. *See* carbon trading
employment: and benefits of local manufacturing, 39–40; and China's science and technology development, 46; and Goldwind, 126, 131, 137, 159, 182; and skilled labor, 42, 126, 137, 159, 167; U.S. green-jobs initiatives, 181
Enercon: and home-country markets, 36; and patents, 97, 98; and wind power industry of India, 150
Energy and Environment Cooperation Initiative (EECI), 172
Energy Conservation Law of the People's Republic of China (2007), 19
energy consumption in China: and energy-efficiency programs, 18–19; and energy intensity, 9; and greenhouse gas emissions, 8; implementing cap on, 18
energy efficiency: China's gains in, 169; China's intensity targets, 18–22; China's policies on, 3; energy savings technologies, 23, 46; and United States, 169; and U.S.–China cooperation, 172, 173, 174, 176, 178, 179, 183; and wind power industry of China, 141
Energy Foundation, 178
energy intensity (ratio of energy consumption to GDP): China's decline in, 8–9, 21; and China's domestic carbon-intensity target, 14–15, 17, 18, 20, 21, 22, 46, 181, 198n64; and China's energy-efficiency programs, 18–22; and energy consumption, 9
energy policy: and China's climate change policy, 12, 13, 14, 15, 16, 18, 168, 169; and China's policy on foreign technology firms, 111–14, 117–18, 119; and China's tax policies, 19–20, 56–57, 69; China's wind energy policy regime, 49–59, 68, 70, 71, 72, 75, 83, 89, 90, 105, 108–9, 110, 111–14, 117–18, 119, 134, 138, 140–41, 143, 162, 163, 165–66, 206n79, 206n82, 207n88, 210–11n127; China's wind energy policy timeline, 50–59, 68–74; India's wind energy policy regime, 162, 163, 165–66; South Korea's wind energy policy regime, 162, 164, 165–66; and U.S.–China cooperation, 183; U.S. policy on climate change, 187; U.S. policy on wind industry, 10, 80, 94, 116, 142, 180–81
Energy Research Center of the Netherlands, 184
Energy Research Institute, 13, 140, 178, 189
energy sector of China: governmental jurisdiction over, 50; and greenhouse gas emissions, 1, 13; R&D investment in China, 48, 49, 205n73; and science and technology development, 46
Energy System Taranto S.p.a., 82
energy technologies: deployment rates of, 21, 64–65. *See also* clean energy technologies; low-carbon energy technology; low-emissions technologies; renewable energy technologies; solar energy technology; wind power technology
Energy Vision 2030 plan of South Korea, 152
Enerwind, 124, 161
engineering capacity of China: and benefits of local manufacturing, 40; and climate change mitigation, 24; government support for, 3, 45, 126, 140, 158, 221n24; history of, 3, 4, 22–25, 44–47; and joint development, 158; and learning networks, 3, 134–35, 159; and overseas expansion, 130; technology acquisition strategies, 158, 162; and wind power development targets, 208n100
Enron Corporation, 94–95
Enron Wind, 82, 93, 94, 95, 97–98, 106
Enron Wind Rotor Production B.V., 151
Enterprise Income Tax Law, 57
Environmental Defense Fund, 177
environmental protection technologies, 23, 129
environmental trends, and China's economic growth, 5–6
Eozen, 124, 162
Ernst and Young, renewable energy "country attractiveness" index, 23
European Union (EU): and Canada, 117; and clean energy collaboration, 168; and intellectual property rights, 98; and Near

European Union (*continued*)
  Zero Emission Coal partnership, 24; science and technology development in, 44, 204*n*56; and Suzlon, 151, 159–60; and wind turbine technology, 28, 29, 75, 159, 163
exports: and benefits of local manufacturing, 39; and China's carbon dioxide emissions, 11, 195*n*30; and China's energy-efficiency program, 19–20; and Chinese wind turbine manufacturers, 64, *64*, 67, 119–20, 138, 139, 141–44, 163, 165; and Goldwind, 130–32, 134–35, 137, 141; and technology transfer, 157; and wind power industry of India, 165; and wind power industry of South Korea, 155, 159, 164

feed-in tariffs: and China's price support for wind energy, 38, 113, 162; China's regional levels of, 55, 72; for China's solar photovoltaics, 182; cost of equipment and raw materials reflected in, 208*n*106; NEA approval for, 74; for renewable electricity in China, 53, 71; and renewable energy technology, 22; and wind concession projects in China, 52; and wind power industry of India, 149; and wind power industry of South Korea, 152–53; and WTO disputes, 117
First Wind, 131
foreign investment: and Chinese Clean Development Mechanism projects, 17; and international trade tensions, 114; and technology transfers, 49; and wind power industry of China, 51; and wind power industry of India, 150
foreign technology firms: Chinese policy environment for, 111–14, 117–18, 119; and Goldwind, 125, 134; and intellectual property rights, 38–39, 108, 119; and international trade tensions, 114–18; and jointly owned enterprises, 2, 4, 51, 58, 118–19; and learning networks within Chinese wind power industry, 3, 62–63, 119; ownership structures of, 112; and technology transfer, 35, 105–11, 112, 114, 115, 118, 119, 148; and wind power industry of China, 57, 58, 61, 62, 67, 70, 73, 75, 105–20, *108*, 210*n*123, 211*n*131; and wind power industry of India, 150, 226*n*11; and wind power industry of South Korea, 156; and wind power technology, 4; as wind turbine manufacturers, 75, 113–14, 150, 226*n*11. *See also specific firms*
fossil fuels: China's reliance on, 18, 24; and U.S.–China cooperation, 172, 185. *See also* coal
Fuhrlander, 104, 135, 157, 161
FutureGen project, 179

Gamesa: history of, 84–85; and technology transfer, 76; timeline of company structure in China, *76*; and Vestas, 36, 81, 85, 202*n*36, 214*n*34; and wind power industry of China, 76, 86, 106; and wind power industry of India, 86, 150
Gamesa Eólica, 81, 85, 202*n*36
Gandasegui, Lopez, 214*n*34
Gao Hu, 140
GaoKe, 224*n*59
Garrad Hassan, 126, 154, 158
GE China Technology Center, 102, 103
GE Drivetrain Technologies, 104
General Electric (GE): and Enron Wind, 95, 97–98; and home-country markets, 36; and intellectual property rights, 97–98, 102, 104; and patents, 97–98, 103–4, 225*n*68; and technology transfers, 98–100, 115; and wind power industry of China, 93–105, 106, 107, 109–10, 112, 115, 116; and wind power industry of India, 150; as wind turbine manufacturer, 32, 82, 93, 94, 96, 104, 112
generation-based incentive (GBI) scheme, 149
Germanischer Lloyd, 132
Germany: and Daewoo, 154; and Enron Wind, 95; and Gamesa, 86; and Goldwind, 131, 159; and innovation, 60; and Samsung, 154; skilled labor force of, 42; and Suzlon, 150, 152, 227*n*17; and Vestas, 81; wind power industry in, 86–87, 96, 114; and wind power industry of China, 51, 86–92, 93, 106, 107, 109, 134; and wind power industry of India, 150; and Windtec, 161; and wind turbine technology development, 28, 29, 31, 38, 152
Ghodawat Energy, 150, 161
Global Wind Energy Council (GWEC), 190*n*1
Global Wind Power, 150, 161
Golden Concord Wind Power, 124–25, 133, 221*n*17
Goldwind: corporate culture of, 121, 137; and deployment of wind turbines, 128–29;

development of wind power technology, 4; expansion of, 138–41; and General Electric, 104; government support for, 121, 143; history of, 122–28, 142–43; and home-country markets, 36, 159; installed wind capacity in China, 122; and joint development, 121, 124–25, 158, 163; and learning networks, 126–27, 132–37, 158, 159; and licensing agreements, 121, 124, 133, 143, 147, 157, 163; and mergers and acquisitions, 121, 133, 157, 158; and outlook of wind power industry of China, 138–42, 144; and overseas markets, 130–32; price of turbines, 132, 223n51; profits and losses of, 123, 220n8; and quality, 130, 132, 138, 228n40; and REpower Systems, 110, 124, 127, 161; and research, development, and demonstration plans, 123, 125–28, 221n20; and technical adaptations for Chinese market, 204n55, 217n103; technology acquisition of, 121, 123–25, 147, 163, 221n18; timeline of company structure in China, 125; and U.S.–China cooperation, 182; and wind farm development, 122, 128–29, 220n5; as wind turbine manufacturer, 112, 121, 122, 125, 127, 128, 129–30, 139, 143, 159

Goldwind University, 126–27
Gore, Al, 172
Goutou, 78, 106, 109
GreenGen, 24
greenhouse gas (GHG) emissions: absolute targets for, 17; and China's climate change policy, 13, 15, 168; and China's domestic carbon-intensity target, 14–15, 20, 21, 22, 198n64; China's emissions challenge, 8–11, 17, 168, 169, 170, 171, 232n48; and China's energy-efficiency program, 20, 169; and China's energy sector, 1, 13; China's footprint, 10; China's monitoring system for, 20, 21; China's per capita emissions, 11; China's reducing of, 183; and data quality and transparency, 15, 21; and developed nations, 14; and developing nations, 15, 16, 166, 196n43; global and local benefits of reduction in, 6; increase in absolute emissions, 21–22; and international technology transfers, 166; South Korea's pledges on, 152; stabilizing, 190–91n2; and United States, 169, 170, 171, 180; and

U.S.–China cooperation, 187; and wind power industry of China, 55; wind power technology, 1–2, 190–91n2. *See also* carbon dioxide emissions; certified emissions reductions (CER)

GreenHunter Energy of Texas, 142
Green New Deal Stimulus Package of South Korea, 153
Greenpeace, 184
Group of 77 (G-77), 13–16
Guangdong Huilai wind concession project, 128
Guangdong Nuclear Wind, 86, 128
Guodian (Longyuan Electric Group) Power Company, 54
Guodian United, 141, 157, 161, 165
Guohua Power Company, 54

Haizhuang. *See* China Shipbuilding Industry Corporation Haizhuang Windpower Equipment Co. Ltd. (CSIC Haizhuang)
Hangzhou Industrial Asset Management Co. Ltd., 101
Hanjin, 153, 154, 158
Hankuk Glass Fiber, 155
Hansen, 151, 226n16
Hansen, H. S., 80
Hansen, Peder, 80
Harakosan Europe BV, 158
Harbin Electric Machinery Company (HEC), 104–5
Harbin Power Equipment Company (HPEC), 99–100, 105
Hewind, 141, 158, 161
high-technology development zones, 44–45
Honeywell, 177
Huachuang, 141
Huadian Power Company, 54
Huaneng Power Company, 24, 54
Huide, 141
hybrid and electric vehicles: and China's Twelfth Five-Year Plan, 23; and U.S.–China cooperation, 174
hydropower: China's capacity in, 22; China's development of, 22, 23; and China's electricity generation, 10, 22; China's manufacturing capacity for, 24; and General Electric, 98, 100–101, 216n80
Hyosung, 153, 154

272   Index

Hyundai Heavy Industries Co. Ltd., 153, 154, 155, 157, 158, 161, 165

Idaswind, 154, 158
IMPSA, 124, 161
India. *See* wind power industry of India
industrial activity: and China's energy-efficiency technology, 18, 19–20; and China's greenhouse gas emissions, 8, 10, 11; China's industrial-restructuring programs, 18; and China's renewable energy technologies, 23; and China's Twelfth Five-Year Plan, 23–24, 23; and China's wind energy policy, 50; and Chinese science and technology development, 46; energy consumption of, 11, 19, 195n29
Inner Mongolia. *See* Inner Mongolia Autonomous Region (IMAR)
Inner Mongolia Autonomous Region (IMAR), 54, 72, 83, 93, 130, 141, 177, 185, 204n55
innovation: and Asian "late industrializing countries," 33, 147; China's capacity for, 24, 119; China's innovation system, 2, 3–4, 43–49, 45, 50, 56, 66–67, 104, 119, 147, 162, 211n130; and China's Twelfth Five-Year Plan, 24; and clean energy development, 180; and Denmark, 60, 211n130; geographically specific hubs of, 34; global model of, 33; goals for, 42; and Goldwind, 143; and learning networks, 34–35, 62–64, 67, 158; and multinational corporations, 33; and technological latecomers, 35–36, 167; and technology transfer, 35–36; and United States, 60, 211n130; and Vestas, 81; and wind turbine manufacturers, 31–36
Inox Wind, 150, 161
integrated gasification combined cycle (IGCC) plants, 24
intellectual property rights (IPR): and Chinese science and technology development, 47–49; enforcement of, 223n53; and foreign technology firms, 38–39, 108, 118, 119; and Gamesa Eólica, 85; and General Electric, 97–98, 102, 104; and Goldwind, 124; and innovation, 2, 3, 67; and joint development, 158; and joint-venture partnerships, 109; and Nordex, 90; and protectionism, 225n68; and technology transfer models, 35, 110,

111, 113, 115, 165–66, 167; and U.S.–China cooperation, 174, 230n15; and wind power industry of India, 146; and wind power industry of South Korea, 146
Intergovernmental Panel on Climate Change, 13
Interim Measures on Renewable Energy Electricity Prices and Cost Sharing Management (2006), 53, 71, 207–8n96
Interim Measures on Revenue Allocation from the Renewable Surcharge (2007), 53, 71, 208n98
Interim Measures on the Management of Special Project Funds for the Industrialization of Wind Power Generation Equipment (2008), 56, 72, 209n113
international cooperation: on climate change, 12–16, 17, 18, 165, 168, 169, 171; and National Energy Commission, 49. *See also* U.S.–China cooperation
International Finance Corporation, 126
international reserve allowances, 180, 231n32
IPCC Special Report on Renewable Energy, 65
ISO 9001 certification, 132
Italy: and Suzlon, 150; and Vestas, 82; and wind turbine technology, 31

Jacobs Energie, 123, 157
Japan: and Canada, 117, 142, 144; and clean energy collaboration, 168; and Nordex, 87; wind power industry in, 162, 228n39; and Windtec, 161; and wind turbine technology, 29, 31, 142, 162
Jeju Island, South Korea, 152, 153, 154
Jiangsu Unipower Wind Power Co. Ltd., 213n19
Jiang Zemin, 172
joint development: and Goldwind, 121, 124–25, 158, 163; and technology transfer models, 3, 35, 135, 147, 158, 160–61, 166
Joint U.S.–China Collaboration on Clean Energy (JUCCCE), 177
joint-venture partnerships: and Chinese wind energy policy, 51, 58, 68, 89, 90, 105, 108, 112; and Chinese wind turbine manufacturers, 51–52, 78–79, 88–92, 99–100, 104–5, 106, 107–11, *108*; and foreign technology firms, 2, 4, 58, 61, 107–11, 112, 117, 215n57; and technology transfer models, 35, 51, 81, 90, 99–100, 106, 109–10, 111, 113, 118–19, 135, 147,

167; and U.S.–China cooperation, 179; and wind power industry of India, 151

Karakosan Europe BV, 155
Kenersys, 150
Kenetech Corporation, 93–94, 95, 97
Kerry, John, 180
Kirk, Ron, 116
Korea Energy Management Corporation (KEMCO), 153
Korean Wind Industry Association, 152
Kvaerner Power Equipment Co. Ltd., 100, 216n86
Kyoto Protocol, 14, 16, 17, 55, 70, 209n108, 209n110

Lagerwey, 161
Lawrence Berkeley Laboratory, 178
leapfrogging. *See* technological leapfrogging
learning networks: access to, 3, 62–64, 134–35; and Chinese wind turbine manufacturers, 63–64; global learning networks, 158–60, 167; and Goldwind, 126–27, 132–37, 158, 159; informal learning networks, 119; and innovation, 34–35, 62–64, 67, 158; and technology acquisition, 158–60; and wind power industry, 147–48; and wind power industry of India, 34–35, 148; and wind power industry of South Korea, 34, 148, 159; and wind turbine manufacturers, 31, 64, 159
licensing agreements: and Goldwind, 110, 121, 124, 133, 143, 147, 157, 163; and Hyundai Heavy Industries, 154; and Suzlon, 151, 159, 160, 164; and technology transfer models, 3, 35, 36, 110, 133, 135, 147, 156–57, 160, 161, 165–66, 167, 224n59
Lieberman, Joseph, 180
Li Junfeng, 140
Li Peng, 172
Little Pringle project, Texas, 154
LM Glasfiber, 150, 154
LM Glasfiber Tianjin, 89
local content requirements: and China's wind industry development, 51–52, 68–70, 101, 139, 165; and Chinese firms, 123; and foreign firms, 89, 91, 92, 101–2, 106, 110–13, 151, 207n87; as protectionist policy, 225n68; and U.S.–China trade disputes, 73, 115–16, 142; and WTO, 57, 113, 115, 142

localization models: definition of localization, 38–39, 43, 204n54, 204n55; and research, development, and demonstration plans, 38, 69; and technology transfer, 43, 105–11, 113, 114, 115; and wind turbine manufacturers, 39–41, 43, 106, 109, 110, 111–12, 113, 115–16, 142, 165, 203n45

Locke, Gary, 57, 181

low-carbon economy: and China's renewable energy resources, 22; China's role in global transition to, 2, 3–4, 191n4; low-carbon pilot provinces and cities, 20–21; and U.S.–China cooperation, 171, 187

low-carbon energy technology: and China's climate change mitigation, 24–25; and China's industry clusters, 165; and Chinese science and technology development, 46; and Clean Development Mechanism, 16; development of, 3–4; and technology transfer, 146, 165–66; and U.S.–China cooperation, 183, 186, 187

low-emissions technologies, 2, 24
low-voltage ride-through (LVRT), 127
Luoyang First Tractor Factory, 51, 215n48

Made, 51, 85–86, 214n37, 215n48
Management Rules on the Administration of Power Generation from Renewable Energy, 55
mandatory market share (MMS), 53
MBB, 94
McKinsey & Company, 184, 232n48
Medium- and Long-Term Plan for Renewable Energy Development in China (2007), 50, 53–54, 59, 71–72
Memorandum of Understanding to Enhance Cooperation on Climate Change, Energy, and Environment (2009), 173
memorandums of understanding (MOUs), 176, 177, 178
mergers and acquisitions (M&A): and barriers to entry, 41; and Goldwind, 121, 133, 157, 158; and technology transfer models, 3, 35, 135, 147, 157–58, 166, 167; and Vensys Energiesysteme GmbH, 147, 157, 158, 163; and wind power industry of South Korea, 155–56

Micon, 76, 77, 161
Ministry of Energy of the People's Republic of China, 50, 206n79
Ministry of Environmental Protection (MEP) of the People's Republic of China, 13, 49, 173
Ministry of Finance (MOF) of the People's Republic of China: and Chinese wind power electricity price, 69; and Chinese wind turbine manufacturers, 56, 57, 72, 74; and climate change policy, 13; and export taxes, 19; and government subsidies, 50, 115, 116, 117, 139; and National Energy Commission of the People's Republic of China, 49; Special Fund, 116; and U.S.– China cooperation, 173
Ministry of Foreign Affairs (MFA) of the People's Republic of China, 12–13, 70, 173
Ministry of Industry and Information Technology (MIIT) of the People's Republic of China, 50, 57, 73, 213n27
Ministry of Land and Resources of the People's Republic of China, 57
Ministry of Machinery of the People's Republic of China, 51
Ministry of New and Renewable Energy (MNRE), India, 149
Ministry of Non-Conventional Energy Sources (MNES), India, 149
Ministry of Science and Technology (MOST) of the People's Republic of China: and climate change policy, 12, 13; and National Energy Commission, 49; and U.S.–China cooperation, 172, 173, 174; and wind energy policy, 206n82; and wind energy R&D expenditures, 56, 68, 69
Minyang: and exports, 141, 142, 165; and home-country markets, 36; R&D expenditures of, 62
Mitsubishi: and patents, 98; as wind turbine manufacturer, 37, 162
multilateralism: China's support for, 169; and stalled negotiations, 171; and U.S.–China cooperation, 117, 186–87
multinational corporations: effect on national innovation systems, 33; and international trade tensions, 115; technology development strategies of, 33; and wind power industry, 94, 105, 106, 107

Nan'ao, 63
National Climate Change Assessment Report of the People's Republic of China, 6, 13
National Climate Change Program of the People's Republic of China, 13
National Conference on Science and Technology of 1978, 44
National Coordination Committee on Climate Change (NCCCC) of the People's Republic of China, 12
National Development and Reform Commission (NDRC) of the People's Republic of China: and carbon-trading programs, 20–21; and climate change policy, 12–13; Department of Climate Change, 12; Energy Research Institute of, 13, 140, 178, 189; establishment of, 206n77; Measures for the Administration of Offshore Wind Power Development, 58, 73; and National Energy Commission, 49–50; Notice on Abolishing the Localization Rate Requirement for Equipment Procurement in Wind Power Projects, 57, 73; Notice on Improving Grid-Connected Wind Power Tariff Policy, 55, 72; Notice on the Relevant Requirements for the Administration of the Construction of Wind Farms, 52; and solar feed-in tariffs, 182; and surcharge on electricity rates, 207–8n96; and U.S.–China cooperation, 172, 173; and wind power industry of China, 52–53, 69, 71, 207n88
National Energy Administration (NEA) of the People's Republic of China, 18, 49, 58, 74, 173, 174
National Energy Bureau of the People's Republic of China, 56
National Energy Commission (NEC) of the People's Republic of China, 49, 206n76, 206n77
National Engineering Research Centers (NERCs), in China, 45
National Guidance Catalogue for Renewable Energy Industry Development, 57, 70
National High Tech R&D Program (863 Program) of China, 44, 56, 68, 69, 71, 126, 185
National Leading Committee on Climate Change, 12
National Natural Science Foundation of China, 45

Index 275

National Program for Medium-to Long-term Scientific and Technological Development (2006–20) of the People's Republic of China, 185
National Renewable Energy Laboratory, 140
National Wind Power Technology Engineering Research Center, 126, 221n24
National Wind Technology Center of the National Renewable Energy Laboratory, 186
Natural Resources Defense Council (NRDC), 177
Near Zero Emission Coal partnership, 24
Nedwind, 77
NEG Micon: history of, 76–78; timeline of company structure in China, 76; and wind power industry of China, 76, 77–79, 106, 109
NEG Micon Goutou Wind Turbine Co. Ltd., 78
NEPC, 161
Netherlands: and solar technology, 184; and Suzlon, 150, 152; windmills of, 32; and wind power industry of China, 77; and wind turbine technology development, 28, 29, 31, 32, 152, 155, 159
New United, 136, 141
Ningxia Electric Power Group, 215n57
Ningxia Tianjing Electric Energy Development Group, 215n57
Ninth Five-Year Plan for Economic and Social Development of the People's Republic of China (Ninth Five-Year Plan), 52, 56, 68
nongovernmental organizations (NGOs): and international climate negotiations, 18; and U.S.–China cooperation, 171, 176, 177–78
Nordex: history of, 86–87; and home-country markets, 36; and joint-venture partnerships, 51; timeline of company structure in China, 88; and wind power industry of China, 87–92, 91, 92, 106, 109–10, 111, 112; as wind turbine manufacturers, 86
Nordtank, 76, 77–78
Norwin, 135, 158, 161, 224n59
Notice on Strengthening the Management of Wind Power Plant Grid Integration and Operation, 58, 74
nuclear power: China's use of, 10, 22, 23; and U.S.–China cooperation, 172

Obama, Barack, 15, 168, 173, 174
offshore wind development in China: administration of, 73; and concession projects, 58, 74; costs of, 40; and General Electric, 103, 104–5; and Goldwind, 124–25, 127, 129, 130, 133, 134, 143; and grid interaction technology, 42; and wind energy policy, 53, 56, 58, 59, 71, 73, 105, 211n128
O'Leary, Hazel, 172
Olympic Games of 2008 (Beijing, China), 128–29, 143
onshore wind development in China: costs of, 40; and Goldwind, 127; and grid interaction technology, 42; and Vestas, 83; and wind energy policy, 58, 59
Optimal Speed Controller (OSC) technology, 77
Organization for Economic Development (OECD), wind energy R&D expenditures in countries of, 29, 29, 31

Pacific Northwest Laboratory, 178
Pakistan, and Goldwind, 131, 132
Papua New Guinea, 15
patents: and Canada, 98; China's policy on, 49, 67, 119; countries leading in wind energy patents, 47; and Enercon, 97, 98; and General Electric, 97–98, 103–4, 225n68. See also intellectual property rights (IPR)
Paulson, Henry, 172
Pedersen, Carsten, 89
Pena, Federico, 172
permanent magnet generators, 43, 123, 127, 132, 204n51
PetroChina, 176, 177
Pioneer Wincon, 150
Pohang University of Science and Technology (POSTECH), 155
Poland, and Goldwind, 131
Portugal: and Goldwind, 131; and Suzlon, 150; wind power industry in, 142
Poulsen, Johannes, 80, 81
power purchase agreements (PPAs), 94
production tax credits (PTCs), 96
Protocol for Cooperation in the Fields of Energy Efficiency and Renewable Energy Technology Development and Utilization (1995), 172

Protocol for Cooperation on a Clean Energy Research Center, 174
Protocol on Cooperation in the Field of Fossil Energy Research and Development (1985), 171
Protocol on Nuclear Physics and Magnetic Fusion (1983), 171
Provisional Management Methods for Wind Power Forecasting, 58, 74
Provisions for Grid-Connected Wind Farm Management (1994), 50–51, 68

Qingdao Huawei Wind Power Co. Ltd., 91–92
Qingdao Institute of Bioenergy and Bioprocess Technology, 176–77

R&D: and barriers to entry, 41; and China's Twelfth Five-Year Plan, 24; and Chinese science and technology development, 45, 46, 47–49, 48, 205n73; and Chinese solar technology expenditures, 183; and Chinese wind energy expenditures, 44, 56, 57, 59, 62, 67, 68, 70, 134, 158; collaboration on, 166; and Denmark, 232n49; and foreign technology firms, 67, 119; and Gamesa, 85, 86; and General Electric, 102, 103, 217n92; and generation of knowledge, 34; and Goldwind, 122, 124, 126, 144, 159; and Hyosung, 154; and Indian wind energy expenditures, 149; and learning networks, 159; and National Energy Commission, 49; and national innovation systems, 33; and South Korean wind energy expenditures, 153, 154, 158; and Suzlon, 150, 151–52, 159, 160, 164, 227n23; and U.S.–China cooperation, 173, 179, 183; and Vestas, 81, 84; wind energy expenditures in OECD countries, 29, 29, 31, 32; and wind turbine technology, 27–29, 32, 37
ReGen Powertech, 124, 150, 161–62
regulatory enforcement, China's system for, 24, 58–59, 165
Renewable Energy Development Fund, 56
Renewable energy development plans, 23, 50, 53–54, 59. *See also Medium- and Long-Term Plan for Renewable Energy Development in China* (2007)
Renewable Energy Law of the People's Republic of China (2005): amendments of 2009, 22, 55–57, 73; and China's low-carbon development strategy, 22; and China's surcharge on electricity rates, 53, 71, 207–8n96, 208n98; provisions of, 53, 70–71
renewable energy resources: China's development goals, 3–4, 22, 53–54, 71–72, 73, 169, 183, 208n100; and China's electricity generation, 10; and China's surcharge on electricity rates, 53, 71, 207–8n96, 208n98; of United States, 4, 169, 183; and wind power resources of South Korea, 152, 162
renewable energy technologies: China's promotion of, 18, 22–24, 23, 53, 183, 187; countries leading in patents, 47, 49; market for, 4; precommercial technologies, 183, 185; transmission infrastructures for, 169; and U.S.–China cooperation, 172, 174, 176, 177, 178, 183. *See also* solar energy technology; wind power technology
Renmin University, 13
REpower Systems: and common knowledge sources, 161; and Gamesa Eólica, 85; and Goldwind, 110, 124, 127, 161; and home-country markets, 36; and licensing agreements, 135, 157; and Suzlon, 151, 157, 158, 161, 164, 226n16; and wind power industry of China, 110, 123, 221n12
research, development, and demonstration (RD&D): and global learning networks, 167; and Goldwind, 123, 125–28, 221n20; and localization of Chinese wind power industry, 38, 69
Ride the Wind Program of China, 51, 68, 89, 111, 215n48
Romax, 154, 158
Rosenzweig, Tim, 131
Rotem, 153, 154, 158
RRB, 82, 150
Russia, and Vensys, 131

Samsung Heavy Industries, 153, 154, 155, 158, 164–65
SBW, 161
science and technology (S&T) metrics, 3, 44–47
Scotland, 86
Second National Communication on Climate Change of the People's Republic of China, 13, 16

Sewind: and common knowledge sources, 161; and exports, 141; and home-country markets, 36; and joint development, 158
Shady Oaks wind farm, Illinois, 130
Shale Gas Resource Initiative, 174
Shanghai Electric, 141
Shenyang Blower Works (Group) Co. Ltd., 157
Shenyang Liming Aero-Engine Group Corporation, 99–100
Shenyang Power Group, 142, 180–81
Shi Pengfei, 117, 140
Siemens: acquisition of Bonus, 76; and home-country market, 36; and wind power industry of China, 109; and wind power industry of India, 150; as wind turbine manufacturers, 94
Sigaard, Svend, 213n20
Sino-Danish Renewable Energy Development Program, 84
Sinovel: and AMSC IP dispute, 117–18; and common knowledge sources, 161; and exports, 141, 164; and General Electric, 104; and home-country markets, 36; and joint development, 158; and licensing agreements, 157, 161; as wind turbine manufacturer, 118, 129, 135, 138, 139, 164
Six Sigma certification system, 101
solar energy technology: and China's Twelfth Five-Year Plan, 23; concentrating solar power, 183, 184–85, 233n53; development of, 166, 183–84, 233n52; and international trade tensions, 117; localization models for, 115–16; and United States, 169, 181, 183, 184–85; and U.S.–China cooperation, 177, 181–82, 183, 184–85
solar hot water heaters, 24
solar photovoltaic technology: China's manufacturing capacity for, 24, 182, 183–84; and U.S.–China cooperation, 177, 181, 183–84
Solyndra, 181
South Africa: and clean energy collaboration, 168; and Goldwind, 130, 132; and Group of 77, 15
South Korea. *See* wind power industry of South Korea
South Korean Ministry of Knowledge Economy, 154

Spain: and Goldwind, 131; and innovation, 60; and Suzlon, 150; and Vestas, 81; wind power industry in, 142; and wind turbine technology, 31, 38, 84–86, 202n36
Specter, Arlen, 180
State Administration of Taxation of the People's Republic of China, 57
State Council of the People's Republic of China: and Clean Development Mechanism, 16, 70; and climate change policy, 12; price control department of, 207n88; and renewable energy technologies, 24, 70–71; and wind turbine manufacturers, 57
State Development and Planning Commission (SDPC) of the People's Republic of China: and climate change, 12; electric grid, 206n82; and renewable energy policy, 206n77; Ride the Wind Program, 51, 68, 89, 111, 215n48; subsidy program for wind power developers, 91
State Economic and Trade Commission (SETC) of the People's Republic of China, 51, 68–69, 206n77
State Electricity Regulatory Commission (SERC) of the People's Republic of China, 50
State Environmental Protection Administration (SEPA) of the People's Republic of China, 12, 14
State Forestry Administration of the People's Republic of China, 173
State Grid Energy Research Institute of the People's Republic of China, 84, 140, 190n1
state-owned enterprises (SOEs), 23, 46
STX Corporation, 153, 154–55, 158
Subsidies and Countervailing Measures (SCM) Agreement, 116. *See also* World Trade Organization (WTO)
Sudwind Energy GmbH (Sudwind), 86, 150–51, 157
Suntech, 182
Su Wei, 12
Suzlon: and global markets, 164–65; and Goldwind, 124; and home-country markets, 36, 159, 160, 164; international headquarters of, 42–43; and learning networks, 159–60; and licensing agreements, 151, 159, 160, 164; and mergers and acquisitions, 157, 158, 164; and R&D, 150, 151–52, 159, 160, 164, 227n23;

278   Index

Suzlon (*continued*)
    and technology transfer, 157, 161; and wind power industry of China, 62, 109, 150, 151, 227*n*19; and wind power industry of India, 149–52, 159, 160
Suzlon Energy Limited Tianjin, 227*n*19
Suzlon Generators, 151
Sweden, 28, 81
SWL, 150

Tacke Windtechnik, 95
Taiwan, 161
Tang Energy, 222*n*40
Tanti, Tulsi, 150
tariffs: and barriers to entry, 43; and energy-efficiency program, 19–20; and wind power industry of China, 52, 53, 55, 57, 69, 74; and wind power industry of India, 149. *See also* feed-in tariffs
taxation in China: carbon taxes, 20; export taxes, 19–20; value-added tax on wind electricity, 56–57, 69
taxation in India, and wind power industry of India, 163
taxation in South Korea, and wind power industry of South Korea, 152
Technical Barriers to Trade Agreement, 43
technological achievement, national pride in, 40
technological leapfrogging: China as example of, 2; and international technology transfers, 166; and technological latecomers, 35, 167; and wind power industry of South Korea, 155, 162, 164
Technology Management Plans (TMPs) of the U.S.–China Clean Energy Research Center, 174, 230*n*15
technology transfer models: and common knowledge sources, 160–62; and developing countries, 147–48; and joint development, 3, 35, 135, 147, 158, 160–61, 166; and joint-venture partnerships, 35, 51, 81, 90, 99–100, 106, 109–10, 111, 113, 118–19, 135, 147, 167; and licensing agreements, 3, 35, 36, 110, 133, 135, 147, 156–57, 160, 161, 165–66, 167, 224*n*59; and mergers and acquisitions, 3, 35, 135, 147, 157–58, 166, 167; risk in, 100; synergies of, 2–3; technology development models, *136–37*, 147; and wind power industry of China, 2, 135, 147, 156; and wind power industry of India, 156; and wind power industry of South Korea, 156
technology transfers: China as recipient of, 2, 64, 75, 78, 79, 90, 98–100; China as source of, 2, 64, 135; and domestic environments, 162; evolving nature of, 1, 4; facilitating international transfers, 166; and foreign investment, 49; and General Electric, 98–100, 115; and Goldwind, 121, 123–25, 133; and innovation, 35–36, 147; and localization models, 43, 105–11, 113, 114, 115; and low-carbon energy technology, 146, 165–66; national security implications of, 114, 218*n*109; networks of, *160*; and partnerships with foreign firms, 35, 145–46; South Korea's strategies for, 155; and Vestas, 81, 83, 85
Tecnometal, 161
telecom industry of China, 23
10 Key Projects Program of China, 19
Tenth Five-Year Plan, 69
Texas, 142, 154, 165, 180–81
Three Gorges Project, 22
Tianjin Climate Exchange (TCX), 21, 198*n*63
Tianjin Economic-Technological Development Area (TEDA), 83
Tianrun Investment, 220*n*5
Top 1,000 Program of China, 18–19
Top 10,000 Program of China, 20
Track II U.S.–China dialogues, 177
trade system: and barriers to entry, 43, 57, 142, 143–44, 225*n*64, 225*n*68; and China's advances in clean energy, 1; and effects of climate change in China, 8; international trade tensions, 114–18, 143; and U.S.–China cooperation, 180, 181, 182. *See also* exports; tariffs; World Trade Organization (WTO)
transparency practices, 21, 62
transportation, and U.S.–China cooperation, 174, 176
Tsinghua-MIT Low Carbon Energy Research Center, 178
Tsinghua University, 13, 84, 119
Turkey, 161
TUV-Nord, 132
Twelfth Five-Year Plan for National Economic and Social Development of the People's Republic of China (Twelfth Five-Year Plan):

and energy-efficiency program, 20; and green and low-carbon development, 2, 191n4; new strategic and emerging industries in, 23–24, 23; and renewable energy resources, 23, 50

Uilk wind farm, Minnesota, 130, 131, 135, 164, 222n40
Unison Corporation, 153, 155, 158, 164
United Kingdom: and Near Zero Emission Coal partnership, 24; wind power industry in, 77; and wind turbine manufacturers, 106; and wind turbine technology, 29, 31
United Nations: and Group of 77, 14; and multilateral forums, 169
United Nations Framework Convention on Climate Change (UNFCCC): and China's climate change policy, 12, 13, 14, 15, 16, 18, 169; and greenhouse gas emissions reports, 20; and Kyoto Protocol, 16. *See also* Conference of the Parties to the United Nations Framework Convention on Climate Change (UNFCCC)
United Power, 36
United States: carbon dioxide emissions of, 9, 10, 15; climate change legislative proposals, 180, 231n32; climate change policy of, 187; coal power plants of, 195n28; and Daewoo, 154, 155; early windmills for electricity generation, 26; and Gamesa, 86; and Goldwind, 130, 131–32, 134, 137, 143, 159, 164, 182, 222n40; greenhouse gas emissions of, 169, 170, 171, 180; and Hyundai, 154, 155; and innovation, 60, 211n130; and learning networks, 148; and Nordex, 87; policy on wind industry, 10, 80, 94, 116, 142, 180–81; renewable energy resources of, 4, 169, 183; Renewable Portfolio Standards, 53; and Samsung, 154, 155, 165; skilled labor force of, 42; and solar energy technology, 169, 181, 183, 184–85; solar resources of, 185; and Suzlon, 150, 151, 165; wind companies in Chinese market, 96; wind power industry in, 65, 80, 81, 82, 100, 105, 113, 183, 186, 233n63; and wind power industry of China, 51, 93–105, 102, 103, 106, 107, 115, 130, 142; wind power technology development in, 93–96; wind turbine manufacturers of, 93, 95–96, 96; and wind turbine technology, 27–28, 29, 31, 34, 52, 65, 75, 163, 212n137

United Steelworkers (USW), 115–17
U.S. Bureau of Export Administration, 114
U.S.–China Advanced Biofuel Forum, 176
U.S.–China Agreement on Cooperation in Science and Technology (S&T Agreement), 171
U.S.–China Clean Energy Forum, 177
U.S.–China Clean Energy Research Center (CERC), 173–74, 177, 179, 230n15
U.S.–China cooperation: areas for expansion, 183–86; barriers to, 178–79; and carbon capture and sequestration (CCS) technologies, 173, 179, 185; and China as global superpower, 169; and clean energy technologies, 4, 117, 168, 169–70, 171, 173, 174, 176, 177, 178, 180–81, 187; and climate change, 4, 171–78, 186–87, 229n10; and competition, 178, 179–82; and concentrating solar power, 183, 184–85; future outlook for, 186–87; and nongovernmental organizations, 171, 176, 177–78; official bilateral cooperation, 171–77, 186, 187; and solar photovoltaic technology, 177, 181, 183–84; timeline of major events, *175–76*; and wind power technology, 183, 185–86
U.S.–China Electric Vehicles Forum, 174
U.S.–China Electric Vehicles Initiative, 174
U.S.–China Energy Cooperation Program (ECP), 174, 176
U.S.–China Energy Efficiency Action Plan, 174
U.S.–China Energy Efficiency Forum, 174, 176
U.S.–China Forum on Environment and Development (1997), 172
U.S.–China Green Energy Council, 177
U.S.–China Green Tech Summit, 177
U.S.–China Joint Commission on Commerce and Trade (JCCT), 116
U.S.–China Joint Statement on Energy Security Cooperation, 173
U.S.–China Presidential Summit (2009), 173, 174, 177
U.S.–China Protocol on Energy Efficiency and Renewable Energy, 179
U.S.–China Renewable Energy Industry Forum, 176, 177
U.S.–China Renewable Energy Partnership, 174, 179

U.S.–China Steering Committee on Clean Energy Science and Technology Cooperation, 174
U.S.–China Strategic and Economic Dialogue (S&ED), 173, 176, 229n10
U.S.–China Strategic Economic Dialogue (SED), 172, 173
U.S.–China Ten-Year Framework for Cooperation on Energy and Environment (TYF), 172–73
U.S. Climate Security Act, 180
U.S. Congress, 142, 179, 180
U.S. Department of Commerce, 172
U.S. Department of Energy (DOE), 172, 174, 176, 185, 195n23
U.S. Department of State, 172, 173, 229n10
U.S. Department of Treasury, 172, 173
U.S. Environmental Protection Agency (EPA), 172, 173
U.S. International Trade Commission, 97, 98
U.S. Low Carbon Economy Act, 180
U.S. Renewable Energy Group, 181
U.S. Trade Representative (USTR), and WTO investigation, 115–17, 209n113, 224n62
U.S. Windpower, 93, 97

Vensys Energiesysteme GmbH: and common knowledge sources, 161; and joint development, 158; and learning networks, 134–35, 159; and licensing agreements, 123–24, 131, 133, 147; and mergers and acquisitions, 147, 157, 158, 163; and R&D, 126, 127
Vermont, wind turbine technology in, 27
Vestas: and Chinese market, 61, 62, 76, 82–84, 109, 112, 122, 213n19, 213n20; and Enron Wind, 96; and Gamesa, 36, 81, 85, 202n36, 214n34; history of, 80–81; and home-country market, 36; and licensing agreements, 36, 202n36; and NEG Micon, 79; and offshore projects, 81, 82, 212n13; and patents, 97; size of turbines, 83–84, 212n12; success of, 31–32, 75–76, 79; and technology transfers, 105–6; timeline of company structure in China, 76; and wind power industry of India, 81, 82, 150; and wind power industry of South Korea, 155
Vestas RRB India Ltd., 82

Wan Gang, 173
Wang Qishan, 229n10
Warner, John, 180
Welch, Jack, 93
Wen Jiabao, 12
Westinghouse Electric Company, 94, 99
wind concession projects of China: and bidding, 54–55, 65–66, 115, 207n87, 207n88, 207n89; and feed-in tariffs, 52; and foreign-owned developers, 210n123; in Huilai, 128; and local content requirements, 69, 101, 111, 139; offshore concessions, 58, 74; in Rudong, 82; wind bases compared to, 72, 129
Wind Energy Group, 77
Windey: and exports, 165; and joint development, 158; and licensing agreements, 157; and REpower, 221n12
wind farms: development of, 28, 50–51, 52, 68, 122, 128–29, 130, 131, 135, 164, 165, 180–81, 220n5, 222n40; performance of, 186
windmills. *See* wind turbines
Wind Power Equipment Manufacturing Industry Access Standards, 57
wind power industry: barriers to entry, 41–43, 145–46; benefits of local manufacturing, 39–41, 106, 109, 203n45; countries leading in, 47; in Denmark, 36, 41, 76, 77–86, 113, 232n49; emerging markets of, 130, 132, 145–46; in Germany, 86–87, 96, 114; goals for innovation, 42; and home-country markets, 36–43; in Japan, 162, 228n39; and multinational corporations, 94, 105, 106, 107; quality control in, 42, 101, 130, 132, 138, 139–40, 142, 144, 145, 152, 163, 165, 227n23; start of local manufacturing, 61; in United Kingdom, 77; in United States, 65, 80, 81, 82, 100, 105, 113, 183, 186, 233n63
wind power industry of China: achievements of, 67; capacity of, 1, 24, 51, 52, 56, 57, 58, 63, 65, 74, 134, 186, 190n1, 210n119, 233n63, 233n64; comparative advantages of, 163–65; development of, 2, 3, 4, 26, 38, 54, 55, 57, 58–59, 63, 66, 67, 111, 146, 146, 163; domestic environment, 162; future viability of, 185–86; and local jurisdiction, 53, 54, 207n93; local market establishment, 50, 51, 60, 61, 82, 101, 211n131, 213n20; outlook of, 138–42; wind installations, 78, 91; wind market shares

Index   281

of, *79*, *92*, *103*, 106, 107, *107*, *108*, 109, 110, 114, 115, 118, 119, 120, 135, 138, 143, *155*, 183, 220n144; and wind power pricing systems, 54–55, 65, 233n64. *See also* Chinese wind turbine manufacturers; feed-in tariffs; offshore wind development in China; onshore wind development in China; tariffs; wind concession projects of China
wind power industry of India: and common knowledge sources, 160–62; comparative advantages of, 163–65; development of, 146, *146*, 148–52; domestic environment, 162; and Gamesa, 86, 150; and Goldwind, 131, 135; and Group of 77, 15; and learning networks, 34–35, 148; and local content requirements, 142; and onshore wind development, 162; policy environment for, 162, 163, 165–66; and tax exemptions, 149; technology acquisition strategies of, 4, 147, 148; and Vestas, 81, 82, 150; wind market shares of, 150, *155*, 164; wind power utilization in, 145; and wind resources, 148, 162; and wind turbine manufacturers, 40, 149–52, 163–66, 227n19
wind power industry of South Korea: capacity of, 152; and common knowledge sources, 160–62; comparative advantages of, 163–65; development of, 146, *146*, 148, 152; domestic environment, 162; growth of, 145; and learning networks, 34, 148, 159; offshore development, 152, 154, 155, 162, 164; onshore development, 152, 153–54, 155, 162; policy environment for, 162, 164, 165–66; technology acquisition strategies of, 4, 147, 148, 159; wind market shares of, *155*; wind resources in, 152, 162; and wind turbine manufacturers, 37, 152, 153–56, 158, 162, 164, 165–66
wind power technology: and barriers to entry, 41; China's development of, 4, 50, 59, 66, 156; cost reductions in, 103, deployment of, 2, 64–65; and foreign technology firms, 4; and greenhouse gas emissions, 1–2, 190–91n2; history of, 26; importing of, 38; India's development of, 156; leaders of, 32; leading world wind markets, *30–31*; offshore development, 28–29; origins of, 3, 4, 145; South Korea's development of, 156; technology transfer networks, *160*, 166;

theoretical physics of, 28; and U.S.–China cooperation, 183, 185–86; U.S. development of, 93–96; wind energy R&D expenditures in OECD countries, *29*. *See also* wind turbines; wind turbine technology
wind resources: in China, 55, 59, 75, 83, 122, 127, 141, 162, 169, 211n128; in India, 148, 162; in South Korea, 152, 162
Windtec, 117, 135, 154, 157, 158, 161, 228n38
wind turbine generators (WTG). *See* wind turbines
wind turbine manufacturers: barriers to entry, 41–43; benefits of local manufacturing, 39–41, 106, 109, 111, 112, 113, 203n45; and domestic environment, 162; and foreign technology firms, 75; global market shares, 37; and home-country markets, 36–43, 75; and leading world wind markets, *30–31*, 31; and learning networks, 31, 64, 159; and licensing agreements, 156–57; and localization models, 39–41, 43, 106, 109, 110, 111–12, 113, 115–16, 142, 165, 203n45; and manufacturing scale, 133–34; and minimum annual demand, 37; and R&D expenditures, 62; and technological latecomers, 35, 36, 167; and technology transfer models, 4; of United States, 93, 95–96, *96*; and wind power industry of India, 40, 149–52, 163–66, 227n19; and wind power industry of South Korea, 37, 152, 153–56, 158, 162, 164, 165–66. *See also* Chinese wind turbine manufacturers; *and specific companies*
wind turbines: components of, 27; costs of, 65–66, *66*, 132, 212n137, 223n51; early windmill designs, 26–27; global market shares, *37*; government-mandated domestic content requirement for, 3, 57; market-specific turbines, 83; mass production of, 28; transportation costs for, 41, 91, 203n48
wind turbine technology: China's development of, 4, 67, 75; China's export of, 64, *64*, 67; and common knowledge sources, 160–62; and direct-drive turbines, 105, 124, 127–28, 134, 217n103; and foreign technology firms, 75; gearbox design, 77, 86, 95, 104, 105, 124, 127, 151, 152, 222n31; and home-country markets, 38; and learning networks, 34; lighter-weight turbines, 128, 222n30; origins

wind turbine technology (*continued*)
of, 26–31; performance of, 186; permanent magnet generators, 43, 123, 127, 132, 204*n*51; sizes of turbines, 28, 60–61, *60*, *61*, 67, 73, 83–84, 86, 87, 90, 91, 109, 133–34, 157, 163, 212*n*12, 213*n*27; South Korea's development of, 155–56; standards for, 28, 140, 165, 186; turbine efficiency, 128, 222*n*31; variable-speed technology, 56, 94, 95, 97; and Vestas, 80, 81

Wind World, 77

Winergy AG, 151

WinWinD, 150

World Bank loans, 46, 126

World Resources Institute, 177

World Trade Organization (WTO): and local content requirements, 142; and military applications of technology, 218*n*109; Subsidies and Countervailing Measures (SCM) Agreement, 116; Technical Barriers to Trade Agreement, 43; and U.S. Trade Representative, 115–17, 209*n*113, 224*n*62; and wind power industry of China, 112–18

World Wildlife Fund for Nature, 178

Wu Gang, 122, 123, 124, 133, 134, 137, 144, 221*n*20

Wu Yi, 172

XEMC, 36, 141

Xi'an Aero Engine Corporation, 51, 88

Xi'an Jiaotong University, 84

Xi'an-Nordex, 51, 88–89, 91, 106

Xie Zhenhua, 14

Xinjiang Agriculture University, 126

Xinjiang Wind Energy Company (XWEC), 122, 123, 125

XJ Group, 157, 161

Yang Jiechi, 12, 14

Yingli, 184

Yinhe, 161

Yituo, 51, 214*n*37

Yu Qingtai, 7, 12, 199*n*79

Zeng Peiyan, 172

Zhang Guobao, 173

Zhejiang Institute of Mechanical and Electrical Engineering, 158

Zhu Rongji, 172

Zhuzhou CSR, 141–42

Zond Systems, 94–95, 215*n*65

Zwolinski, Steve, 98

GPSR Authorized Representative: Easy Access System Europe, Mustamäe tee
50, 10621 Tallinn, Estonia, gpsr.requests@easproject.com

www.ingramcontent.com/pod-product-compliance
Lightning Source LLC
Chambersburg PA
CBHW021356290426
44108CB00010B/261